Biopolymers in Food Colloids: Thermodynamics and Molecular Interactions

Biopolymers in Food Colloids: Thermodynamics and Molecular Interactions

By

Maria Semenova
Eric Dickinson

Edited by

Elena Burlakova
Gennady Zaikov

BRILL

LEIDEN • BOSTON
2010

This book is printed on acid-free paper.

ISBN 978 90 04 17186 2

Printed and bound in Great Britain by
CPI Antony Rowe, Chippenham and Eastbourne

Preface

> "The fascination of a growing science lies in the work of the pioneers at the very borderland of the unknown, but to reach this frontier one must pass over well travelled roads; of these one of the safest and surest is the broad highway of thermodynamics." (Lewis, G.N., Randall, M. (1923). *Thermodynamics and the Free Energy of Chemical Substances*, New York: McGraw-Hill.)

The term 'food colloids' can be applied to all edible multi-phase systems such as foams, gels, dispersions and emulsions. Therefore, most manufactured foodstuffs can be classified as food colloids, and some natural ones also (notably milk). One of the key features of such systems is that they require the addition of a combination of surface-active molecules and thickeners for control of their texture and shelf-life. To achieve the requirements of consumers and food technologists, various combinations of proteins and polysaccharides are routinely used. The structures formed by these biopolymers in the bulk aqueous phase and at the surface of droplets and bubbles determine the long-term stability and rheological properties of food colloids. These structures are determined by the nature of the various kinds of biopolymer–biopolymer interactions, as well as by the interactions of the biopolymers with other food ingredients such as low-molecular-weight surfactants (emulsifiers).

There is a growing aspiration to take the subject of food colloids into a new era addressing more directly the issues relevant to the formulation of foods with greater potential health benefit. Such functional foods are envisaged as contributing significantly to reducing the risk of various age-related diseases and to enhancing human health through improved diet and eating habits. To achieve this objective using the existing set of food ingredients will require a greater depth of understanding and control of biopolymer interactions, since these interactions underpin the application of modern nanoscience and nanotechnology to the design of novel biological structures. The challenge is to combine an appealing texture, taste, flavour and shelf-life with the enrichment of food products with more healthy ingredients. This has to involve the development of new delivery technologies for controlled release and improved bioavailability.

The authors of this book have a scientific perspective that is strongly influenced by thermodynamics, and a conviction that the thermodynamic approach still has continuing relevance and importance. Our aim is to demonstrate how an understanding of the thermodynamic basis of the interactions of food biopolymers in aqueous solution and at interfaces can be used by food scientists and engineers to design colloidal systems with novel or improved characteristics — physico-chemical, nutritional, or sensory. The book is arranged in four parts.

In Part One we present some trends and developments in the area of food colloids as perceived from the point of view of the thermodynamics and molecular interactions of biopolymers. Chapter one describes the significance of biopolymer interactions in the context of nanoscience and nanotechnology. It is suggested that the production of nanoscale particles and structures for use in food colloids should first of all rely on an in-depth understanding of the thermodynamically driven interactions and self-assembly processes involving biopolymers. This forms the basis of the nanotechnological 'bottom-up' approach. Chapter two discusses applications of biopolymers for micro- and nano-encapsulation of bioactive food ingredients. We present examples of the main classes of low-molecular-weight bioactive molecules which are currently considered to be helpful for human well-being and which can be therefore used as food supplements as well as active components in skin-care applications. Research developments in the area of nanoparticle delivery systems for micro-nutrients and nutraceuticals are mentioned. In Chapter three we discuss the use of molecular thermodynamics in providing both qualitative and quantitative relationships to describe fundamental phenomena occurring in the equilibrium state of systems containing biopolymers.

Part Two outlines the fundamental principles and practices underlying the study of biopolymer interactions. Chapter four characterizes the different kinds of intermolecular forces that can occur between biopolymers in bulk aqueous media, including the interfacial region. Chapter five sets out the thermodynamic parameters that can describe these interactions quantitatively, together with the experimental methods available for their determination.

Part Three describes a range of important specific examples of the interactions of individual biopolymers in the bulk aqueous medium of food colloids. Chapter six is devoted to the subject of the self-assembly of food biopolymers, and how this self-assembly is affected by conditions such as pH, ionic strength, divalent ions, cosolutes, *etc*. It is indicated how biopolymer self-assembly can form the basis of the 'bottom-up' nano-biotechnological approach, which attempts to mimic Nature in the creation of new and varied structures with potential applications. It is explained how self-assembled biopolymer nanostructures can account for various aspects of the stability, rheology and microstructure of emulsions and foams. Chapter seven shows how physical and covalent interactions between different kinds of biopolymers can affect the properties of food colloids. The significance of the strength and character (net attractive or net repulsive) of the biopolymer–biopolymer interaction in relation to colloid stability is discussed. The role of protein–polysaccharide electrostatic complexes and covalent conjugates in the formulation of 'intelligent' nanoscale encapsulation systems is highlighted.

In Part Four (Chapter eight) we focus on the interactions of mixed systems of surface-active biopolymers (proteins and polysaccharides) and surface-active lipids (surfactants/emulsifiers) at oil–water and air–water interfaces. We describe how these interactions affect mechanisms controlling the behaviour of colloidal systems containing mixed ingredients. We show how the properties of biopolymer-based adsorption layers are affected by an interplay of phenomena which include self-association, complexation, phase separation, and competitive displacement.

We anticipate that this book could be of interest to scientists, technologists and engineers working in food-related areas in industry and academia. For the benefit of the less specialized reader, we have tried wherever possible to illustrate concepts and phenomena with schematic diagrams and images from microscopy. The coverage is not meant to be exhaustive. Food science is a broad subject, and the various examples highlighted in this volume necessarily reflect the interests of the authors. It is therefore not surprising that there is particular emphasis on milk proteins as interfacial functional ingredients, and on light scattering as an experimental technique for studying biopolymers in solution. In order to provide authenticity and some degree of authority, many of the specific examples are taken from the published work of the authors' own research groups. Without the skill, insight and dedication of our research students and collaborators, this book would not have been possible. The names of all the people whom we have had the great pleasure to collaborate with over many years are to be found in the bibliography sections. We are most grateful to you all.

Maria G. Semenova (Moscow)
Eric Dickinson (Leeds)

ACKNOWLEDGEMENTS

Grateful thanks are due to authors and the following copyright holders for granting permission to make use of published material: Elsevier (Figs. 1.1, 2.1, 2.2, 2.3, 3.1, 3.2, 3.3, 3.4, 3.6, 6.1, 6.3, 6.4, 6.6, 6.9, 6.10, 6.13b, 6.14, 6.15, 6.17, 6.21, 6.22, 7.1, 7.2, 7.3, 7.4, 7.7, 7.8, 7.9, 7.13, 7.14, 7.16, 7.17, 7.18, 7.19, 7.20, 7.21, 7.22, 7.23, 7.25, 7.26, 7.28, 8.1, 8.3, 8.4, 8.6, 8.7, 8.8, 8.10, 8.11); Wiley Inter-Science (Fig. 1.2); American Chemical Society (Figs. 6.8, 6.18, 6.19, 7.11, 7.12, 8.2, 8.9, 8.13, 8.14); Royal Society of Chemistry (Figs. 6.13a,c,d, 6.20, 8.5, 8.15).

Contents

Part Four — BIOPOLYMER INTERACTIONS AT THE INTERFACES IN FOOD COLLOIDS

PART ONE

TRENDS AND DEVELOPMENTS

CHAPTER ONE

SIGNIFICANCE OF BIOPOLYMER INTERACTIONS IN CONTEXT OF NANOSCIENCE AND NANOTECHNOLOGY

We begin by attempting to put the subject of biopolymer interactions into some kind of technological and sociological context. Our observations of the current commercial scene would suggest that the main areas of focus of ongoing research and development in the food industry are:

(i) optimizing ingredient formulation to improve the consumer appeal of food, such as its colour, flavour, texture or consistency;
(ii) extending the shelf-life of foods by improving stability;
(iii) controlling the encapsulation and release of flavours and nutrients;
(iv) enhancing the absorption of nutrients or nutraceuticals;
(v) developing functional foods, which target nutrition for different lifestyles and for an ageing population as well as enabling consumers to modify their diets, depending on their own nutritional needs or tastes; and
(vi) removing undesirable molecules from foods.

Food in the future will not only be a good source of basic nutrients with good sensory appeal and safety, but it will also contribute effectively to the well-being and health of everybody individually. This individualized approach to foods will increasingly become more relevant as the relationships between genes, diet and health are further established (Sanguansri and Augustin, 2006). A number of new processes and materials derived from nanotechnology can potentially provide answers to many of these needs because they offer the ability to control and manipulate the properties of substances at the molecular level (Chen *et al.*, 2006; Sanguansri and Augustin, 2006; Weiss *et al.*, 2006; Bouwmeester *et al.*, 2007; Chaudhry *et al.*, 2008). Nanotechnology applications for the food sector are still relatively new compared to those in the biomedical area, but there have been rapid developments in recent years as the relationship between food and health becomes clearer and more health claims are allowed for foods (Ransley *et al.*, 2001). For example, various kinds of nanoparticles have good potential for incorporation into food colloids as nutrient delivery systems (nanocapsules, nanocontainers, nanogels, *etc.*) built from self-assembled lipid monomers or peptides (Taylor *et al.*, 2005; Chen *et al.*, 2006; Graveland-Bikker and de Kruif, 2006; Mozafari *et al.*, 2006; Sanguansri and Augustin, 2006; Bouwmeester *et al.*, 2007; Chaudhry *et al.*, 2008).

Nanoencapsulation of bioactive compounds, having positive impact on human health, will be increasingly incorporated in novel formulations of food products. The aim will be to enhance the bioavailability of food ingredients, in order to deliver compounds at the right site and at the right time (Gouin, 2004; Acosta, 2009). These new food products, containing nanoencapsulates loaded with bioactive compounds, will be designed to provide a healthy and/or low calorie diet, whilst still maintaining the possibility to enjoy the food. Such novel applications will therefore contribute to the role of foods in preventive healthcare (Kampers, 2007). Consequently there is a tendency towards a blurring of the boundaries between the food and pharmaceutical domains — and also between the food and cosmetics domains. It is indeed in these non-food industries that significant advances have already been achieved in the application of nanotechnology. In response to such rapid developments, food scientists will need to examine how nanoscience and nanotechnology may lead to fundamental changes in the food supply, and to find the most appropriate ways in which value can be added to natural ingredients and traditional foods (Chen *et al.*, 2006; Sanguansri and Augustin, 2006). Certainly it can be anticipated that more universal 'nanofood' products will appear on the world markets within the next few years as a result of the global organization of international food companies (Chaudhry *et al.*, 2008). Whether the consumer is considered by the industry to be ready yet for these scientific developments is a matter for conjecture. It is well known that the senior management of one multinational food company (Nestlé) strongly recommends its scientist employees not to make use of the apparently taboo word 'nano-' in their technical publications!

One question that is sometimes asked is whether 'nanoscience' is really the same as old-fashioned 'colloid science'. In the broadest sense, the subjects of nanoscience and nanotechnology are concerned with understanding, creating and applying atomic or molecular level manipulation and assembly to organic and inorganic matter in ways consistent with the basic physical laws. In particular, nanotechnology is defined as the engineering of matter over the length scale from 1 to 1000 nanometres (a nanometre is one billionth of a metre), and nanoscience is the knowledge and understanding of how to manipulate and characterize matter on the nanoscale. Both these terms embrace phenomena whose effects can be observed at the macro-scale, but can only be understood and mastered when looking and operating at the 'nano' level. The length scale of nanoscience is formally the same as the one covered by colloid science; but it is now commonly accepted that the nanoscale extends up to ~100 nm, rather than to ~1 μm as for the traditionally defined colloid.

The field of nanobiotechnology, as defined in the work of Sletmoen *et al.* (2008), is the part of nanotechnology dealing with and inspired by biology. It can therefore be described as a multi-disciplinary area lying at the interface between engineering, biology and the physical sciences. The area includes the exploitation of biomaterials, and devices or methodologies based on biomaterials, that are characterized by having dimensions of their functional components on the scale between molecular lengths and a few hundred nanometres. In this context, the most promising applications are those that mimick biological assembly processes, insofar as they suggest biologically inspired ways of fabricating adaptable, switchable or responsive materials, where the interactions between the nanoparticles can be controlled or modulated reversibly (Min *et al.*, 2008). On this basis the field of nanobiotechnology can provide the methods and framework for building new food materials and developing new food products.

It is generally recognized that the successful implementation of nanoscale research and technology represents a great intellectual and scientific challenge. The emergence of novel properties of organic and inorganic matter at the 'nano' level differentiates them from bulk materials, and it allows the introduction and development of new applications of materials that can be exploited by industry (Sanguansri and Augustin, 2006). In principle, all kinds of organic and inorganic matter can act as nanoparticles having one or more spatial dimensions of the order of 100 nm or less (Min *et al.*, 2008). It has to be emphasized, however, that the size at which particles display substantially different properties from those of the bulk substance varies from one material to another, and in some cases it can be much larger than 100 nm.

Nanoparticles can be spherical, tubular, or irregularly shaped; they can exist in fused, aggregated or agglomerated forms; and they can be compositionally homogeneous or heterogeneous (Graveland-Bikker and de Kruif, 2006; Graveland-Bikker *et al.*, 2006; Bouwmeester *et al.*, 2007; Letchford and Burt, 2007; Tiede *et al.*, 2008; Min *et al.*, 2008). Hence it is clear that size is not the only critical factor to consider. It is necessary that nanoparticles be characterized as completely as possible; for example, the total surface area may be highly relevant, as well as other characteristics like flexibility and shape (Bouwmeester *et al.*, 2007; Min *et al.*, 2008). In a food product, it is possible that a bigger surface area could enhance texture and mouthfeel by increasing water-binding capacity, flavour release, bioavailability, *etc.* (Sanguansri and Augustin, 2006). It seems likely to turn out that one of the major complicating factors in nanoparticle characterization is the strong dependence of the physicochemical characteristics of nanoparticles on the matrix in which they are dispersed. There is therefore a need to characterize the properties of nanoparticles *in situ* in the food matrix (Oberdorster *et al.*, 2005; Powers *et al.*, 2006). Experimental techniques such as atomic force microscopy (AFM) can be very useful in this context by generating new information on the sizes, shapes and interactions of biopolymers and nanoparticles in food systems (Morris, 2006).

In relation to food, however, a cautious approach is always required. Not much is yet known about the relationship between the physico-chemical characteristics of nanoparticles and their behaviour in the human body. In particular, the potential effects of the presence of nanoparticles through out the gastrointestinal route are largely unknown. This is an important issue that needs further research. More generally, reliable knowledge on the actual or potential toxicity of nanoparticles is limited. What seems likely to be the case, however, is that the potential toxicity and physiological behaviour of nanoparticles will be affected by a wide range of factors, including particle number density and mass concentration, surface area, net electrical charge, chemical composition and reactivity, particle size and size distribution, the state of aggregation, as well as aggregate structure and shape (Tiede *et al.*, 2008). Identification of the potential risks of any imminent applications of specific nanoparticles in foods will be a matter for urgent investigation (Chau *et al.*, 2007; Walsh *et al.*, 2008).

Engineered nanoparticles can be prepared in two ways: 'top-down' by breaking apart conventional bulk substances, or 'bottom-up' by building up structures from the molecular scale. There also is a growing trend to combine the 'top-down' and 'bottom-up' approaches to produce more sophisticated nanoparticle systems (Horn and Rieger, 2001).

The 'top-down' approach involves size reduction by the application of three main types of force — compression, impact and shear. In the case of colloids, the small entities produced are subsequently kinetically stabilized against coalescence with the assistance of ingredients such as emulsifiers and stabilizers (Dickinson, 2003a). In this approach the ultimate particle size is dependent on factors such as the number of passes through the device (microfluidization), the time of emulsification (ultrasonics), the energy dissipation rate (homogenization pressure or shear-rate), the type and pore size of any membranes, the concentrations of emulsifiers and stabilizers, the dispersed phase volume fraction, the charge on the particles, and so on. To date, the 'top-down' approach is the one that has been mainly involved in commercial scale production of nanomaterials. For example, the approach has been used to produce submicron liposomes for the delivery of ferrous sulfate, ascorbic acid, and other poorly absorbed hydrophilic compounds (Vuillemard, 1991;

Kosaraju *et al.*, 2006), as well as to encapsulate probiotic cultures (Feijoo *et al.*, 1997) and to make vitamin E nanoparticles (100 nm) stabilized by a starch coating suitable for fortified beverages (Chen and Wagner, 2004). And, by using microfluidization followed by solvent evaporation, Tan and Nakajima (2005) have reported preparation of β-carotene nanodispersions (60–140 nm).

The more challenging 'bottom-up' approach relies first of all on the self-assembling properties of molecules or particles. These properties define their susceptibilities to spontaneous ordering processes that occur under strong thermodynamic control, *i.e.*, under balancing of the attractive and repulsive forces between them, in order to build supramolecular structures, microstructures, and the higher hierarchical structures of functional materials. Secondly, when thermodynamics no longer dictates the self-assembly pathway (or does so only after a very long time), particles can be ordered by directed assembly, using the externally applied forces, such as flow-induced viscous and lubrication forces, compressive or shear stresses, gravity, magnetic and electric fields (Min *et al.*, 2008). Future use of the 'bottom-up' approach within industry is expected to increase (Förster and Konrad, 2003; Sanguansri and Augustin, 2006, Acosta, 2009).

In the 'bottom-up' approach, a large variety of ordered nano-, micro- and macrostructures may be obtained by changing the balance of all the attractive and repulsive forces between the structure-forming molecules or particles. This can be achieved by altering the environmental conditions (temperature, pH, ionic strength, presence of specific substances or ions) and the concentration of molecules/particles in the system (Min *et al.*, 2008). As this takes place, the interrelated processes of formation and stabilization are both important considerations in the production of nanoparticles. In addition, as particles grow in size a number of intrinsic properties change, some qualitatively, others quantitatively; some affect the equilibrium (thermodynamic) properties, and others affect the non-equilibrium (dynamic) properties such as relaxation times.

In a clear exposition of the physical concepts involved, Min and co-authors (2008) distinguish the separate influences of particle size on equilibrium effects and dynamic effects. Regarding equilibrium effects, they postulate that, as particle size exceeds the range of the interparticle forces, the phase diagram will change qualitatively. This change therefore depends on the types of forces acting between the particles — in other words, whether they are short-ranged (attractive van der Waals forces between identical materials) or long-ranged (repulsive electrostatic 'double-layer' forces) relative to the particle size. Another thermodynamic scaling effect concerns polydispersity of soft self-assembling particles such as micelles. With increasing size, the polydispersity in size (radius *R* or aggregation number *N*) increases in absolute terms, but decreases relative to *R* or *N*. That is, the normalized distribution becomes narrower with increasing size. Turning to dynamic effects, firstly there is the linear increase in the van der Waals adhesion or binding energy between particles with increasing radius. Because of this, such systems can show 'ageing effects' such as deformational creep. Secondly, there is more rapid surface and bulk diffusion in the case of nanoparticles. For example, if a surface molecule takes one year to diffuse to the other side of a 10 μm colloidal particle, it would take only 30 s to diffuse to the other side of an equivalent 10 nm nanoparticle. Furthermore, for 10 nm spheres in water at room temperature, Min *et al.* (2008) have noted that the rotational relaxation time is around 1 μs; this is some 10^9 times faster than for 10 μm spheres. For these reasons, it is argued that most nanoparticle systems are non-equilibrium systems: to understand and control them, one needs to appreciate the strength and range of the diverse interaction forces and energies amongst their constituent atoms, as well as amongst the particles themselves.

Nature itself makes extensive use of the self-assembly strategy to grow nanoscale structures using the 'bottom-up' approach. Probably one of the more spectacular examples of electrostatically driven molecular assembly processes in Nature is the wrapping of DNA around histones to form nucleosomes, the building blocks of chromosomes (Sletmoen *et al.*, 2008). Another impressive example of Nature's self-assembly strategy, relating specifically to food colloids, is the self-association of the major milk protein, casein. This archetypal association colloid, the so-called 'casein micelle', supplies calcium ions and protein to the human diet in a readily assimilated functional form (Holt, 1992; Horne, 1998, 2002, 2003, 2006; Dickinson, 2006a).

The ability to utilise Nature's strategies to build artificial molecular devices, with the possibility to impart responsive and tunable properties typical of amphiphilic self-assemblies, is likely to have very significant impact on nanoscience and soft matter science over the next decade (Menger, 2002; Niemeyer, 2002; Berti, 2006; Graveland-Bikker and de Kruif, 2006; Graveland-Bikker *et al.*, 2006; Manski *et al.*, 2007; Min *et al.*, 2008). One prominent example of the successful use of this approach is the ability to template bone nanostructure by assembling a set of peptide amphiphiles into cylindrical fibres. Apparently the observed cylindrical micelles are the result of a peculiar directional electrostatic interaction between head groups of the peptide amphiphiles, which possesses an attractive character, unlike that found with most conventional micelles (Hartgerink *et al.*, 2001; Niece *et al.*, 2003; Tsonchev *et al.*, 2004).

In the case of food science and food processing, the main concepts of nanoscience and nanotechnology provide a sound framework for further developing our understanding of the interactions and self-assembly behaviour of the basic food components (proteins, polysaccharides, lipids) into nano/microstructures under well-controlled and extreme environments (pH, ionic strength, heat, shear, high pressure, drying, freezing, *etc.*). These interactions and self-assembly processes subsequently control food hierarchical structures, and ultimately food texture, food rheology, and food functionality on the macroscopic scale. One may presume that such developments could be especially promising in the field of food colloids, where the primary structure-building elements are nanosized molecules (biopolymers, emulsifiers and lipids) along with colloidal-sized particles (emulsion droplets, air bubbles, lipid micelles or biopolymer aggregates). These are all contained in a liquid dispersion medium and hence are responsive to environmental variations. Therefore, the application of nanoscience and nanoengineering to food colloids is concerned with understanding and manipulating the molecular and colloidal interactions involving all these entities for the purpose of developing innovative food products of outstanding quality (Dickinson, 2003a, 2004, 2006a; McClements, 2006; Sanguansri and Augustin, 2006; Morris, 2005, 2006; van der Linden, 2006; Semenova, 2007). According to this interpretation, both the 'top-down' and the 'bottom-up' approaches of nanoscience and nanotechnology have actually been used successfully already by food scientists for rather a long time (Dickinson, 2003a, 2006a; Veerman *et al.*, 2003; Bolder *et al.*, 2006; Graveland-Bikker and de Kruif, 2006; Graveland-Bikker *et al.*, 2006; Manski *et al.*, 2007).

Table 1.1 summarizes the types of processes for generating nanoparticles that are currently involved in the 'top-down' and 'bottom-up' approaches (or their mixed variants). It should be noted that many of these processes have their origin and primary applications predominantly in the pharmaceutical industry; so far the commercial food applications of nanotechnology are still in their infancy. For each of the processes mentioned in Table 1.1, the kinds of nanoparticles involved are listed along with a brief indication of their characteristic properties and their approximate particle dimensions. Also presented are some recent literature sources on these various topics (mainly review-type articles) where the interested reader can obtain further background material.

Table 1.1 Examples of processes involved in nanotechnological 'top-down' and 'bottom-up' approaches.

Type of process	Type of nanoparticles	References
	'top-down' approach	
High-pressure homogenization and microfluidization (including principle of the rapid expansion of supercritical solutions of mostly hydrophobic compounds)	Fine emulsion droplets, liposomes and nanoparticles: 40–200 nm	Vuillemard, 1991; Feijoo *et al.*, 1997; Chen and Wagner, 2004; Tan and Nakajima, 2005; Kosaraju *et al.*, 2006; Acosta, 2009
Ultrasound emulsification	Fine emulsion droplets: 40–200 nm	Abismail *et al.*, 1999; Behrend *et al.*, 2000; Shukla and Halpern, 2005
Membrane and microchannel emulsification	Fine emulsion droplets: 50–1000 nm (with a narrow size distribution range)	Joscelyne and Trägårdh, 1999, 2000; Kobayashi and Nakajima, 2006
Electrified coaxial liquid jets	Fine emulsion droplets with electrostatic stabilization of droplets: 10 nm–10 μm	Loscertales *et al.*, 2002; Sanguansri and Augustin, 2006
Shear-induced anisotropic structure formation	Worm-like micelles, bundle-like and string-like structures in dense biopolymer and colloidal systems	Manski *et al.*, 2007

Table 1.1 *Continued*

	'bottom-up' approach	
Self-assembly of polymers in the bulk	Polymer micelles, polymerosomes, gelled macromolecules, nano-tubes, protein fibres/tapes produced by aggregation at low pH, controlled release vehicles, smart delivery systems: 50–500 nm	Förster and Konrad, 2003; Sanguansri and Augustin, 2006; Dickinson 2006a; Graveland-Bikker and de Kruif, 2006; van der Linden, 2006
Electrostatic self-assembly of polymers in the bulk	Polyelectrolyte capsules; controlled release vehicles, smart delivery systems: 50–500 nm	Anal *et al.*, 2008
Layer-by-layer electrostatic deposition of biopolymers at interface of emulsion droplets	Polyelectrolyte capsules; microparticles with novel functionality and triggered release: multiple layers around oil droplets exhibiting improved stability towards environmental stresses (temperature, ionic strength, pH, freezing, dehydration)	McClements, 2005, 2006; Sanguansri and Augustin, 2006
Self-assembly of colloidal particles onto emulsion droplets	Colloidosomes, controlled release vehicles: 50–500 nm	Dinsmore *et al.*, 2002; Sanguansri and Augustin, 2006;
Phase separation into water-in-water emulsions	Spheres, teardrops, fibres, trapped in kinetically stable state by changing solution or environmental conditions (temperature, pH, ionic composition, solvent quality) so that one or both phases thickens or gels	Norton and Frith, 2001; McClements, 2006

Table 1.1 *Continued*

	'bottom-up' approach	
Spontaneous emulsification and solvent diffusion method	Solid lipid nanoparticles: 20–80 nm	Horn and Rieger, 2001; Cui *et al.*, 2006; Ribeiro *et al.*, 2008
Self-microemulsifying method	Fine emulsion droplets: < 100 nm	Pouton, 2000; Cui *et al.*, 2005
Lipid micellization	Vesicles for encapsulation, controlled release, functional incorporation of proteins and mimicking of biological membranes	Singh *et al.*, 1995; Taylor *et al.*, 2005; Sagalowicz *et al.*, 2006; Mozafari *et al.*, 2006
	mixed variants of 'top-down' and 'bottom-up' approaches	
Pressure cycling: high hydrostatic pressure (*e.g.*, 500 Mpa) induces disintegration of casein micelles and reassociation on pressure reduction	Hydrophobic casein particles, formed under pressure, reassociate into smaller and more irregularly shaped aggregates	Dickinson 2006a
Lipohilic solvent method: melting a lipid matrix (*e.g.*, saturated fatty acids) and dissolving hydrophobic bioactive ingredients in hot melt, which is emulsified and then spray congealed	Solid lipid nanoparticles containing wide range of hydrophobic bioactive ingredients	Müller *et al.*, 2000; Üner, 2006; Ma *et al.*, 2007; Acosta, 2009

Against this background, the design and modification of structures and interactions of the major functional and nutritional components of food systems assumes a new significance. Biopolymers are especially important in this respect for a number of reasons. Firstly, the individual protein and polysaccharide molecules are of nanoscale dimensions. Secondly, owing to the special features of their molecular structures (a large number of both polar and non-polar functional groups), they are prone to self-assembly in aqueous media or at interfaces. This self-association is the result of weak non-specific physical bond formation (biopolymer–biopolymer or biopolymer–solute[1]) in response to changes in pH, temperature, *etc.*, or addition of ions, enzymes, *etc.* A third reason for the importance of biopolymers is that they comprise the nanoscale building blocks of particles, aggregates, fibres, complexes and networks commonly found in food gels and dispersions (Veerman *et al.*, 2003; Murray and Ettelaie, 2004; Dickinson, 2004, 2006a,b; Graveland-Bikker *et al.*, 2006; van der Linden, 2006; Morris, 2005, 2006; Weiss *et al.*, 2006; Bolder *et al.*, 2006; Zhang *et al.*, 2006; Manski *et al.*, 2007; Huppertz and de Kruif, 2008). Table 1.2 illustrates schematically some examples of nanoparticles based on biopolymers, lipids and surfactants which can be present in the dispersion medium and at interfaces in food colloids.

Various kinds of intermolecular non-covalent physical interactions can contribute to the self-assembly of biopolymers, emulsifiers and lipids. These include hydrogen bonding, attractive electrostatic forces (between opposite charges), van der Waals forces, and hydrophobic forces between non-polar groups in water (McClements, 2006; Morris, 2006; Anal *et al.*, 2008; Min *et al.*, 2008). Individual intermolecular energies are rather small in magnitude, but their combination in regular assembly can make the nanoscale structure stable. Additionally, we have repulsive electrostatic forces and excluded volume interactions that may induce mutual biopolymer segregation and phase separation in binary biopolymer systems and in mixtures of biopolymers with surfactants (Tolstoguzov, 2002; Semenova, 2007; Dickinson, 2003b; McClements, 2006). Finally, of course, we recognize that, in any biopolymer-based aqueous solution or colloidal dispersion, most of the system is occupied by the water molecules. The molecular organization of water structure contributes predominantly to the free energy of the whole system, including an important contribution from the so-called hydration force between molecules/particles, the precise nature of which is still not properly understood (Israelachvili and Wennerström, 1996; Tolstoguzov, 2002).

---

[1] The solute may itself be amphiphilic like a surface-active lipid (emulsifier) or just a simple hydrophilic or hydrophobic compound.

Table 1.2 Examples of the types of nanoparticles that can potentially be present in food colloids.

Types of nano-particles	Putative structure		Substances forming nanoparticles
	Homogeneous composition **(A)**	Heterogeneous composition **(B)**	
Individual molecules or colloidal particles			
Compact ellipsoidal or spherical particles			**(A)** globular proteins; **(B)** complexes of proteins with amphiphilic compounds
Random coils or flexible polymers		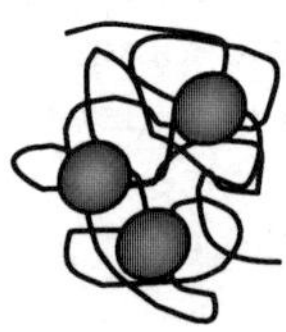	**(A)** polysaccharides or disordered proteins; **(B)** complexes of polysaccharides or disordered proteins with amphiphilic compounds; inclusion complexes in helical part of biopolymer coils
Colloidal particles stabilized by single or multiple biopolymer and/or surfactant adsorbed layers		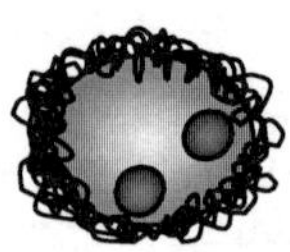	**(A)** oil/water droplets/bubbles; **(B)** oil/water droplets with bioactive compounds dissolved in the core

Table 1.2 *Continued*

Types of nano-particles	Putative structure		Substances forming nanoparticles
	Homogeneous composition **(A)**	Heterogeneous composition **(B)**	
	Self-assembled aggregates		
Micellar protein assemblies	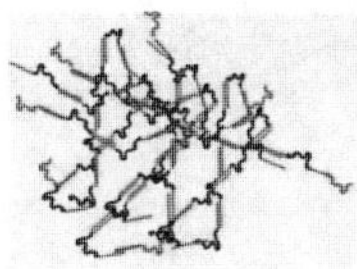	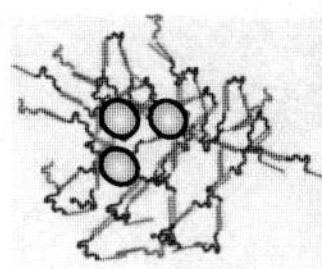	**(A)** native micellar casein, sodium caseinate, β-casein; **(B)** complexes of proteins with amphiphilic compounds
Nanogel particles	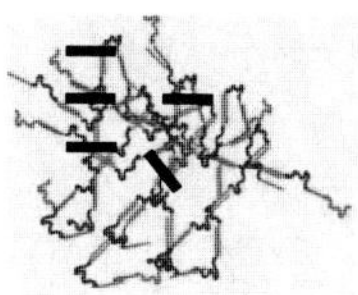	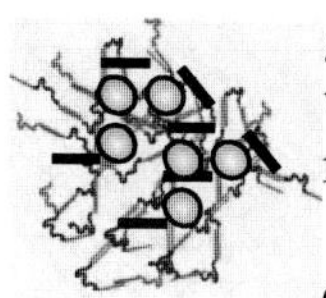	**(A)** Caseins cross-linked by the enzyme trans-glutaminase within the casein micelle; **(B)** complexes of cross-linked proteins with amphiphilic compounds
Protein–polysaccharide electrostatic complexes		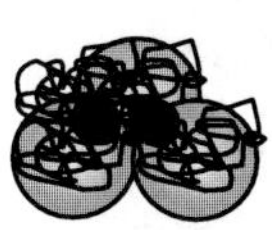	**(A)** Oppositely charged proteins and poly-saacharides; **(B)** protein–polysaccharide complex with amphiphilic com-pounds

Table 1.2 *Continued*

Types of nano-particles	Putative structure		Substances forming nanoparticles
	Homogeneous composition **(A)**	Heterogeneous composition **(B)**	
Protein–polysaccharide covalent conjugates			**(A)** Covalently linked proteins and polysaccharides; **(B)** conjugate with amphiphilic compounds
Protein nanotubes	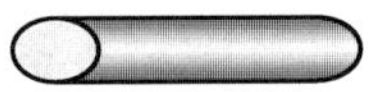		**(A)** hydrolyzed α-lactalbumin; **(B)** complexes of protein nanotubes with amphiphilic compounds
Protein nano-fibres		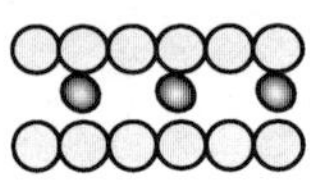	**(A)** α_{s1}-casein and globular proteins in their molten state; **(B)** aggregates of protein nano-fibres bridged by bioactive species
Micelles of lipids and surfactants	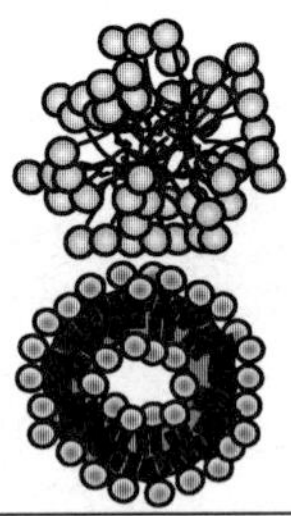	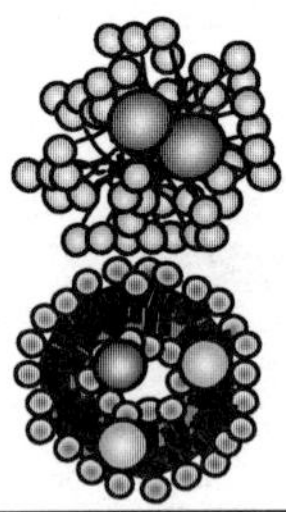	**(A)** surfactants, phospholipids; **(B)** surfactant micelles and phospholipid liposomes loaded with bioactive compounds

Why do particular nanoscale structures actually form? The answer in simple terms is that the formation is a slow step-by-step process, like the gradual descent into a valley from a mountain. That is, rather than requiring large amounts of energy for their creation, it appears that self-assembled nanoscale structures are formed through a series of optimized processes that utilize the tendency of a system to minimize its overall free energy, thereby minimizing any required activation energies. Therefore the production of specific nanoscale structures for food science and technology must rely first and foremost on an in-depth understanding of the thermodynamically driven self-assembly processes in multicomponent food systems (Manski *et al.*, 2007; Morris, 2006). We require an understanding of both the structural and thermodynamic characteristics of the final self-assembled nanostructures as well as the molecular mechanisms of their formation and degradation. Once this information is available, it can be used to design, stabilize and manipulate nanostructures rationally so as to enhance functionality under the conditions encountered in food manufacture and in product usage by consumers (Murray and Ettelaie, 2004; Graveland-Bikker and de Kruif, 2006; Morris, 2006; Weiss *et al.*, 2006; Manski *et al.*, 2007). Probably most relevant for the case of food colloids is the understanding and development of novel interactions between proteins and polysaccharides, and also between biopolymers and emulsifiers/lipids. Controlling the interactions of particles, droplets and bubbles covered by these mixed ingredients could make possible tailor-made improvements in existing products as well as leading to novel food system formulations (Murray and Ettelaie, 2004; McClements, 2006; Morris, 2006; Dickinson, 2003a,b, 2006a). Table 1.3 sets out the potential impact of the different kinds of physical biopolymer interactions in relation to the phenomena making a major contribution to the nanostructuring of food colloids.

One of the more spectacular examples of the development of novel interactions for nanostructuring of food systems is the self-assembly of partially hydrolysed molecules of α-lactalbumin at neutral pH in the presence of appropriate cations (Ca^{2+}, Mn^{2+}, Zn^{2+}, Cu^{2+} or Al^{3+}). These ordered nanostructures possess enhanced functionality for thickening, gelation and encapsulation, as compared to the individual protein molecules or their disordered aggregates. The molecules assemble into rather stiff nanotubes with a cavity diameter of 16 nm and a length of a few micrometres (Figure 1.1). The specific ion size and its preferred ligand coordination number seem to play a key mechanistic role. But hydrolysis is needed to make the α-lactalbumin prone to self-assembly.

Table 1.3 Potential impact of different kinds of biopolymer interactions affecting phenomena relevant to the nanostructuring of food colloids.

Interactions	Phenomena	References
Reduction in protein–protein repulsion due to lowering of net charge or screening of charge, as triggered by change in pH or ionic strength, or addition of specific ions	(i) Change in osmotic pressure in the aqueous phase and hence in negative depth of the Gibbs free energy of depletion interactions; (ii) loss of electrostatic stabilization of the droplets in emulsions; (iii) increase in viscoelasticity of concentrated protein-stabilized emulsions (*i.e.*, liquid-like emulsions converted into soft solid-like texture); (iv) reduction in creaming stability of dilute emulsions	Dickinson *et al.*, 2001; Dickinson 2006a; Semenova *et al.*, 1999, 2005
Temperature-triggered self-assembly of surface active protein due to strengthening of hydrogen bonding interactions as temperature is lowered or hydrophobic interactions as temperature is raised	Temperature-sensitive emulsion rheology and gelation	Dickinson 2006a; Eliot and Dickinson, 2003
Sucrose-triggered disintegration of the assembled nanostructures of sodium caseinate at pH ≥ p*I*	(i) Increase in degree of interconnectivity of protein gel microstructure upon lowering of pH; (ii) enhancement in stiffness and homogeneity of the protein gel	Belyakova *et al.*, 2003; Dickinson, 2006a

Table 1.3 *Continued*

Interactions	Phenomena	References
Biopolymer–surfactant and/or biopolymer–biopolymer attraction in adsorption layers due to opposite electrical charges of the interacting molecules	(i) Increase in stability of nanostructured multilayer emulsions against environmental stresses; (ii) formation of smart delivery systems with functional component trapped within core of multilayer emulsion to be released in response to specific trigger (pH, ionic strength, temperature); (iii) formation of double emulsions, allowing separation of reactive components into different phases, and protecting substance trapped within inner phase droplets for ultimate release to a specific body site	Benichou *et al.*, 2002; de Kruif *et al.*, 2004; McClements, 2005, 2006; Weiss *et al.*, 2006; Dickinson, 2006b
Self-assembly and stratification of surface-active proteins and nanoparticles at surface of bubbles, as driven by hydrophobic interactions	(i) Enhanced foam coalescence stability due to modified interfacial structure/rheology of adsorbed films; (ii) change in rate and extent of shrinkage of bubbles due to disproportionation	Murray and Ettelaie, 2004
Self-assembly of hydroysed α-lactalbumin into nanotubes in presence of specific cations	(i) Efficient and responsive thickening and gelling agent, with low protein concentration providing rheological enhancement with reversible gel formation due to physical interactions; (ii) ability to encapsulate specific molecules via manipulation of nanotube internal structural properties; (iii) controlled release via controlled disassembly	Graveland-Bikker *et al.*, 2006; Graveland-Bikker and de Kruif, 2006

Table 1.3 *Continued*

Interactions	Phenomena	References
Self-assembly of globular proteins into nano-fibres under specific conditions of pH, ionic strength and thermal treatment	Efficient gelation at low concentration due to nano-fibre elongated structure	Veerman *et al.*, 2003; van der Linden, 2006
Thermodynamically unfavourable interactions from excluded volume effects and electrostatic repulsion between unlike biopolymers	(i) Liquid–liquid segregative separation (continuous phase); (ii) large increase in protein surface activity at oil–water interface and protein loading on emulsion droplets; (iii) decrease in size of droplets during emulsification; (iv) increase in surface shear viscosity of protein layer; (v) intensification of depletion flocculation of emulsions leading to poorer stability of the dilute emulsions with respect to creaming; (vi) increase in viscoelastisity of concentrated emulsions	Dickinson and Semenova, 1992; Dickinson, 2003b; 2006b; Dickinson *et al.*, 1998; Semenova *et al.*, 1999; Semenova, 2007
Thermodynamically favourable interactions arising from complexation or entropy-driven association between unlike biopolymers	(i) Increase in thermodynamic stability with respect to phase separation of mixed solutions; (ii) marked reduction in protein surface activity; (iii) substantial increase in surface shear viscosity of adsorbed protein layer; (iv) marked increase in viscoelasticity of emulsions of high oil volume fraction, and rapid serum separation of emulsions of low oil volume fraction, in both cases due to bridging flocculation	Dickinson and Semenova, 1992; Dickinson, 2003b, 2006b; Dickinson *et al.*, 1998; Semenova *et al.*, 1999; Semenova, 2007

Another controlling factor is that the nanotubes can only be formed within a rather narrow ion concentration window. These α-lactalbumin nano-tubes have been found to consist of several hydrolysis products with molar mass ranging from 10 to 14 kDa. This mixture of hydrolysis products self-assembles into helical structures via a nucleation and growth mechanism. The minimum protein concentration to form nano-tubes of α-lactalbumin is reported to be ~20 g/l (50 °C, 75 mM Tris buffer, pH 7.5, 2 mol Ca^{2+} per mol protein) (Graveland-Bikker and de Kruif, 2006). Below this protein concentration, fibrillar and/or random aggregates are obtained (Otte *et al.*, 2005). Cross-linking with either transglutaminase or glutaraldehyde increases the stability of the α-lact-albumin nanotubes (Graveland-Bikker and de Kruif, 2006).

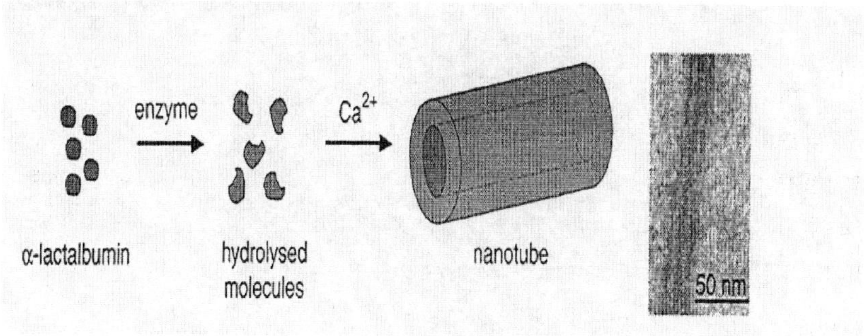

Figure 1.1 Schematic representation of self-assembly of partially hydrolysed α-lactalbumin into nanotubes in the presence of Ca^{2+}. The image at the extreme right shows an electron micrograph of the nanotubes (75 mM Tris buffer, pH = 7.5, 2 mole Ca^{2+} per mole α-lactalbumin; prepared by negative staining with 3% uranyl acetate for 1 min). Reproduced from Graveland-Bikker and de Kruif (2006) with permission.

Another prominent example of food biopolymer nanostructuring is the self-assembly of globular proteins — like β-lactoglobulin, bovine serum albumin, ovalbumin, soy and pea proteins — into nano-fibrils of up to a micrometre size in aqueous media (Figure 1.2). These nanoscale structures vary significantly in contour length, stiffness, aggregation mechanism, and reversibility/irreversibility of structure formation, depending on the thermodynamic conditions, which in turn are influenced by pH, temperature treatment, ionic strength, protein concentration and the presence of specific ions (*e.g.*, Ca^{2+}).

example 2

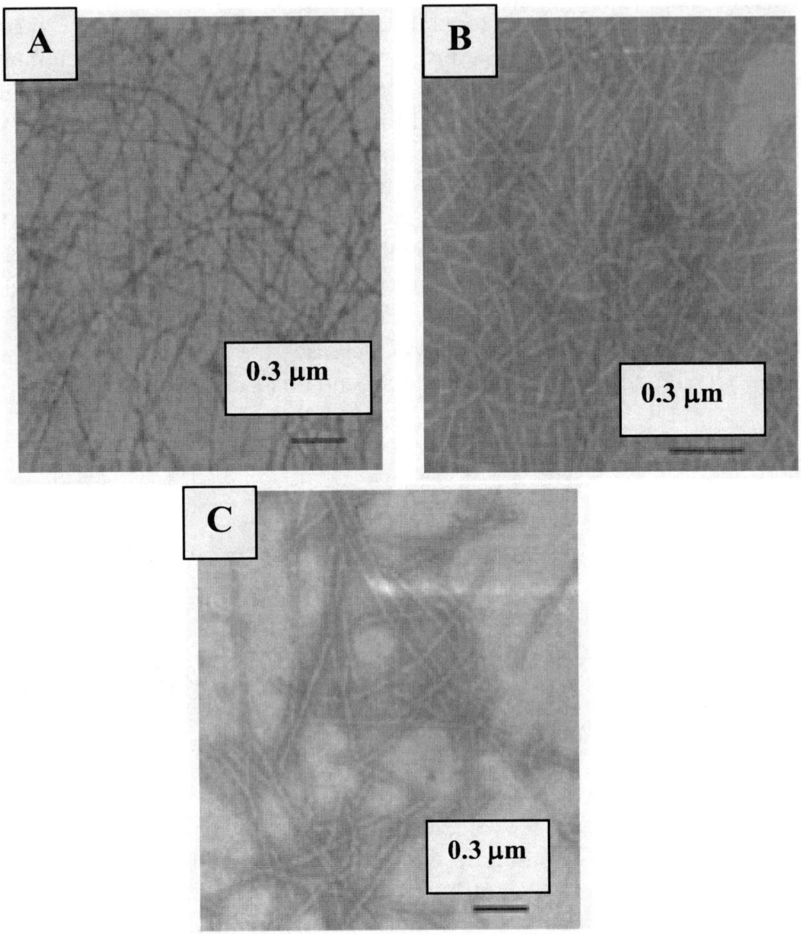

Figure 1.2 TEM micrographs of β-lactoglobulin at pH = 2: after heating at 80 °C for 10 h (image A); after pH was then adjusted to pH = 7 (image B) or pH = 8 (image C). Reproduced from Veerman *et al.* (2003) with permission.

Protein-based nano-fibrils have been found to be very efficient gelling agents, insofar as a protein gel can be formed even at very low protein concentrations (*e.g.*, 0.07 wt%). Applying an external shear flow can further increase the gel-forming ability of these protein nano-fibres due to their preferred orientation along the direction of the flow. The robustness of the fibres towards dilution and other physical treatments is considered a particular advantage for food processing applications (Veerman *et al.*, 2003; van der Linden, 2006).

Much research activity continues to be directed towards finding novel self-assembling biopolymers to form nanostructures for different applications. One such example is N-alkyl-N-dimethyl/N-alkyl-N-trimethyl chitosan,[2] which is capable of forming polymeric micelles in water with an average particle diameter ranging from 36 to 218 nm. Both the length of the alkyl group and the degree of N-trimethylation affect the size of the polymeric micelles (Zhang *et al.*, 2006). Additional research is required to investigate further the possible utilization of these chitosan-derived nanomaterials in controlled release and targeted delivery of bioactive compounds. In the medical field, it has already been shown that N-alkyl-O-sulfate chitosan is able to form self-assembling nanomicelles that can be used as a vehicle for solubilization and slow release of paclitaxel, a hydrophobic anticancer compound (Zhang *et al.*, 2004).

There is growing interest in research into various kinds of protein–polysaccharide complexes. This is because it is widely recognized that a greater insight into the multiscale structure of protein–polysaccharide complexes and the underlying mechanisms of soluble/insoluble complex formation is necessary for understanding properly how to design functional ingredients for the structuring of edible nanoparticles. These biopolymer complexes, which are themselves nanoparticles, can also provide protective multilayer films around solid nanoparticles or liquid nanodroplets. The potential applications of these systems includes the encapsulation of health beneficial nutrients, the protection of reactive compounds against chemical or biochemical attack, and the triggering of the targeted release of specific ingredients during the processes of digestion (de Kruif *et al.*, 2004; McClements, 2005, 2006; Dickinson, 2007, 2008; Turgeon *et al.*, 2007; Anal *et al.*, 2008). It is now well established that either soluble or insoluble electrostatic complexes can be formed, depending on the solution pH. The structural parameters of the complexes (size, net charge, shape, *etc.*) are determined by the specific biopolymer conformations, the protein-to-polysaccharide ratio, and the solution conditions (pH, ionic strength) (de Kruif *et al.*, 2004; Guzey and

[2] Chitosan is a cationic polymer of glucosamine, which is itself a deacetylated product of chitin (p*K* value between 6.3 and 7.0). It possesses a unique cationic nature relative to other neutral or negatively charged polysaccharides. Because of its net positive charge chitosan is able to form complexes with anionic proteins and other compounds. It has become extremely popular with biomedical researchers because is a cheap non-toxic natural polymer which is compatible with living tissue. Moreover, it posess antifungal and antimicrobial activities, since the protonated amino group (NH_3^+) in chitosan can bind to anionic sites on bacterial and fungal cell wall surfaces.

McClements, 2006; Anal *et al.*, 2008). Different combinations of proteins and polysaccharides (*e.g.*, β-lactoglobulin + pectin, carrageenan or alginate; casein + pectin) have been investigated within the context of multilayer emulsion stabilization (Guzey and McClements, 2006). It seems that the main technical challenge associated with the utilization such complex formation for 'layer-by-layer' emulsion stabilization is the avoidance of bridging flocculation (McClements, 2005, 2006).

The weak physical forces that hold together self-assembled nanoparticles are, of course, susceptible to disruption under the influence of thermodynamic and/or mechanical stresses. Hence some workers have investigated ways to reinforce nanoscale structures via covalent bonding. For instance, improved stability of protein nanoparticles, in particular, casein micelles, can be achieved by enzymatic cross-linking with the enzyme transglutaminase, which forms bonds between protein-bound glutamine and lysine residues. By this means native casein micelles can be converted from semi-reversible association colloids into permanent nanogel particles (Huppertz and de Kruif, 2008).

Looking realistically into the future, we can envisage nanotechnology and nanoscience providing the tools for developing far more precise and effective methods for the manipulation of food biopolymers and their supramolecular assemblies. These advances should provide tailor-made improvements in the quality and texture of many food products, with improvement in product shelf-life and freshness, along with maintenance of safety. In summary, these technological advances will contribute to the well-being of individual consumers in several ways (Bolder *et al.*, 2006; Graveland-Bikker *et al.*, 2006; Dickinson, 2004, 2006a; Morris, 2005, 2006; Weiss *et al.*, 2006):

(i) new food processing methods at nanoscale and microscale levels;
(ii) new functional materials and formulations, based on well-defined molecular components and their self-assembled nanostructures;
(iii) efficient controlled delivery and release systems;
(iv) intelligent contaminant detection; and
(v) novel and responsive packaging materials.

In the case of smart delivery systems, the release of the encapsulated substances from nanoscale structures would be triggered by exposure to the appropriate environmental conditions (pH, ionic strength, temperature) or by the application of an external stress (high pressure, electric field, ultrasound). In practice, one may reasonably envisage a range of specific applications being developed to control the bioaccessibility of individual antioxidants, vitamins or nutraceuticals at well-defined positions along the gastrointestinal route (Sanguansri and Augustin, 2006).

In the next chapter of this book we attempt to outline the potential of nanotechnology in the area of delivery of nutraceuticals from 'functional foods', the ultimate aim of this strategy being to reduce the risks of chronic diseases in the general population. One of the nanotechnological approaches is the nano- or micro-encapsulation of bioactive compounds within small vesicles of submicron diameter. These structures have the potential to protect bioactive compounds against degradation/oxidation and from undesirable interactions with other food components during food processing and storage. Nanocapsules can improve the stability and solubility of a bioactive substance and therefore increase its bioavailability and delivery to cells and tissues in a targeted fashion, as in the phrase "in the right place and at the right time" (Gouin 2004; Taylor *et al.*, 2005; Lalush *et al.*, 2005; Chen *et al.*, 2006; Sanguansri and Augustin, 2006; Bouwmeester *et al.*, 2007; Letchford and Burt, 2007; Semenova *et al.*, 2008).

BIBLIOGRAPHY

Abismail, B., Canselier, J.P., Wilhelm, A.M., Delmas, H., Gourdon, C. (1999). Emulsification by ultrasound: drop size distribution and stability. *Ultrasonics Sonochemistry*, **6**, 75–83.

Acosta, E. (2009). Bioavailability of nanoparticles in nutrient and nutraceutical delivery. *Current Opinion in Colloid and Interface Science*, **14**, 3–5.

Anal, A.K., Tobiassen, A., Flanagan, J., Singh, H. (2008). Preparation and characterization of nanoparticles formed by chitosan–caseinate interactions. *Colloids and Surfaces B: Biointerfaces*, **64**, 104–110.

Behrend, O., Ax, K., Schubert, H. (2000). Influence of continuous phase viscosity on emulsification of ultrasound. *Ultrasonics Sonochemistry*, **7**, 77–85.

Belyakova, L.E., Antipova, A.S., Semenova, M.G., Dickinson, E., Matia-Merino, L., Tsapkina, E.N. (2003). Effect of sucrose on molecular and interaction parameters of sodium caseinate in aqueous solution: relationship to protein gelation. *Colloids and Surfaces B: Biointerfaces*, **31**, 31–46.

Benichou, A., Aserin, A., Garti, N. (2002). Protein–polysaccharide interactions for stabilization of food emulsions. *Journal of Dispersion Science and Technology*, **23**, 93–123.

Berti, D. (2006). Self assembly of biologically inspired amphiphiles. *Current Opinion in Colloid and Interface Science*, **11**, 74–78.

Bolder, S.G., Hendrickx, H., Sagis, L.M.C., van der Linden, E. (2006). Ca^{2+}-induced cold-set gelation of whey protein isolate fibrils. *Applied Rheology*, **16,** 258– 264.

Bouwmeester, H., Dekkers, S., Noordam, M., Hagens, W., Bulder, A., de Heer, C., ten Voorde, S., Wijnhoven, S., Sips, A. (2007). Health impact of nanotechnologies in food production. Report 2007, a co-production of RIKILT and RIVM (Internet).

Chau, C.F., Wu, S.H., Yen, G.C. (2007). The development of regulations for food nanotechnology. *Trends in Food Science and Technology*, **18**, 269–280.

Chaudhry, Q., Scotter, M., Blackburn, J., Ross, B., Boxall, A., Castle, L., Aitken, R., Watkins, R. (2008). Applications and implications of nanotechnologies for the food sector. *Food Additives and Contaminants*, **25**, 241–258.

Chen, C.C., Wagner, G. (2004). Vitamin E nanoparticle for beverage applications. *Chemical Engineering Research and Design*, **82**, 1432–1437.

Chen, H., Weiss, J., Shahidi, F. (2006). Nanotechnology in nutraceuticals and functional foods. *Food Technology*, **60**(3), 30–36.

Cui, F., Shi, K., Zhang, L., Tao, A., Kawashima, Y. (2006). Biodegradable nanoparticles loaded with insulin–phospholipid complex for oral delivery: preparation, *in vitro* characterization and *in vivo* evaluation. *Journal of Controlled Release*, **114**, 242–250.

Cui, S., Zhao, C., Chen, D., He, Z. (2005). Self-microemulsifying drug delivery systems (SMEDDS) for improving *in vitro* dissolution and oral absorption of *Pueraria lobata* isoflavone. *Drug Development and Industrial Pharmacy*, **31**, 349–356.

de Kruif, C.G., Weinbreck, F., de Vries, R. (2004). Complex coacervation of proteins and anionic polysaccharides. *Current Opinion in Colloid and Interface Science*, **9**, 340–349.

Dickinson, E. (2003a). Food colloids… drifting into the age of nanoscience. Editorial overview. *Current Opinion in Colloid and Interface Science*, **8**, 346–348.

Dickinson, E. (2003b). Hydrocolloids at interfaces and the influence on the properties of dispersed systems. *Food Hydrocolloids*, **17**, 25–39.

Dickinson, E. (2004). Food colloids: the practical application of protein nanoscience in extreme environments. Editorial overview. *Current Opinion in Colloid and Interface Science*, **9**, 295–297.

Dickinson, E. (2006a). Structure formation in casein-based gels, foams, and emulsions. *Colloids and Surfaces A: Physicochemical* and *Engineering Aspects*, **288**, 3–11.

Dickinson, E. (2006b). Colloid science of mixed ingredients. *Soft Matter*, **2**, 642–652.

Dickinson, E. (2007). Food colloids... Editorial overview. How do interactions of ingredients control structure, stability and rheology? *Current Opinion in Colloid and Interface Science*, **12,** 155–157.

Dickinson, E. (2008). Interfacial structure and stability of food emulsions as affected by protein–polysaccharide interactions. *Soft Matter*, **4**, 932–942.

Dickinson, E., Semenova, M.G. (1992). Emulsifying behaviour of protein in the presence of polysaccharide under conditions of thermodynamic incompatibility. *Journal of the Chemical Society, Faraday Transactions*, **88**, 849–854.

Dickinson, E., Semenova, M.G., Antipova, A.S., Pelan, E. (1998). Effect of high-methoxy pectin on properties of casein-stabilized emulsions. *Food Hydrocolloids*, **12**, 425–432.

Dickinson, E., Semenova, M.G., Belyakova, L.E., Antipova, A.S., Il'in, M.M., Tsapkina, E.N., Ritzoulis, C. (2001). Analysis of light scattering data on the calcium ion sensitivity of caseinate solution thermodynamics: relationship to emulsion flocculation. *Journal of Colloid and Interface Science*, **239**, 87–97.

Dinsmore, A.D., Hsu, M.F., Nikolaides, M.G., Marquez, M., Bausch, A.R., Weitz, D.A. (2002). Colloidosomes: selectively permeable capsules composed of colloidal particles. *Science*, **298**, 1006–1009.

Eliot, C., Dickinson, E. (2003). Thermoreversible gelation of caseinate stabilized emulsions at around body temperature. *International Dairy Journal*, **13**, 679– 684.

Feijoo, S.C., Hayes, W.W., Watson, C.E., Martin, J.H. (1997). Effects of Microfluidizer® technology on *Bacillus licheniformis* spores in ice cream mix. *Journal of Dairy Science*, **80**, 2184–2187.

Förster, S., Konrad, M. (2003). From self-organising polymers to nano- and biomaterials. *Journal of Materials Chemistry*, **13**, 2671–2688.

Gouin, S. (2004). Microencapsulation: industrial appraisal of existing technologies and trends. *Trends in Food Science and Technology*, **15**, 330–347.

Graveland-Bikker, J.F., de Kruif, C.G. (2006). Unique milk protein-based nanotubes: food and nanotechnology meet. *Trends in Food Science and Technology*, **17**, 196–203.

Graveland-Bikker, J.F., Schaap, I.A.T., Schmidt, C.F., de Kruif, C.G. (2006). Structural and mechanical study of self-assembling protein nanotubes. *Nano Letters*, **6**, 616–621.

Guzey, D., McClements, D.J. (2006). Formation, stability and properties of multilayer emulsions for application in the food industry. *Advances in Colloid and Interface Science*, **128–130**, 227–248.

Hartgerink, J.D., Beniash, E., Stupp, S.I. (2001). Self-assembly and mineralization of peptide–amphiphile nanofibers. *Science*, **294**, 1684–1688.

Holt, C. (1992). Structure and stability of the bovine casein micelle. In Anfinsen, C.B., Edsall, J.D., Richards, F.R., Eisenberg, D.S. (Eds). *Advances in Protein Chemistry*, San Diego: Academic Press, vol. 43, pp. 63–151.

Horn, D., Rieger, J. (2001). Organic nanoparticles in the aqueous phase — theory, experiment, and use. *Angewandte Chemie, International Edition*, **40**, 4330–4361.

Horne, D.S. (1998). Casein interactions: casting light on the black boxes, the structure in dairy products. *International Dairy Journal*, **8**, 171–177.

Horne, D.S. (2002). Casein structure, self-assembly and gelation. *Current Opinion in Colloid and Interface Science*, **7**, 456–461.

Horne, D.S. (2003). Casein micelles as hard spheres: limitations of the model in acidified gel formation. *Colloids and Surfaces A: Physicochemical* and *Engineering Aspects*, **213**, 255–263.

Horne, D.S. (2006). Casein micelle structure: models and muddles. *Current Opinion in Colloid and Interface Science*, **11**, 148–153.

Huppertz, T., de Kruif, C.G. (2008). Structure and stability of nanogel particles prepared by internal cross-linking of casein micelles. *International Dairy Journal*, **18**, 556–565.

Israelachvili, J., Wennerström, H. (1996). Role of hydration and water structure in biological and colloidal interactions. *Nature*, **379**, 219–225.

Joscelyne, S.M., Trägårdh, G. (1999). Food emulsions using membrane emulsification: conditions for producing small droplets. *Journal of Food Engineering*, **39**, 59–64.

Joscelyne, S.M., Trägårdh, G. (2000). Membrane emulsification — a literature review. *Journal of Membrane Science*, **169**, 107–117.

Kampers, F. (2007). Micro- and nanotechnologies for food and nutrition in preventative healthcare. *Food Science and Technology*, **21**, 20–23.

Kobayashi, I., Nakajima, M. (2006). Generation and multiphase flow of emulsions in microchannels. In Kockmann, N. (Ed.). *Advanced Micro and Nanosystems 5: Micro Process Engineering*, Weinheim: Wiley, pp. 149–171.

Kosaraju, S.L., Tran, C., Lawrence, A. (2006). Liposomal delivery systems for encapsulation of ferrous sulfate: preparation and characterization. *Journal of Liposome Research*, **16**, 347–358.

Lalush, I., Bar, H., Zakaria, I., Eichler, S., Shimoni, E. (2005). Utilization of amylose–lipid complexes as molecular nanocapsules for conjugated linoleic acid. *Biomacromolecules*, **6**, 121–130.

Letchford, K., Burt, H. (2007). A review of the formation and classification of amphiphilic block copolymer nanoparticulate structures: micelles, nanospheres, nanocapsules and polymersomes. *European Journal of Pharmaceutics and Biopharmaceutics*, **65**, 259–269.

Loscertales, I.G., Barrero, A., Guerrero, I., Cortijo, R., Marquez, M., Ganan-Calvo, A.M. (2002). Micro/nano encapsulation via electrified coaxial liquid jets. *Science*, **295**, 1695–1698.

Ma, Q., Xia, Q., Lu, Y., Hao, X., Gu, N., Lin, X., Luo, D. (2007). Preparation of tea polyphenol-loaded solid lipid nanoparticles based on the phase behaviour of hot microemulsions. *Diffusion and Defect Data B: Solid State Phenomena*, **121–123**, 705–708.

Manski, J.M., van der Goot, A.J., Boom, R.M. (2007). Advances in structure formation of anisotropic protein-rich foods through novel processing concepts. *Trends in Food Science and Technology*, **18**, 546–557.

√McClements, D.J. (2005). Theoretical analysis of factors affecting the formation and stability of multilayered colloidal dispersions. *Langmuir*, **21**, 9777–9785.

√McClements, D.J. (2006). Non-covalent interactions between proteins and polysaccharides. *Biotechnology Advances*, **24**, 621–625.

Menger, F.M. (2002). Supramolecular chemistry and self-assembly. *Proceedings of the National Academy of Sciences USA*, **99**, 4818–4822.

Min, Y., Akbulut, M., Kristiansen, K., Golan, Y., Israelachvili, J. (2008). The role of interparticle and external forces in nanoparticle assembly. *Nature Materials*, **7**, 527–538.

Morris, V.J. (2005). Is nanotechnology going to change the future of food technology? *International Review of Food Science and Technology*, **3**, 16–18.

Morris, V.J. (2006). Nanotechnology and its future in new product development. *Journal of the Institute of Food Science and Technology*, **20**, 15–17.

Mozafari, M.R, Flanagan, J., Matia-Merino, L., Awati, A., Omri, A., Suntres, Z.E., Singh, H. (2006). Recent trends in the lipid-based nanoencapsulation of antioxidants and their role in foods. *Journal of the Science of Food and Agriculture*, **86**, 2038–2045.

Müller, R.H., Mäder, K., Gohla, S. (2000). Solid lipid nanoparticles (SLN) for controlled drug delivery — a review of the state of the art. *European Journal of Pharmaceutics and Biopharmaceutics*, **50**, 161–177.

Murray, B.S., Ettelaie, R. (2004). Foam stability: proteins and nanoparticles. *Current Opinion in Colloid and Interface Science*, **9**, 314–320.

Niece, K.L., Hartgerink, J.D., Donners, J.J.J.M., Stupp, S.I. (2003). Self-assembly combining two bioactive peptide-amphiphile molecules into nanofibers by electrostatic attraction. *Journal of the American Chemical Society*, **125**, 7146– 7147.

Niemeyer, C.M. (2002). Self-assembled nanostructures based on DNA: towards the development of nanobiotechnology. *Current Opinion in Chemical Biology*, **4**, 609–618.

Norton, I.T, Frith, W.J. (2001). Microstructure design in mixed biopolymer composites. *Food Hydrocolloids*, **15**, 543–553.

Oberdorster, G., Maynard, A., Donaldson, K., Castranova, V., Fitzpatrick, J., Ausman, K., Carter, J., Karn, B., Kreyling, W., Lai, D., Olin. S., Monteiro-Riviere, N., Warheit, D., Yang, H. (2005). Principles for characterizing the potential human health effects from exposure to nanomaterials: elements of a screening strategy. *Particle and Fibre Toxicology 2*, article no. 8.

Otte, J., Ipsen, R., Bauer, R., Bjerrum, M.J., Waninge, R. (2005). Formation of amyloid-like fibrils upon limited proteolysis of bovine α-lactalbumin. *International Dairy Journal*, **15**, 219–229.

Powers, K.W., Brown, S.C., Krishna, V.B., Wasdo, S.C., Moudgil, B.M., Roberts, S.M. (2006). Research strategies for safety evaluation of nanomaterials. Part VI. Characterization of nanoscale particles for toxicological evaluation. *Toxicological Sciences*, **90**, 296–303.

Pouton, C.W. (2000). Lipid formulations for oral administration of drugs: non-emulsifying, self-emulsifying, and self-microemulsifying drug delivery systems. *European Journal of Pharmaceutical Sciences*, **11**, S93–S98.

Ransley, J.K., Donnelly, J.K., Read, N.W. (Eds). (2001). *Food and Nutritional Supplements. Their Role in Health and Disease*. Berlin/Heidelberg: Springer-Verlag.

Ribeiro, H.S., Chu, B.S., Ichikawa, S., Nakajima, M. (2008). Preparation of nanodispersions containing β-carotene by solvent displacement method. *Food Hydrocolloids*, **22**, 12–17.

Sagalowicz, L., Leser, M.E., Watzke, H.J., Michel, M. (2006). Monoglyceride self-assembly structures as delivery vehicles. *Trends in Food Science and Technology*, **17**, 204–214.

Sanguansri, P., Augustin, M.A. (2006). Nanoscale materials development — a food industry perspective. *Trends in Food Science and Technology*, **17**, 547–556.

Semenova, M.G. (2007). Thermodynamic analysis of the impact of molecular interactions on the functionality of food biopolymers in solution and in colloidal systems. *Food Hydrocolloids*, **21**, 23–45.

Semenova, M.G., Antipova, A.S., Belyakova, L.E., Dickinson, E., Brown, R., Pelan, E., Norton, I. (1999). Effect of pectinate on properties of oil-in-water emulsions stabilized by α_{s1}-casein and β-casein. In Dickinson, E., Rodriguez Patino, J.M. (Eds). *Food Emulsions and Foams: Interfaces, Interactions and Stability*, Cambridge, UK: Royal Society of Chemistry, pp. 163–175.

Semenova, M.G., Belyakova, L.E., Dickinson, E., Eliot-Laizé, C., Polikarpov, Yu.N. (2005). Caseinate interactions in solution and in emulsions: effect of temperature, pH and calcium ions. In Dickinson, E. (Ed.). *Food Colloids: Interactions, Microstructure and Processing*, Cambridge, UK: Royal Society of Chemistry, pp. 209–217.

Semenova M.G., Belyakova, L.E., Polikarpov, Yu.N., Antipova, A.S., Anokhina, M.S. (2008). Utilization of sodium caseinate nanoparticles as molecular nanocontainers for delivery of bioactive lipids to food systems: relationship to the retention and controlled release of phospholipids in the simulated digestion conditions. In Williams, P.A., Phillips, G.O. (Eds). *Gums and Stabilisers for the Food Industry 14*, Cambridge, UK: Royal Society of Chemistry, pp. 326–333.

Shukla, T.P., Halpern, G.J. (2005). Emulsified liquid shortening compositions comprising dietary fibre gel, water and lipid. US Patent No. 2005/0064068A1.

Singh, A., Markowitz, M., Chow, G.M. (1995). Materials fabrication via polymerisable self-organized membranes: an overview. *Nanostructured Materials*, **5**, 141–153.

Sletmoen, M., Maurstad, G., Stokke, B.T. (2008). Potentials of bionanotechnology in the study and manufacturing of self-assembled biopolymer complexes and gels. *Food Hydrocolloids*, **22**, 2–11.

Tan, C.P., Nakajima, M. (2005). β-Carotene nanodispersions: preparation, characterization and stability evaluation. *Food Chemistry*, **92**, 661–671.

Taylor, T.M, Davidson, P.M., Bruce, B.D., Weiss, J. (2005). Liposomal nanocapsules in food science and agriculture. *Critical Reviews in Food Science and Nutrition*, **45**, 587–605.

Tiede, K., Boxall, A.B.A., Tear, S.P., Lewis, J., David, H., Hassellov, M. (2008). Detection and characterization of engineered nanoparticles in food and the environment. *Food Additives and Contaminants*, **25**, 795–821.

Tolstoguzov, V. (2002). Thermodynamic aspects of biopolymer functionality in biological systems, foods, and beverages. *Critical Reviews in Biotechnology*, **22**, 89–174.

Tsonchev, S., Schatz, G.C., Ratner, M.A. (2004). Electrostatically-directed self-assembly of cylindrical peptide amphiphile nanostructures. *Journal of Physical Chemistry B*, **108**, 8817–8822.

Turgeon, S.L., Schmitt, C., Sanchez, C. (2007). Protein–polysaccharide complexes and coacervates. *Current Opinion in Colloid and Interface Science*, **12**, 166–178.

Üner M. (2006). Preparation, characterization and physico-chemical properties of solid lipid nanoparticles (SLN) and nanostructured lipid carriers (NLC): their benefits as colloidal drug carrier systems. *Pharmazie*, **61**, 375–386.

van der Linden, E. (2006). Innovations with protein nano-fibres. *World's Poultry Science Journal*, **62**, 439–442.

Veerman, C., Sagis, L.M.C., van der Linden, E. (2003). Gels at extremely low weight fractions formed by irreversible self-assembly of proteins. *Macromolecular Bioscience*, **3**, 243–247.

Vuillemard, J.C. (1991). Recent advances in the large-scale production of lipid vesicles for use in food products: microfluidization. *Journal of Microencapsulation*, **8**, 547–562.

Walsh, S., Balbus, J.M., Denison, R., Florini, K. (2008). Nanotechnology: getting it right the first time. *Journal of Cleaner Production*, **16**, 1018–1020.
Weiss, J., Takhistov, P., McClements, D.J. (2006). Functional materials in food nanotechnology. *Journal of Food Science*, **71**, 107–115.

Zhang, C., Ping, Q.E., Zhang, H. (2004). Self-assembly and characterization of paclitaxel-loaded N-octyl-O-sulfate chitosan micellar system. *Colloids and Surfaces B: Biointerfaces*, **39**, 69–75.
Zhang, C., Ding, Y., Ping, Q.E., Yu, L.L. (2006). Novel chitosan-derived nanomaterials and their micelle-forming properties. *Journal of Agricultural and Food Chemistry*, **54**, 8409–8416.

CHAPTER TWO

APPLICATIONS OF BIOPOLYMERS FOR MICRO- AND NANO-ENCAPSULATION OF BIOACTIVE FOOD INGREDIENTS

1. *Bioactive Food Ingredients*

Bioactive food ingredients, or so-called nutraceuticals, are a new class of food supplements with biological added value, containing prophylactic and therapeutic agents for the improvement of human health. The active agents are antioxidants (enzymatic or non-enzymatic), different kinds of vitamins, essential fatty acids, phospholipids, minerals, phytochemicals, *etc.*, which are considered to contribute towards prevention of the development of various diseases like cancer, diabetes, cardiovascular disease, circulatory disorders, skin conditions, immune dysfunction, autoimmune diseases and neurodegenerative disorders. It has been commonly stated that these bioactive compounds offer major advantages over existing drugs because they exhibit multifunctional benefits for well-being, in contrast to the use of drugs for specific diseases, which often have harmful side effects (Omenn *et al.*, 1996; Edge *et al.*, 1997; Gibbs *et al.*, 1999; Albanes, 1999; Ransley *et al.*, 2001; Erhardt *et al.*, 2003; Ficarra *et al.*, 2004; Ruxton *et al.*, 2004, 2005; Shahidi and Miraliakbari, 2005; Ratnam *et al.*, 2006). It has also been suggested that more than two-thirds of human cancers arising from mutations in multiple genes could be prevented by a change in lifestyle, including dietary modification. In addition, it is recognized that the human antioxidant defence system is incomplete without dietary antioxidants (Ransley *et al.*, 2001; Ficarra *et al.*, 2004; Tur *et al.*, 2005; Kaur *et al.*, 2007).

Let us briefly mention here some of the reported benefits of specific neutraceuticals. Firstly, it has been demonstrated that appreciable protection against UV-induced epidermal damage can be obtained as a result of long-term oral administration (a period of 3 months) of a combination of the antioxidants ascorbic acid (vitamin C) and D-α-tocopherol (vitamin E) in human volunteers (Placzek *et al.*, 2005). Several potential health benefits of the carotenoids have been reported: for example, lutein and zeaxanthin, which are found at high concentrations in the human eye, have been postulated to be beneficial in relation to age-related macular degeneration and cataracts (Stringham and Hammond, 2005), and lycopene, which is found at high concentrations in tomatoes, is considered to reduce the risk of prostate cancer (Basu and Imrhan, 2007).

In the case of essential fatty acids, a recent study surveying the relationship between long chain ω-3 fatty acids and disease in 38 different countries has found an inverse relationship between long chain ω-3 availability and cardiovascular disease mortality in both men and women (Hibbeln *et al.*, 2006). Dietary long chain ω-3 fatty acids are also beneficial in relation to inflammatory diseases. For example, they can apparently benefit patients with morning stiffness, joint pain, grip strength, and can reduce the need for anti-inflammatory drugs in patients suffering from rheumatoid arthritis (Cleland *et al.*, 1988; Kremer *et al.*, 1985; Geusens *et al.*, 1994). In addition, intake of phytosterols has been reported to reduce low density lipoprotein (LDL) in humans by inhibiting the absorption of dietary cholesterol (Wong, 2001; Ostlund, 2004), *i.e.*, an intake of 1.6 g phytosterols per day results in ~10 % reduction in LDL cholesterol (Hallikainen *et al.*, 2000). Also it appears that the intestinal absorption of phytosterols is very low: so dietary phytosterols do not have adverse effects on health. Also demonstrated to exhibit specific biological activities are certain food proteins (including soy, dairy, fish, and meat proteins), peptides arising from protein digestion (*e.g.*, casein phosphopeptides, calcium binding peptides, *etc.*), various amino acids (tryptophan, tyrosine, leucine, isoleucine, valine), and polysaccharides in the form of dietary fibre (including non-digestible carbohydrates and lignin) (Playne *et al.*, 2003; Fernstrom, 2005; Redgwell and Fischer, 2005; McClements *et al.*, 2009).

At the same time, it is necessary to realize, as Faulks and Southon (2008) have noted recently, that an increased intake of nutraceuticals via foods may not simply confer specific health benefits. In addition, there may be a more general effect of stimulating changes in homeostasis, which may cumulatively confer health benefits. Furthermore, some other factors, such as individual human variation, gender, age and lifestyle, must also be considered, both in relation to personalized nutrition and in order to identify requirements of specific subgroups for 'healthy eating'.

Table 2.1 presents examples of the main classes of low-molecular-weight bioactive molecules which are currently considered to be helpful for human well-being and which can be therefore used as food supplements as well as active components in skin-care applications (Ratnam *et al.*, 2006; McClements *et al.*, 2009). The required physicochemical properties of effective bioactive compounds, which should be considered in the formulation of the prophylactic and therapeutic dietary supplements at their desired oral dosages, are described in the scientific literature. These properties are: (i) solubility in aqueous media; (ii) permeability through the gastrointestinal tract and cell membranes; (iii) physical stability; and (iv) bioavailability.

Table 2.1 Bioactive molecules considered beneficial for human health and therefore which are useful as potential food ingredients. (Bioactive molecules labelled * are not produced in the human body, and so need to be supplied in a healthy human diet.)

Bioactive molecule	Structural formula
(I) Non-enzymatic antioxidants	
1. Vitamins	
vitamin A – retinol (oil-soluble)	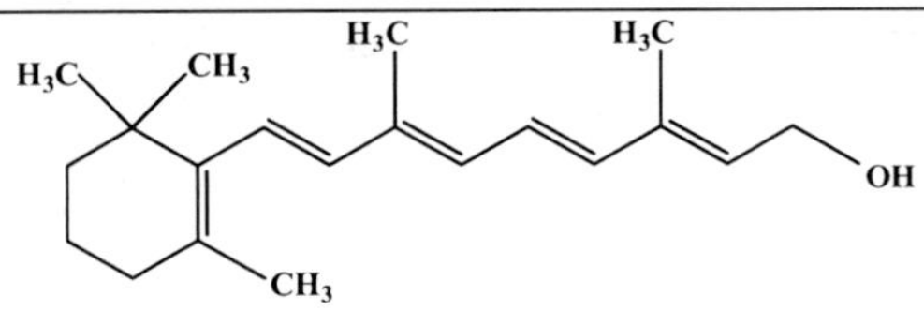
vitamin C – ascorbic acid (water-soluble)*	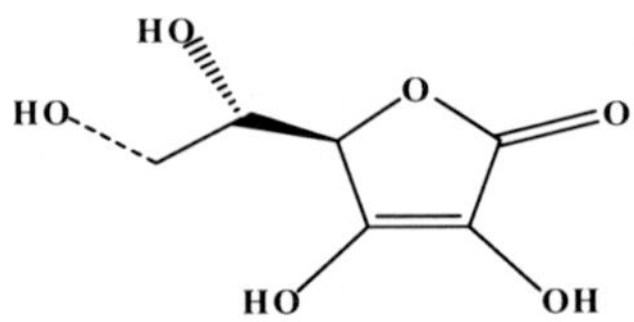
vitamin E – α-tocopherol, (oil-soluble)*	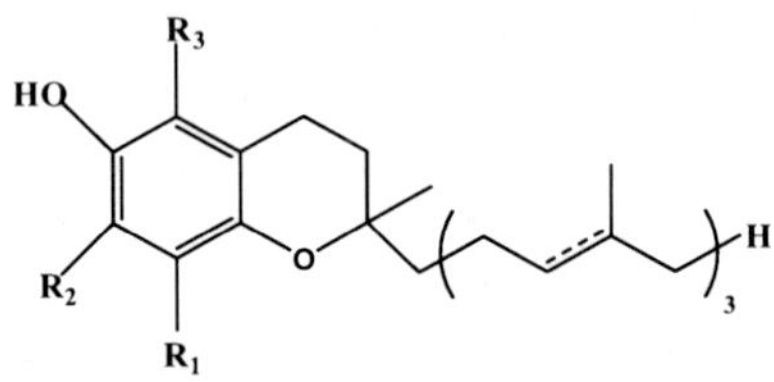 **α-tocopherol, $R_1 = R_2 = R_3 = CH_3$** **α-tocotrienol, $R_1 = R_2 = R_3 = CH_3$** **γ-tocopherol, $R_1 = R_2 = CH_3$; $R_3 = H$** **γ-tocotrienol, $R_1 = R_2 = CH_3$; $R_3 = H$** **β-tocopherol, $R_1 = R_3 = CH_3$; $R_2 = H$** **β-tocotrienol, $R_1 = R_3 = CH_3$; $R_2 = H$** **δ-tocopherol, $R_1 = R_2 = R_3 = H$** **δ-tocotrienol, $R_1 = R_2 = R_3 = H$**
vitamin K – phylloquinone (oil-soluble)	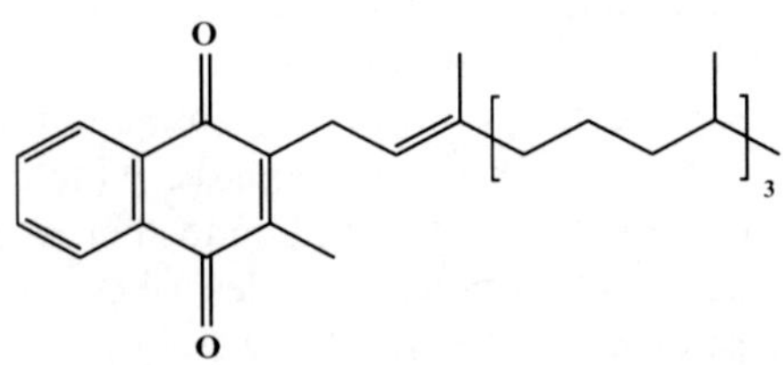

Table 2.1 *Continued*

Bioactive molecule	Structural formula
	(I) Non-enzymatic antioxidants
	2. Minerals*
	Zn, Se Organic and inorganic salts of these minerals
	3. Carotenoids*
β-carotene	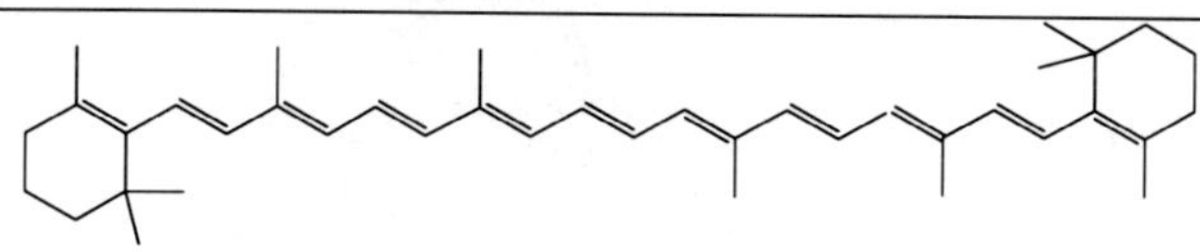
lycopene	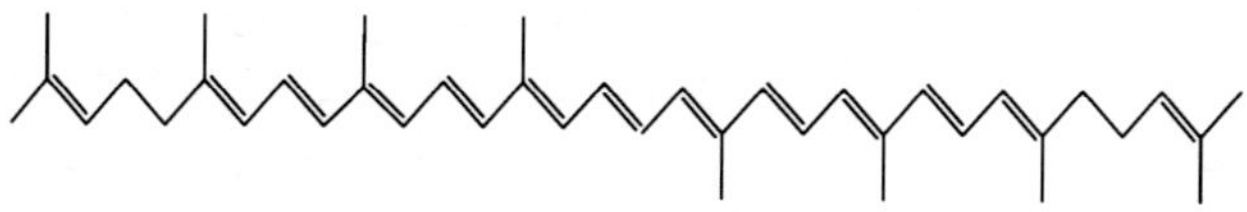
lutein	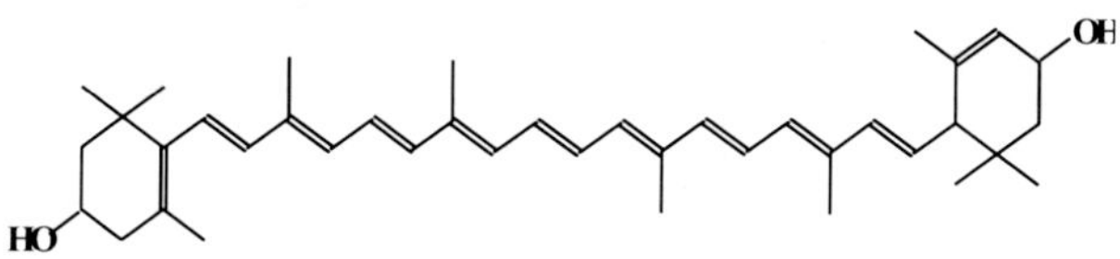
zeaxanthin	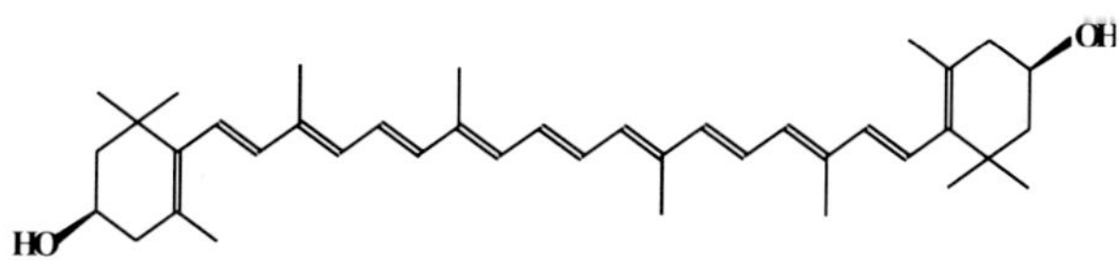
astaxanthin	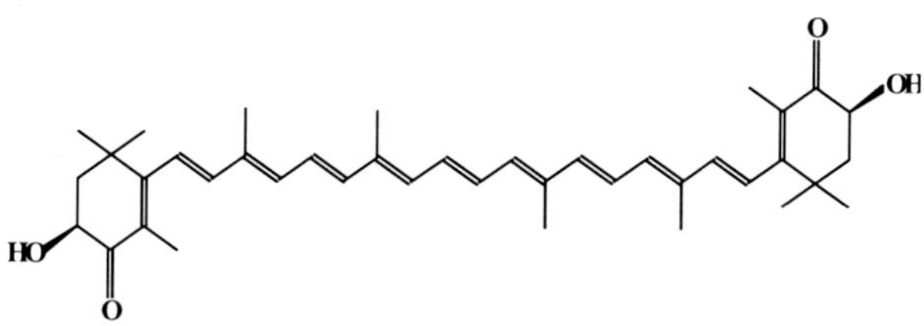

Table 2.1 *Continued*

Bioactive molecule	Structural formula
	(I) Non-enzymatic antioxidants
	4. Organosulfur compounds*
allyl methyl sulphide	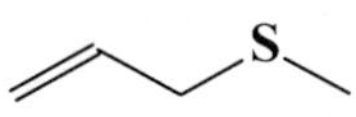
indoles	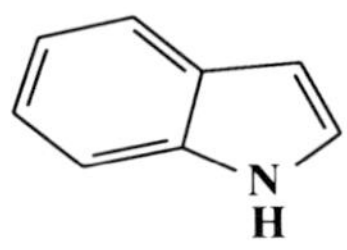
	5. Low-molecular-weight antioxidants
glutathione	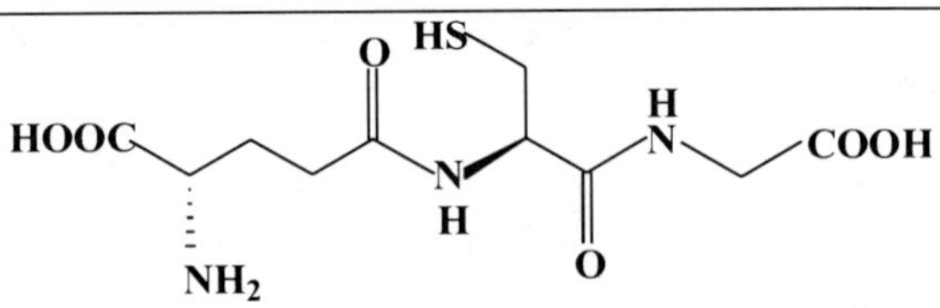
uric acid	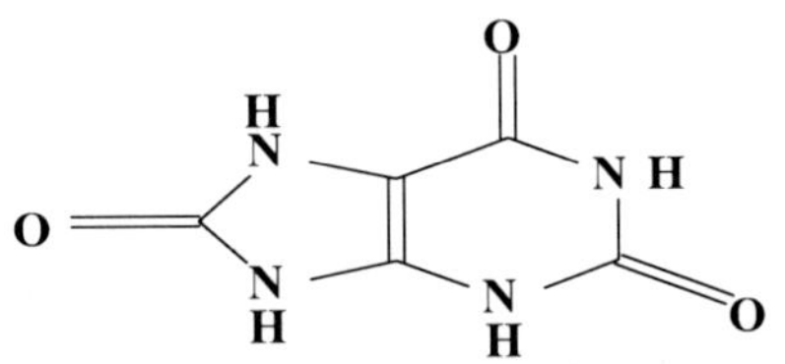
	6. Antioxidant cofactor*
coenzyme Q_{10} (ubiquinone)	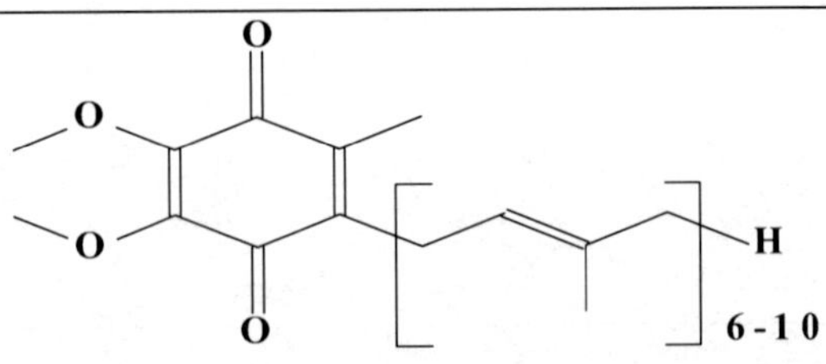

Table 2.1 *Continued*

Bioactive molecule	Structural formula
(I) Non-enzymatic antioxidants	
7. Polyphenols*	
A. Flavonoids	
Flavonols:	
quercetin	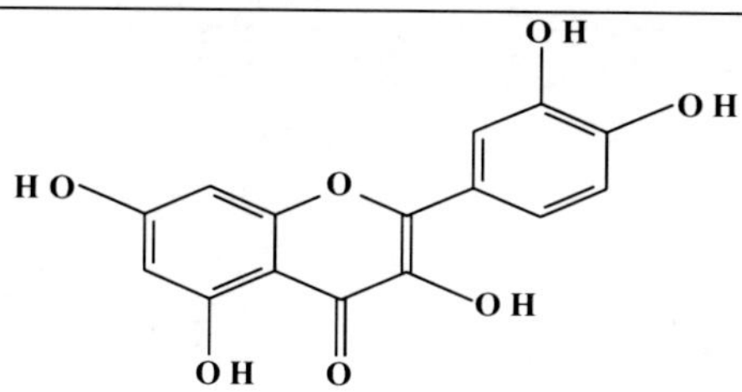
kaempferol	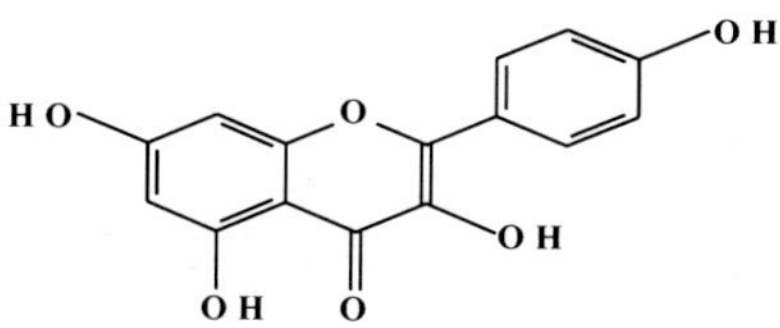
Flavanols:	
epicatechin	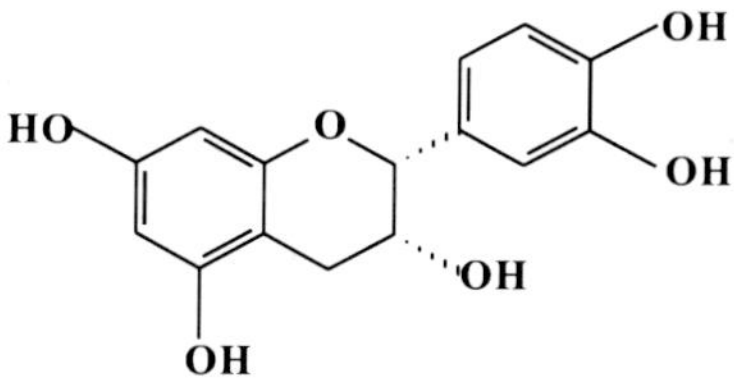
epigallocatechin gallate	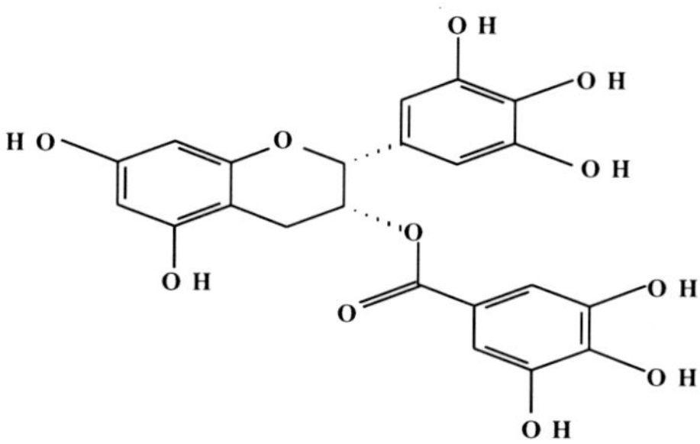

Table 2.1 *Continued*

Bioactive molecule	Structural formula
(I) Non-enzymatic antioxidants	
7. Polyphenols*	
A. Flavonoids	
Flavanones: hesperidin	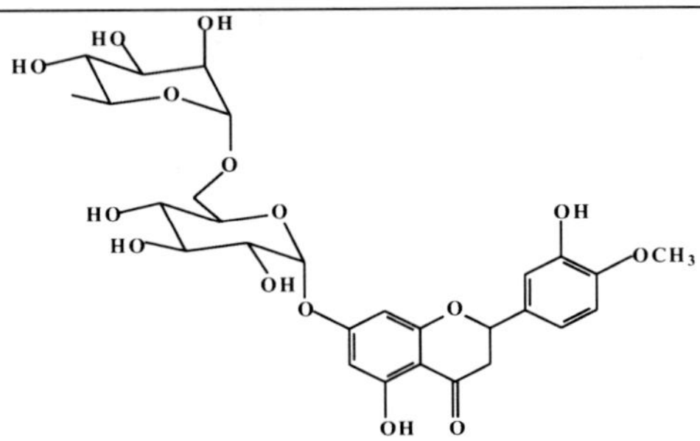
Isoflavanoids: genistein	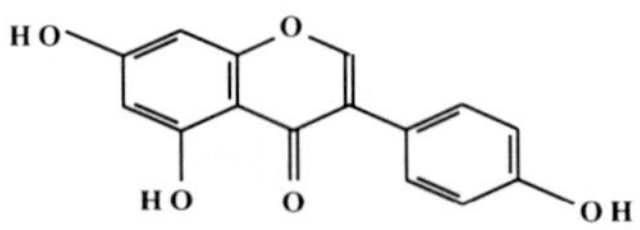
Anthocyanidins: cyanidin	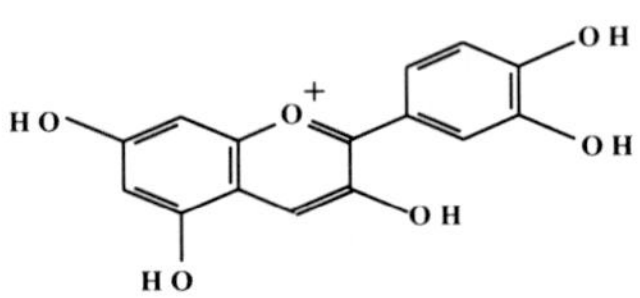
pelargonidin	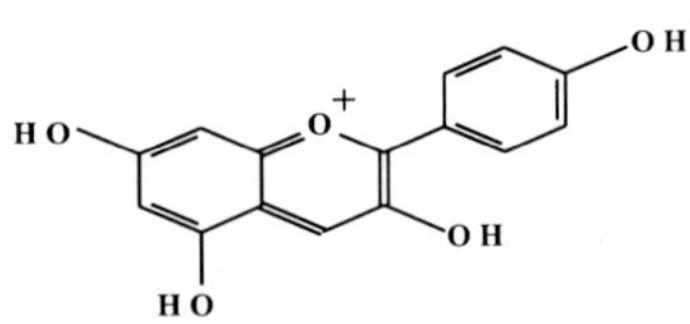
Flavones: chrysin	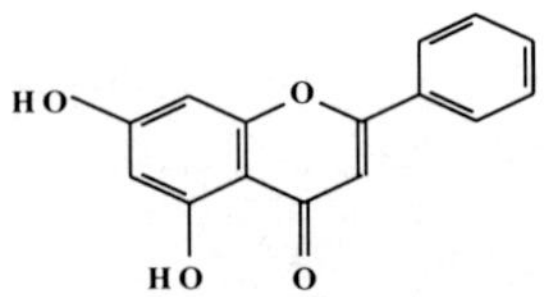

Table 2.1 *Continued*

Bioactive molecule	Structural formula
(I) Non-enzymatic antioxidants	
7. Polyphenols*	
B. Phenolic acids	
Hydroxycinnamic acids:	
ferulic acid	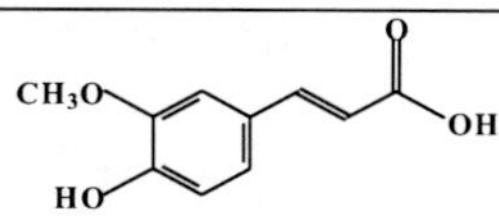
p-coumaric acid	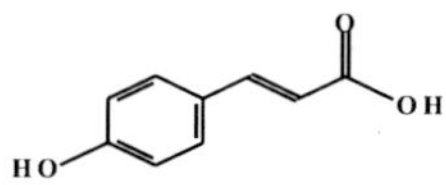
Hydroxybenzoic acids:	
gallic acid	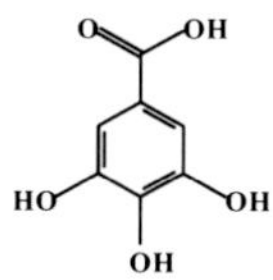
ellagic acid	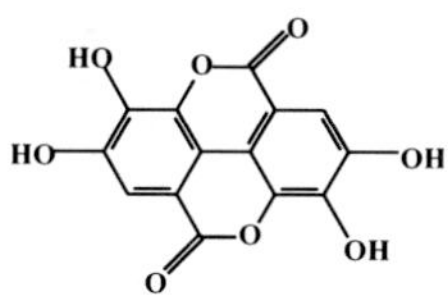
(II) Minerals*	
Ca, Mg, K, Fe, Mn, Cr, Cu	
(III) Essential fatty acids*	
ω-3 Fatty acids:	
α-linolenic acid (18:3)	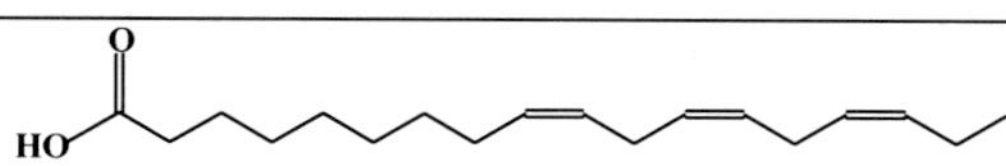
eicosapentaenoic acid (20:5)	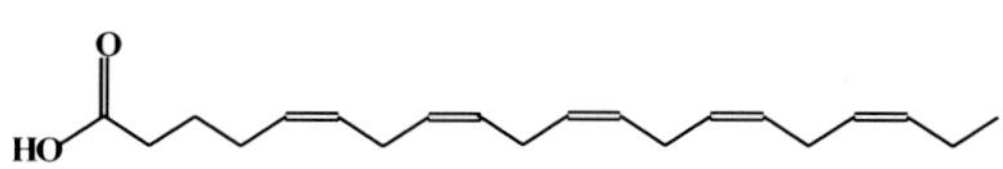
docosahexaenoic acid (22:6)	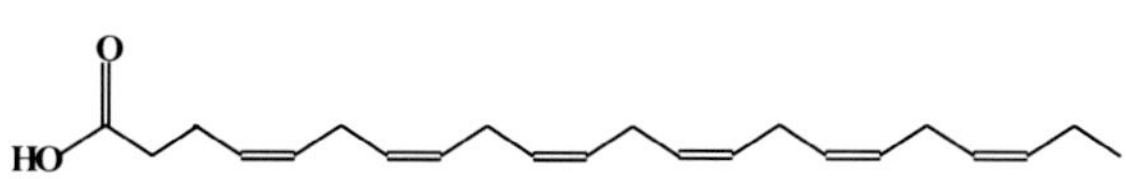

Table 2.1 *Continued*

Bioactive molecule	Structural formula
	(III) Essential fatty acids*
ω-6 Fatty acids: linoleic acid (18:2)	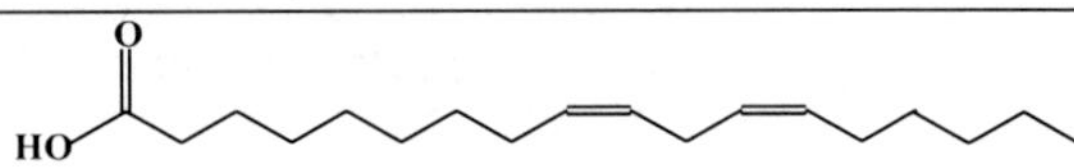
γ-linolenic acid (18:3)	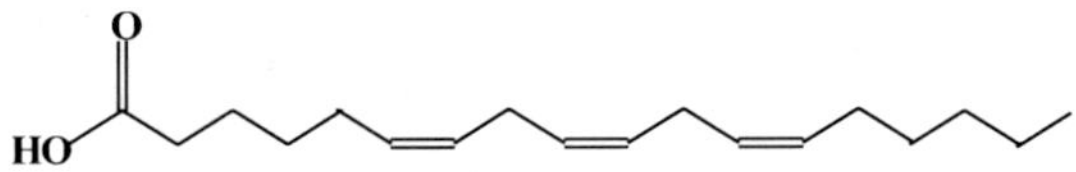
arachidonic acid (20:4)	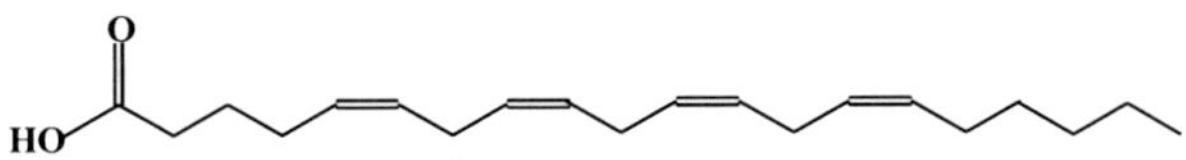
	(IV) Phospholipids
phosphatidylcholine (lecithin)	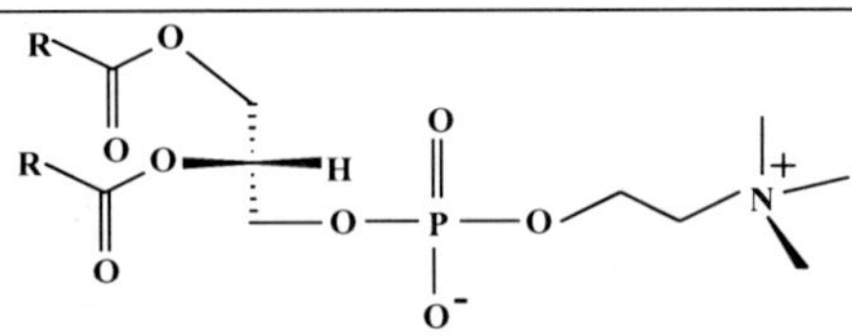
phosphatidyl-ethanolamine	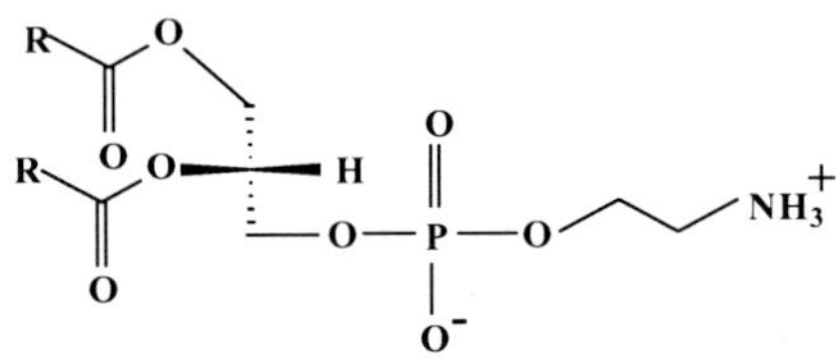
phosphatidyl-inositol	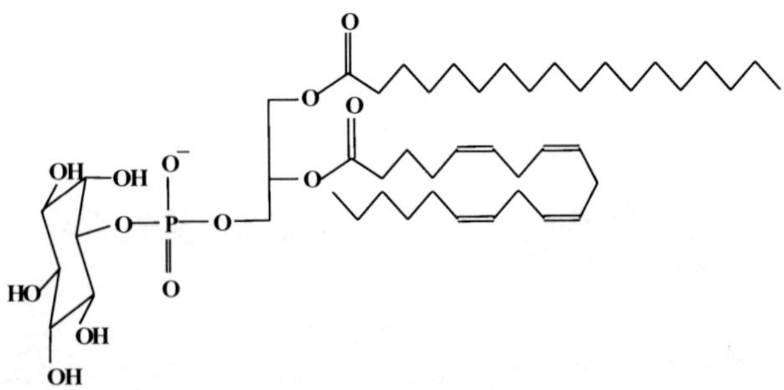

Table 2.1 *Continued*

Bioactive molecule	Structural formula
(IV) Phospholipids	
phosphatidylserine	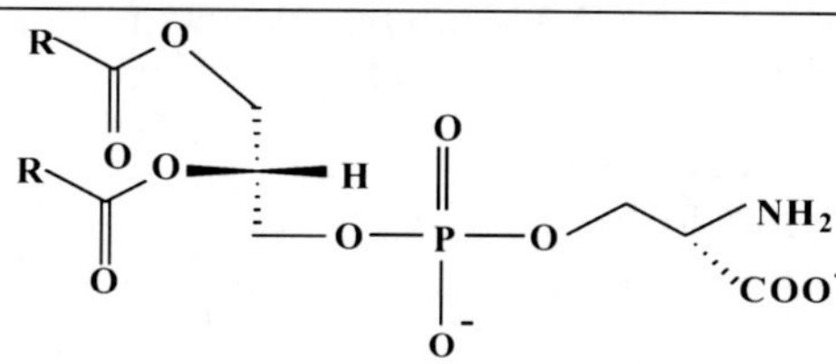
(V) Vitamins and vitamin-like substances	
vitamin B_1 – thiamine (water-soluble)	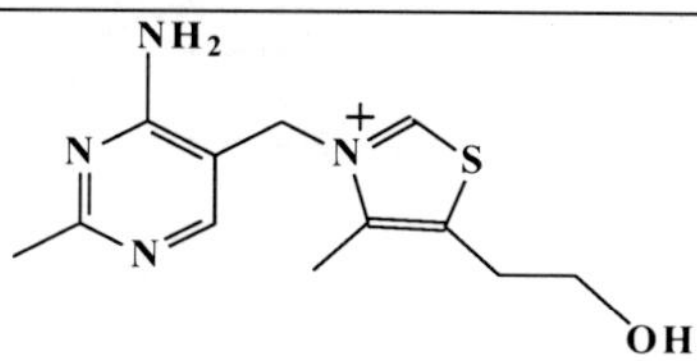
vitamin B_2 – riboflavin (water-soluble)	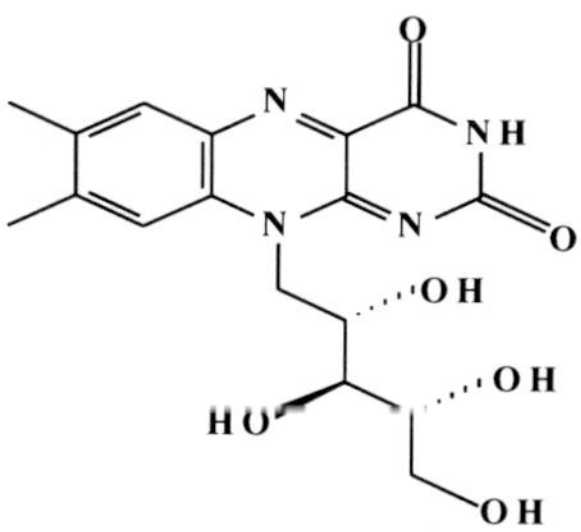
vitamin B_3 – niacin (water-soluble)	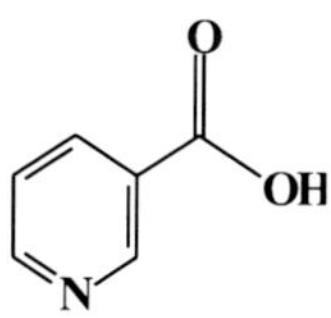
vitamin B_5 – pantothenic acid (water-soluble)	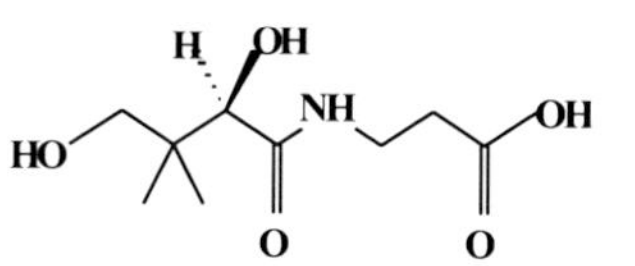

Table 2.1 *Continued*

Bioactive molecule	Structural formula
(V) Vitamins and vitamin-like substances	
vitamin B_6 – pyridoxine (water-soluble)	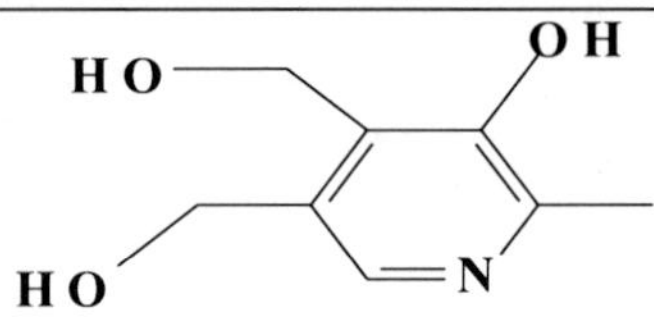
vitamin B_7 – biotin (water-soluble)	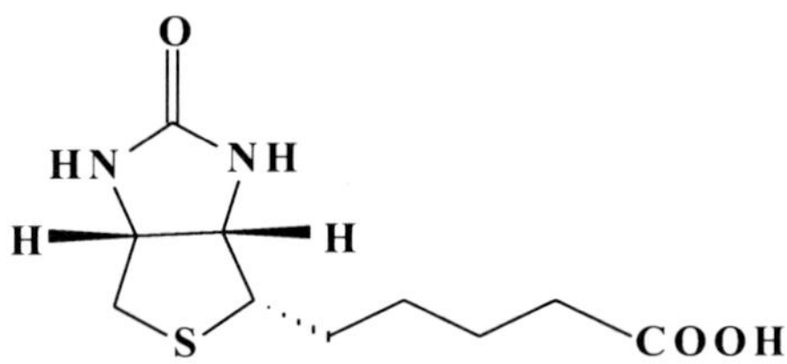
vitamin B_9 – folic acid (water soluble)	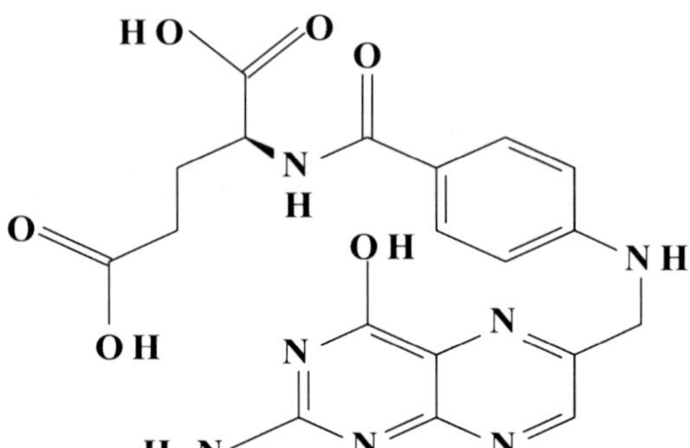
lipoic acid	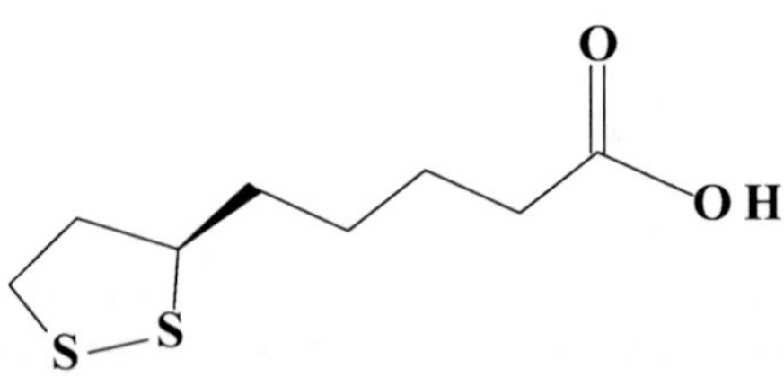

Table 2.1 *Continued*

Bioactive molecule	Structural formula
	(V) Vitamins and vitamin-like substances
vitamin B_{12} – cyanocobalamin (water-soluble)	**R=(CN,OH,CH_3, Deoxyadenosyl)** 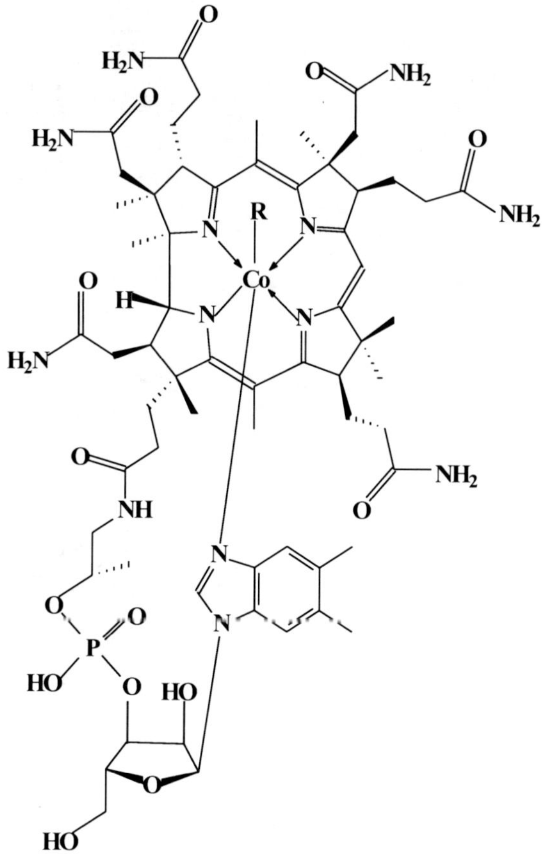

Table 2.1 *Continued*

Bioactive molecule	Structural formula
(V) Vitamins and vitamin-like substances	
vitamin D_2–ergocalciferol (oil-soluble)	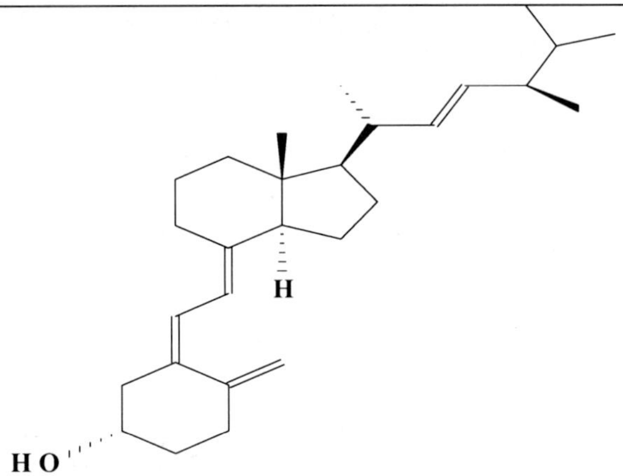
cholecalciferol (oil-soluble)	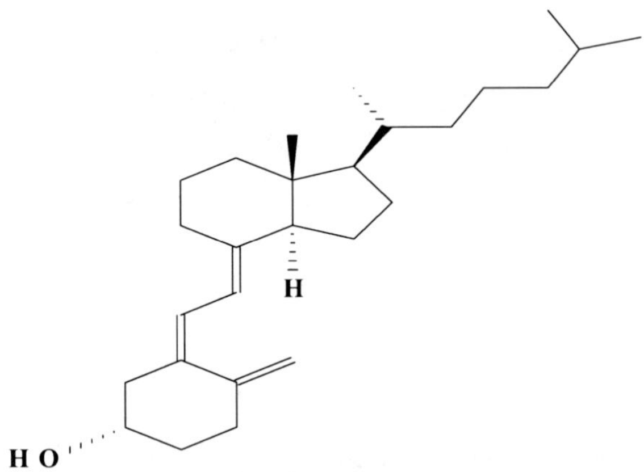
(VI) Terpenoids	
isopentenyl pyrophosphate	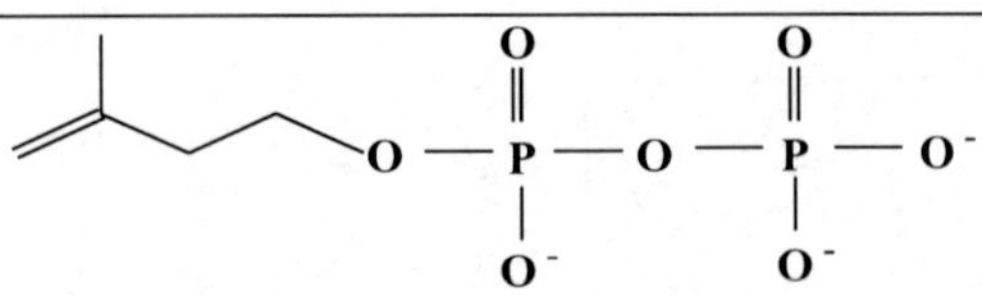

Table 2.1 *Continued*

Bioactive molecule	Structural formula
	(VII) Phytosterols
β-sitosterol	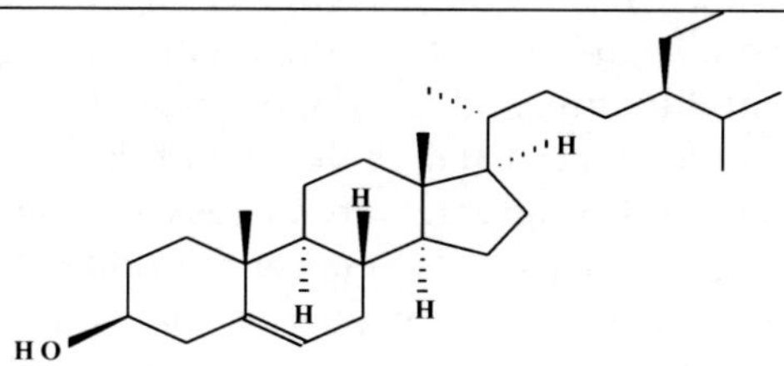
egosterol	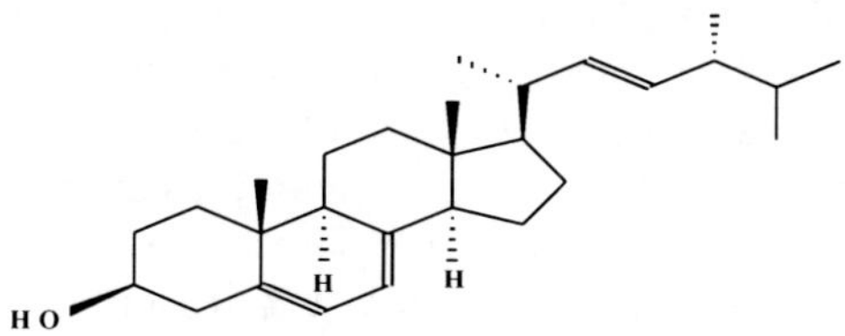
	(VIII) Alkaloids
caffeine	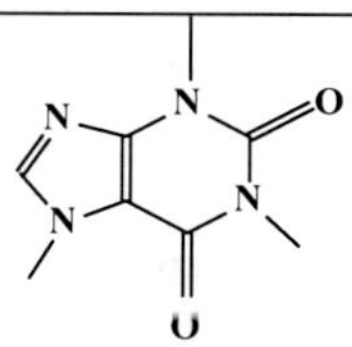
theobromin	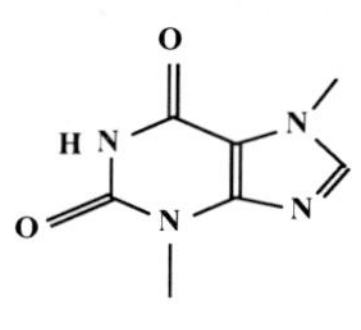
theophyllin	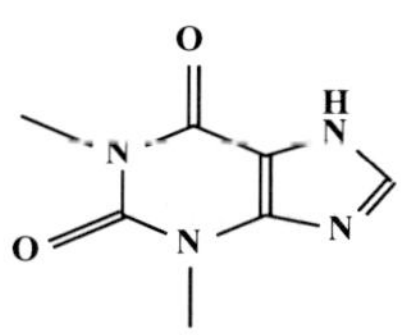

Let us try to clarify the meaning and significance of this concept of 'bioavailability'. Formally, the term bioavailability is defined as the rate and extent to which the active ingredient is (i) liberated from the food matrix, (ii) absorbed by the body, and (iii) made available at the site of action (Ratnam *et al.*, 2006; Acosta, 2009). However, on the basis of various experimental studies, it has been suggested that, if a bioactive food component spends sufficient time within the gastrointestinal tract, then the rate of digestion and absorption might be less important than its total extent. On the other hand, if the rate of digestion or absorption can be slowed down sufficiently, this may cause a reduction in the bioavailability of the component (McClements *et al.*, 2008).

A detailed insight into the biopharmaceutical properties of the list of food supplements in Table 2.1 is beyond the scope of this chapter (and the authors of this book). What we can simply say here is that, according to biopharmaceutical classification, it seems that the two key parameters affecting the bioavailability of most of these bioactive compounds are (a) the solubility in the aqueous medium and (b) the membrane permeability (Amidon *et al.*, 1995; Horter and Dressman, 2001; Hecq *et al.*, 2005; Lipinski *et al.*, 2001). Furthermore it is important to note that, in order properly to understand and control bioavailability issues, it is necessary to assess carefully the combination of physico-chemical processes that take place during gastrointestinal digestion *in vivo*: dissolution in gastrointestinal fluids, enzymatic reactions, precipitation, emulsification (and emulsion breaking), solubilization into bile micelles (essential for delivery into cells), and absorption (active and passive) (McClements *et al.*, 2008; Velikov and Pelan, 2008; Singh *et al.*, 2009).

Some viable framework for quantifying the main physico-chemical contributions to bioavailability is clearly desirable. For the case of lipids, for example, the bioavailability (F) has been represented by a simple equation (Versantvoort *et al.*, 2004; McClements *et al.*, 2008):

$$F = F_{\mathrm{B}} \times F_{\mathrm{T}} \times F_{\mathrm{M}} \quad . \tag{2.1}$$

In this equation, F_{B} is defined as the bioaccessibility coefficient, *i.e.*, fraction of lipid released from the food matrix into juices of the gastrointestinal tract; F_{T} is defined as the transport coefficient, *i.e.*, fraction of the released lipid component transported across the intestinal epithelium; and F_{M} is the fraction of the lipid component that reaches the systemic circulation without being metabolized. It is usually important to measure the concentration of a bioactive component at a particular location to establish its efficacy. Thus, the concentration–time profile of a specific bioactive component at a particular site of action may be helpful in the assessment of its beneficial or adverse affects on human health and wellness (McClements *et al.*, 2008).

With the exception of water-soluble vitamins and isoflavones, most of the bioactive molecules listed in Table 2.1 are predominantly lipophilic. Therefore they are usually insoluble in water or in aqueous body fluids, resulting in inefficient permeability, low bioavailability and poor efficacy *in vivo* (Ratnam *et al.*, 2006; Spernath and Aserin, 2006). In fact, it has been revealed that a high-fat diet results in better bioavailability of lipophilic bioactive compounds like, for example, carotenoids (Brown *et al.*, 2004), coenzyme Q_{10} (Weber *et al.*, 1997a,b), lycopene (Boileau *et al.*, 1999, 2002) and quercetin (Lesser *et al.*, 2004). However, a high-fat diet certainly cannot be recommended for most people in highly industrialized countries because it commonly leads to obesity, diabetes and cardiovascular diseases. On the contrary, there is an increasing demand for zero-fat or low-fat foods. Therefore, the poor solubility in water of these lipophilic bioactive compounds poses a challenge in terms of their effective delivery. On the other hand, highly soluble nutraceuticals (vitamins, isoflavones) are quickly eliminated from the body — for example, 73% of ascorbic acid can be removed from the body in less than 24 hours (Levine *et al.*, 1996). For such water-soluble compounds, sustained release and prolonged activity are desirable, and so it is required to design the appropriate delivery systems to achieve this objective. Besides, many bioactive compounds have a lot of unsaturated bonds in their molecules, and they need to be protected against oxidation and degradation under extreme environmental conditions (high temperature, light exposure, pH, ionic strength, *etc.*) in order to maintain their physical and chemical stability, as well as their biological activity, during food processing, product storage, and digestion in the gastrointestinal tract following oral administration. Moreover, the surface activity of some bioactive molecules, like the phytosterols, makes them especially susceptible to oxidation, since they can accumulate at oil–water interfaces where oxidative stress is high (Cercaci *et al.*, 2007). Nanoengineering that typically increases the total surface area of the nutraceutical delivery vehicles, due to their smaller size, can also cause an increase in the oxidation. An enchanced extent of oxidation can lower the taste quality of a food product, because of the perception of off-flavours and odours. In addition, some essential micronutrients, like the transition metals iron and copper, are themselves catalysts in the oxidation process (Sergent *et al.*, 1999; McClements and Decker, 2000), and consequently it is important to minimize their concentration in ionic form. Another relevant factor is that some of these bioactive substances (*e.g.*, carotenoids) are crystalline at room temperature in their pure form.

The crystalline nature of these lipids provides challenges in the manufacture of certain types of food products. To address successfully these sometimes conflicting and demanding objectives, it is clear that specific encapsulation processes must be designed to formulate pure bioactive substances into their appropriate dosage forms and also to improve their overall efficacy and safety (Augustin and Hemar, 2009; McClements *et al.*, 2008, 2009).

2. *Encapsulation Processes and Delivery Systems*

Encapsulation processes have their research roots in the pharmaceutical industry. They were developed over forty years ago for enhancing the stability of formulations and controlling the release of drugs. Nowadays, these processes involve the coating, incorporation, entrapment, absorption or dispersion of one or more pure compounds within small capsules/vesicles/carriers of micrometre or nanometre dimensions. Various synthetic polymers (polyethyleneglycol, polymethylmethacrilate, polylacticcoglycolic acid, polyvinyl alcohol, poly-ε-caprolactone, polyisobutylcyanoacrylate, poly(styrene)–poly(ethylene oxide) block copolymers, *etc.*) have been employed as the surrounding wall or barrier, with the encapsulated drugs forming the core of the capsules (Kwon *et al.*, 2002; Langer and Peppas, 2003; Dziubla *et al.*, 2005; Shea *et al.*, 2005; Yoncheva *et al.*, 2005; des Rieux *et al.*, 2006; Augustin and Hemar, 2009).

Whilst it is known that micro- and nano-capsules do exhibit enhanced delivery over the more conventional dosage forms such as tablets, the detailed mechanisms by which this occurs are not yet well understood (Gibbs *et al.*, 1999; Ratnam *et al.*, 2006; Acosta, 2009). More recently, the nanoscale dimensions of such delivery systems, especially below 500 nm (Acosta, 2009), has enabled the promise of tackling problems of low oral bioavailability or inefficient delivery of poorly water-soluble drugs. The improvement in diffusive transport with the use of nanoparticles allows crossing of biological barriers, leading to an increase in cellular uptake (Moghimi *et al.*, 2001; Kreuter, 2001; Nishioka and Yoshino, 2001; Merisko-Liversidge *et al.*, 2003; McClements *et al.*, 2008).

More generally, it is commonly suggested nowadays that reducing the size of encapsulates to the nanoscale offers increased technical opportunities because of enhancements in the following factors: (i) the apparent solubility of the active ingredients; (ii) the rate of mass transfer; (iii) the gastrointestinal retention time in the mucus covering the intestinal epithelium; (iv) the rate of release (due to the large surface area); and (v) the direct uptake of particles by the intestinal epithelium (Jani *et al.*, 1990; Horn and Rieger, 2001; Hussain *et al.*, 2001; des Rieux *et al.*, 2006; Chen *et al.*, 2006b; Medina *et al.*, 2007; Lai *et al.*, 2007). Additional modifications of surface properties (*e.g.*, coatings or biomolecular flags), using polysaccharides (dextran, chemically modified starch, alginates, pectins, chitosan, carrageenan, *etc.*) as target-recognition groups, can increase the bioadhesion of the bioactive compounds towards the mucus (mucin) or can facilitate their targeted delivery (Charych *et al.*, 1996; Qaqish and Amiji, 1999; Takeuchi *et al.*, 2001; Vandamme *et al.*, 2002; des Rieux *et al.*, 2006; Tiede *et al.*, 2008; Acosta, 2009). In particular, the incorporation of chitosan, a weak cationic polysaccharide, introduces hydrophilic groups on the surface of a particle, which can then establish weak ionic interactions with the negatively charged mucin layer that coats the surface of enterocytes (which represent 90–95% of epithelial cells of the intestine) (des Rieux *et al.*, 2006). Polyethylene glycol coating has also been used to increase the hydrophilicity of the surface of nanoparticles (Yoncheva *et al.*, 2005; des Rieux *et al.*, 2006). And it has been found that an increase in the adhesion of nanoparticles to the mucin layer can achieved with the use of cationic coatings produced with synthetic polymers or cationic surfactants such as alkyl trimethylammonium salts (Sakuma *et al.*, 2002; des Rieux *et al.*, 2006; Lamprecht *et al.*, 2006). It may be noted, however, that if the interaction between cationic nanoparticles and the mucin proteins is too strong, the particles remain stuck to the mucin and so do not permeate through the epithelial tissue (Hussain *et al.*, 2001).

There is considerable current interest in protecting certain compounds against release in the stomach or intestine (Singh *et al.*, 2009). A large number of polysaccharides (dextran, chitosan, chemically modified starch, alginates, pectins, carrageenan, xanthan, *etc.*), and their combinations or chemical derivatives, are resistant to gastric and intestinal bacteria, but they are specifically hydrolysed by colonic bacteria. This makes these polysaccharides useful as coating and matrix materials for specific delivery of bioactive molecules to the colon (Macleod *et al.*, 1999; Vandamme *et al.*, 2002).

Many new delivery systems are now being formulated in the form of dispersions of colloidal particles with different shapes and morphologies (homogeneous spheres, core–shell-type particles, hollow particles, *etc.*) (Velikov and Pelan, 2008; McClements *et al.*, 2008, 2009; Augustin and Hemar, 2009). For example, it has been established that nanoemulsions with droplet sizes in the range 50–400 nm can be used successfully as drug delivery systems for oral, parenteral and transdermal administration (Benita, 1998; Sarker, 2005). One of the advantages of nanoemulsions over thermodynamically stable microemulsions is that a lower surfactant concentration is needed for their stabilization (Velikov and Pelan, 2008).

Polymeric nanomicelles are recognized as potential passive targeting carriers for anticancer agents. This is because of a number of favourable characteristics of their properties: (i) their ability to trap hydrophobic substances in the hydrophobic core and to facilitate good solubility in aqueous media owing to their hydrophilic surface; (ii) their permanence and high mechanical strength; and (iii) their relatively small particle size (Yokoyama *et al.*, 1990; Kataoka *et al.*, 1993; Zhang *et al.*, 2006). Most especially, the nanoscale dimension of polymeric micelles is recognized as a crucial condition for their effective drug delivery function *in vivo*.

Marked advances in pharmaceutical applications of encapsulation and in the development of cost-effective preparation techniques and materials have allowed a significant increase in use of encapsulation processes in the food industry. Research activities in the area of nanoparticle delivery systems for micronutrients and nutraceuticals which can be incorporated into food products have increased almost exponentially during the past decade (Ransley *et al.*, 2001; McClements *et al.*, 2008, 2009; Faulks and Southon, 2008; Augustin and Hemar, 2009). Moreover, current market projections for these technologies suggest a multifold increase in their commercial potential over the next decade, as functional foods become increasingly popular among consumers as a result of increased awareness of bioactive food ingredients and their impact on human health and physiological functions. Nowadays, it seems like the citizen is constantly being encouraged to address medical issues like, for example, cardiovascular health and obesity through choice of food products rather than with drugs. Nevertheless, the design of functional foods involving bioactive ingredients remains a big technological challenge (Ransley *et al.*, 2001; McClements *et al.*, 2008, 2009; Faulks and Southon, 2008; Augustin and Hemar, 2009). One of the main difficulties is to replace some of the polymers and surfactants used in the pharmaceutical industry with food-grade alternatives (Acosta, 2009). Furthermore, the incorporation of nutraceuticals in the design of functional foods brings with it enormous technological challenges in the maintenance of the physico-chemical stability, appearance, texture, flavour and taste of the products, along with the required high extent of bioavailability of the nutraceuticals (Velikov and Pelan, 2008). Listed below is the set of necessary requirements for an efficient edible delivery system as set out by McClements and co-workers in their recent review article (2009).

(i) The delivery system should be fabricated from food-grade (GRAS) ingredients using easily implemented processing operations.

(ii) The delivery system should be capable of being economically manufactured using inexpensive ingredients. Ultimately, the benefits gained from encapsulating the functional component (*e.g.*, an improved shelf-life, enhanced marketability, novel functionality) should outweigh additional costs associated with encapsulation.

(iii) The delivery system should efficiently encapsulate an appreciable amount of the functional component in a form that is easily incorporated into food systems, *i.e.*, it should have:
(a) a high loading capacity (LC), defined as the mass of encapsulated material per unit mass of carrier material ($LC = M_E/M_C$);
(b) a high loading efficiency (LE), defined as the ability of the delivery system to retain encapsulated material over time ($LE = 100 \times M_E(t)/M_E(0)$, where $M_E(t)$ and $M_E(0)$ are the masses of encapsulated material at time t and initially, respectively).

(iv) Depending on the application, the delivery system may have the capability to protect the functional component from chemical degradation (*e.g.*, oxidation, hydrolysis) keeping it in its active state.

(v) Depending on the application, the delivery system may have the capability to release the functional component at a particular site of action, at a controlled rate and/or in response to a specific environmental trigger (*e.g.*, pH, ionic strength or temperature). If so, this should involve a high level of delivery efficiency (DE), as defined by $DE = 100 \times M_E(D)/M_E(I)$, where $M_E(I)$ and $M_E(D)$ are masses of encapsulated material in the initial delivery system and delivered to the active site, respectively.

(vi) The delivery system should be compatible with the food matrix that surrounds it.

(vii) The delivery system should be resistant to the various kinds of environmental stresses that a food typically experiences during its production, storage, transport, and consumption.

(viii) The delivery system should have the capability of maintaining the bioactivity of the functional component within the human body prior to its being delivered to the desired site of action, e.g., resisting the high acidity and enzyme activity of the stomach.

Nowadays a wide variety of food ingredients are already produced in an encapsulated form. These comprise artificial sweeteners (aspartame), flavouring agents such as oils or spices (with desirable flavour but possibly undesirable odour), natural colorants (*e.g.*, β-carotene, turmeric), preservatives, acids (citric, lactic and ascorbic), bases, buffers, enzymes, lactic acid bacteria, and some antioxidants (Kirby, 1991; Gibbs *et al.*, 1999; Chen *et al.*, 2006b; Ubbink and Krüger, 2006; Augustin and Hemar, 2009). It is well established that, as a result of their encapsulation, these ingredients can keep their stability and functionality under moderately extreme conditions (heat, acidic/basic pH, moisture) of prolonged storage and processing in the period prior to release or targeted delivery.

Table 2.2 lists the characteristics of the major encapsulation processes and the most widely used wall/carrier materials in the food industry. We take note of the key roles of food-grade biopolymers (proteins and polysaccharides) and lipids (fat) in the encapsulation processes developed for food manufacture. The primary role of the biopolymers follows from the special features of their biological origin, their molecular structures, their physico-chemical properties, and their specific functions in certain food applications. In particular, because of their biological origin, food biopolymers show good biocompatibility, biodegradability and non-toxicity in any combinations. And, of course, because they also have basic nutritional and structural functions in the human diet, this makes them highly desirable ingredients in food product manufacture.

3. Biopolymer-Based Delivery Vehicle Ingredients

Biopolymers are convenient as delivery vehicle ingredients because of their generally amphiphilic nature. This amphiphilicity provides a high solubility in aqueous media, and the simultaneous availability of many charged, polar and hydrophobic functional groups for engaging in different types of interactions (electrostatic, hydrophobic and hydrogen bonding) with other kinds of hydrophilic and lipophilic substances. At the molecular level, these interactions can occur in the interior and at the surface of biopolymer molecules, and they can be manipulated by varying the solution conditions (pH, ionic strength, temperature) or by the application of an external stress (*e.g.*, high pressure, electric field, ultrasound). Moreover, biopolymers possess various conformations, namely, globular, random coil, helical, worm-like and tubular, which can be favourable for biopolymer encapsulating abilities, such as the coating, incorporation, entrapment, or absorption of different bioactive molecules. Therefore, all these features of biopolymers, along with the nanometre scale of both their molecules and their soluble aggregates, presents a great potential for utilization of novel biopolymers or biopolymer assemblies in the development of smart nanosized natural delivery systems for controlling the release of nutraceuticals via the oral route. Moreover, there is one further clear advantage of using food-grade biopolymers for the formulation of delivery systems — their abundant availability from renewable natural sources.

Table 2.2 Encapsulation processes and wall/carrier materials developed for the food industry.

Encapsulation process	Wall/carrier material	Advantages (references)
Spray drying (using a high air velocity at 180–360 °C) (most common technique, being relatively inexpensive and technically straightforward)	Modified starch, dextrin, maltodextrins, sugar beet pectin, and gum arabic ('gold standard' ingredient for microencapsulation)	Efficient retention and inhibition of oxidation of flavours and bioactive molecules; long shelf-life of microcapsules (Gibbs *et al.*, 1999; Bhandari *et al.*, 1992; Gouin, 2004; Vega and Roos, 2006; Ubbink and Krüger, 2006; Gharsallaoui *et al.*, 2007; Drusch, 2007; Shaw *et al.*, 2007; Augustin and Hemar, 2009)
Spray chilling, spray cooling (using a high air velocity at 30–120 °C) or freeze drying	Vegetable oil (possibly hydrogenated or fractionated); chitosan, maltodextrin, whey proteins	Frozen liquids, heat-sensitive materials and those not soluble in the usual solvents can be encapsulated (Gibbs *et al.*, 1999; Gouin, 2004; Klaypradit and Huang, 2008; Augustin and Hemar, 2009)
Encapsulation based on supercritical (SC) fluids (usually carbon dioxide)	Beeswax, hydrogenated vegetable oil derivatives, paraffin wax, stearic acid or stearoyl acohol, gelatin, whey proteins, cellulose, *etc.*	Main advantage here is the absence of water and the very mild temperature (< 30 °C) throughout the process); encapsulation of sensitive materials (enzymes, volatile flavours, chemically sensitive ingredients) can benefit from SC fluid-based spray drying (Gouin, 2004)

Table 2.2 *Continued*

Encapsulation process	Wall/carrier material	Advantages (references)
Extrusion at high temperature into a dehydrating liquid (isopropyl alcohol); syringe-extrusion into a calcium chloride solution	Maltodextrins with different dextrose equivalent; starches and starch derivatives; gelatin, alginate	Encapsulated material is totally isolated by the wall material, and any core is washed from the outside; shelf-life up to at least two years (Gibbs *et al.*, 1999; Gouin, 2004; Augustin and Hemar, 2009)
Fluidized bed coating: solid particles (50–500 μm) suspended in temperature- and humidity-controlled chamber of high-velocity air where coating is atomized	Hydrogenated vegetable oil, fatty acids, emulsifiers, starches, gums, maltodextrins	Method allows isolation of antagonistic components, like iron and ascorbic acid, in nutritional mixes in small tablets (Gibbs *et al.*, 1999; Gouin, 2004)
Inclusion complexation with biopolymers (nanocapsules with oil interior having average size of ~ 300 nm)	β-Cyclodextrins; chemically modified cyclodextrins, including fatty acid chains grafted to hydroxyl groups	High aqueous solubility and stability (> 5 months) of bioactive ingredients (vitamins A, E and K) resulting in improved bioavailability; moisture and temperature conditions of the mouth allow release of bound material (Skiba *et al.*, 1996; Memiolu *et al.*, 2002; Gamberini and Ficarra, 2002; Bangalore *et al.*, 2005)

Table 2.2 *Continued*

Encapsulation process	Wall/carrier material	Advantages (references)
Formation of multiple emulsions (W/O/W, O/W/O, and O/W/W, where W = water and O = oil)	Whey protein–polysaccharide complexes and covalent conjugates	Efficient storage carriers for sustained release of hydrophilic and lipophilic molecules; protection of sensitive molecules from aggressive environmental conditions in gastrointestinal tract (Benichou *et al.*, 2002, 2004, 2007; Shima *et al.*, 2004; McClements *et al.*, 2009)
Electrostatic complexation or coacervation in bulk aqueous phase and at droplet surface in (macro)emulsions (size > 0.5 μm)	Gelatin; gum arabic; β-lactoglobulin; low-methoxyl pectin; SDS–chitosan, SDS–chitosan–pectin; whey protein–chitosan	Hydrophilic or hydrophobic monolayer or multilayer coatings can be used to microencapsulate hydrophobic or hydrophilic substances; controlled and targeted release in response to environmental conditions (temperature, pH, pressure or chemical reactions) (Schmitt *et al.*, 1998; Gibbs *et al.*, 1999; Dickinson, 2004; de Kruif *et al.*, 2004; Zhu *et al.*, 2005; Guzey and McClements, 2006; Shaw *et al.*, 2007; McClements *et al.*, 2008, 2009; Livney, 2008; Zimet and Livney, 2009; Grigoriev and Miller, 2009)

Table 2.2 *Continued*

Encapsulation process	Wall/carrier material	Advantages (references)
Emulsification (by high-energy homogenization or ultrasonic atomizer) of encapsulated material with low-molecular-weight surfactants (Tweens) or polymeric emulsifiers (proteins, polysaccharides; covalent conjugates of proteins with polysaccharides) into nanoemulsions (40–500 nm); entrapping fat-soluble and water-soluble bioactive ingredients in emulsion gels	Gelatin, soy protein, milk proteins, modified starch, dextran, maltodextrin	Good physical stability against creaming (sedimentation); transparency and appealing colour, good oxidative stability, digestion rate and bioavailability (Chaiyasit *et al.*, 2000; Silvestre *et al.*, 2000; Horn and Rieger, 2001; Hogan *et al.*, 2001; Spernath *et al.*, 2002; Amar *et al.*, 2003; Chen and Wagner, 2004; Mun *et al.*, 2007; Jafari and Bhandari, 2007;Velikov and Pelan, 2008; Klaypradit and Huang, 2008; Yuan *et al.*, 2008; Grigoriev and Miller, 2009; Augustin and Hemar, 2009); following oral administration, increase in bioavailability of poorly soluble and/or poorly permeable lipophilic bioactive compounds as compared with powder (Kommuru *et al.*, 2001)

Table 2.2 *Continued*

Encapsulation process	Wall/carrier material	Advantages (references)
Loading in liposomes (25–1000 nm)	Phospholipids	Simultaneous delivery of both hydrophilic and hydrophobic bioactive molecules, owing to the unique structure of liposomes, in which an aqueous volume is entirely enclosed by a membrane composed of double-chained lipid molecules; responsiveness to changes in lipid composition, size, surface charge and method of preparation; controlled enzyme release in cheese-making for reducing ripening time and preventing spoilage by bacteria (Kirby, 1991); prevention of oxidation of unsaturated lipids (Haynes *et al.*, 1992); improvement of solubility, stability and penetration through the plasma membrane of cells of bioactive ingredients; dramatic increase in cellular specific activity under the action of bioactive ingredients (Sinha *et al.*, 2001; Minko *et al.*, 2002; Ratnam *et al.*, 2006; Kaur *et al.*, 2007)

Table 2.2 *Continued*

Encapsulation process	Wall/carrier material	Advantages (references)
Entrapping of bioactive ingredients by polymer matrix in gel or microgel particles; heat-induced or cold-induced aggregation and gelation of globular proteins (microcapsules of 5–5000 μm)	Chitosan, cross-linked chitosan; alginate gels (in presence of calcium ions); starch; egg white, soybean and whey proteins	Slow-release formulation of bioactive ingredients; the maintenance of active forms of bioactive molecules; large size microgel particles (200–5000 μm) can entrap large molecules (> 5 kDa) (King, 1983; Gibbs *et al.*, 1999; Kumar *et al.*, 2004; Desai and Park, 2005; Renard *et al.*, 2002; Chen *et al.*, 2006a; McClements *et al.*, 2008, 2009; Augustin and Hemar, 2009)
Incorporation of bioactive ingredients into lipid (micellar) or polymeric nanoparticles (1–1000 nm)	(See Tables 1.1 and 1.2 in chapter one)	Improvement in aqueous solubility, stability, permeability and bioavailability of ingredients; protection against oxidation and gastrointestinal degradation (Bhardwaj *et al.*, 2005; Chen *et al.*, 2006a; Semo *et al.*, 2007; McClements *et al.*, 2008, 2009; Augustin and Hemar, 2009); provision of sustained release; lymphatic absorption of filled nanoparticles via gut-associated lymphoid tissue that avoids first-pass metabolism in liver (Delie, 1998; Hussain *et al.*, 2001; Ratnam *et al.*, 2006)

Nature itself gives us a spectacular example of a biopolymer-based delivery system in the form of the native casein micelle of mammalian milk (Lemay *et al.*, 2007). This is primarily a colloidal delivery system for calcium, where the micronutrient is in the form of calcium phosphate, which does not give a bitter taste, and which provides good bioavailability owing to its colloidal size, amorphous state and quick dissolution in gastric conditions (pH ~ 1–2). Nevertheless, the casein micelle structure is unique: there are no other readily available natural delivery systems for most nutraceuticals. Therefore some new designs are clearly required (Velikov and Pelan, 2008; McClements *et al.*, 2008, 2009).

3.1. *Protein Gels and Gel Particles*

It is already fairly well demonstrated how different kinds of biopolymer interactions could be used for the development of delivery systems for a number of nutraceuticals. Amongst the most important kinds of transformations are the heat-induced and cold-induced attractive interactions of globular proteins (egg white, soybean and whey proteins), which can cause aggregation and gelation into networks. Thus, thermal aggregates and gels with different structures can be produced by the predominantly hydrophobic attractions between unfolded molecules of globular proteins (Lefèvre and Subirade, 2000). In turn, the cold-induced aggregation and gelation of globular proteins can be achieved by electrostatic binding of bivalent ions (Ca^{2+}, Fe^{2+}) to pre-heated (*i.e.*, pre-denatured) protein globules (Roff and Foegeding, 1996; Veerman *et al.*, 2003; Maltais *et al.*, 2005), followed by strengthening of hydrophobic attractions between the partly neutralized and unfolded protein molecules. The high surface area of such interconnecting networks forms an efficient framework for the physico-chemical entrapment of nutraceuticals by the highly porous gel or the protein gel particles. Furthemore, a fundamental advantage of such a nutraceutical carrier system is that the protein network itself, formed in the bulk of the food structure, can stabilize the soft solid-like texture that is the highly desirable characteristic of many semi-solid food products (Chen *et al.*, 2006a).

In the case of cold-induced aggregation and gelation, two different types of gel microstructure, namely 'filamentous' and 'particulate' (Figure 2.1), have been obtained by adding different concentrations of a ferrous salt to solutions of pre-denatured β-lactoglobulin (the major whey protein). This substantial difference in microstructure turns out to have a major impact on the iron delivery, due to the different sensitivities of the structures to the relevant environmental conditions, such as pH and the presence of digestive enzymes. In particular, the filamentous gel microstructure has shown greater promise as a protein-based matrix for transporting iron and promoting its effective absorption *in vivo* (Remondetto *et al.*, 2002, 2003, 2004).

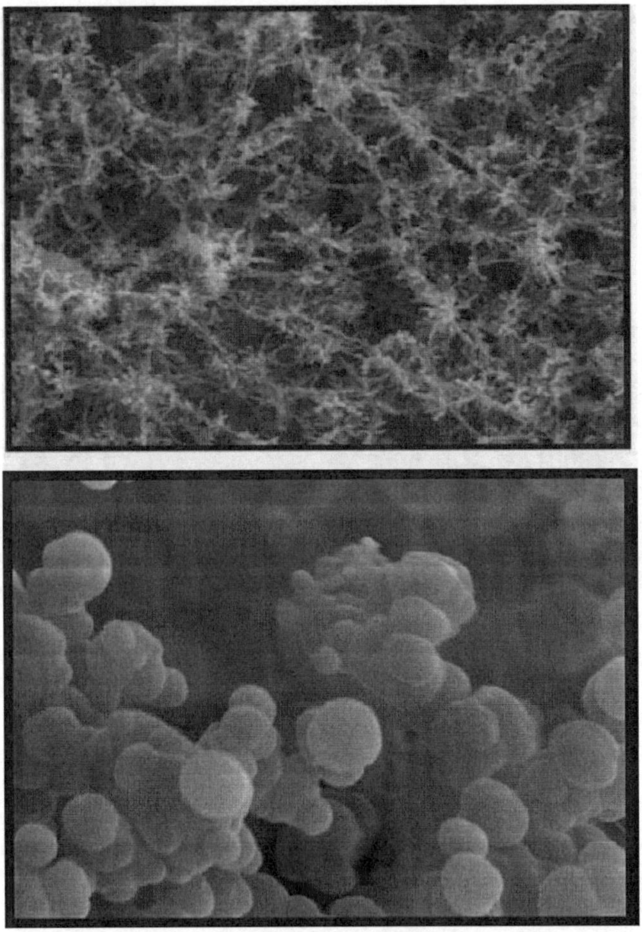

Figure 2.1 Scanning electron micrographs of Fe^{2+}-induced cold-set gels of β-lactoglobulin: filamentous gel (top) and particulate gel (bottom). The gel microstructure depends on the iron/protein ratio. At low iron/protein ratios, a homogeneous filamentous network is obtained, whereas at high iron/protein ratios an aggregated particle gel is produced. Reproduced from Chen *et al.* (2006a) with permission.

3.2. *Biopolymers as Emulsifying Wall Material for Encapsulation*

Another characteristic property of many biopolymers (proteins, modified starch, chitosan, *etc.*) which is useful for the encapsulation of bioactive molecules is their ability to adsorb at the oil–water interface and to form adsorbed layers that are capable of stabilizing oil-in-water (O/W) emulsions against coalescence (see Table 2.2). It is worthwhile to note here that the formation of an emulsion is one of the key steps in the encapsulation of hydrophobic nutraceuticals by the most common technique used nowadays in the food industry (spray-drying). The adsorption of amphiphilic biopolymers at the oil–water interface involves the attachment of their hydrophobic groups to the surface of the oil phase (or even their slight penetration into it), whilst their hydrophilic parts protrude into the aqueous phase providing a bulky interfacial layer.

Owing to their high level of intrinsic hydrophobicity, proteins play the primary role in the formation of adsorbed protective layers around emulsion droplets and in colloidal stabilization of food O/W emulsions (Dickinson, 2001, 2006; Keogh *et al.*, 2001; Kagami *et al.*, 2003; McClements, 2004; Damodaran, 2005; McClements *et al.*, 2008). The nature of the protein conformation seems to be a controlling factor in the encapsulating ability of food proteins following adsorption on emulsion droplets containing hydrophobic nutraceuticals (see section 1 in chapter eight for more details). Thus, for example, in a test of a series of proteins as potential storage carriers for β-carotene nanoparticles (size 17 nm), it was shown (Chu *et al.*, 2007) that the milk protein ingredient sodium caseinate, containing individual caseins having largely random-coil conformations, was the most efficient amongst all the ingredients studied (including whey protein concentrate, whey protein isolate, and whey protein hydrolysate). Moreover, the biopolymeric structure of the sodium caseinate and whey protein adsorbed layers, as compared with a surfactant layer made from a conventional low-molecular-weight surfactant (Tween 20), allowed a higher rate of lipid hydrolysis in the emulsion droplets. This behaviour was attributed (Mun *et al.*, 2007) to the greater permeability of enzyme through the protein adsorbed layer, thereby allowing the lipase to come into close proximity to the lipid substrates. (See, however, the recent reservations about the experimental validity of this interpretation by Singh *et al.* (2009).)

Turning now to the use of a surface-active polysaccharide, it has been shown that vitamin E nanoparticles can be produced by ultra-high-pressure homogenization, and then stabilized against physical instability (flocculation, coagulation, coalescence, ringing, creaming) by means of encapsulation with starch octenyl succinate (Chen and Wagner, 2004).

According to these authors, this encapsulation allows, on the one hand, the conversion of vitamin E nanoparticles into a better handling powder by spray-drying, and, on the other, the provision of optical clarity and long-term shelf-life (over a 6-month study period) to low-fat beverages prepared by dispersing the powdered form of the vitamin into water or apple juice (see Figure 2.2) (Chen and Wagner, 2004).

Figure 2.2 Illustration of the optical clarity of (a) vitamin E 50% (from DSM Nutritional Products (Kaiseraugst, Switzerland), average particle size 350 nm), (b) vitamin E 15% (from DSM Nutritional Products (Kaiseraugst, Switzerland), a nanoparticle product, average particle size 100 nm), and (c) vitamin E 17% (competitor product, average particle size 730 nm). Reproduced from Chen and Wagner (2004) with permission.

It has been proposed that sugar beet pectin (containing up to 10.6 % of associated protein) can be considered as a promising alternative to gum arabic (with ~ 2 % protein) when used as emulsifying wall material for microencapsulation of hydrophobic nutraceuticals (Drusch, 2007). Figure 2.3 shows an example of spray-dried pectin-based microcapsules containing fish oil. This type of edible oil is an excellent dietary source of essential polyunsaturated fatty acids, such as eicosapentaenoic and docosahexaenoic, which have been recognized to show cardioprotective effects, including anti-arrhythmic and anti-inflammatory effects, lowered blood pressure, and improved endothelial function (Balk *et al.*, 2006).

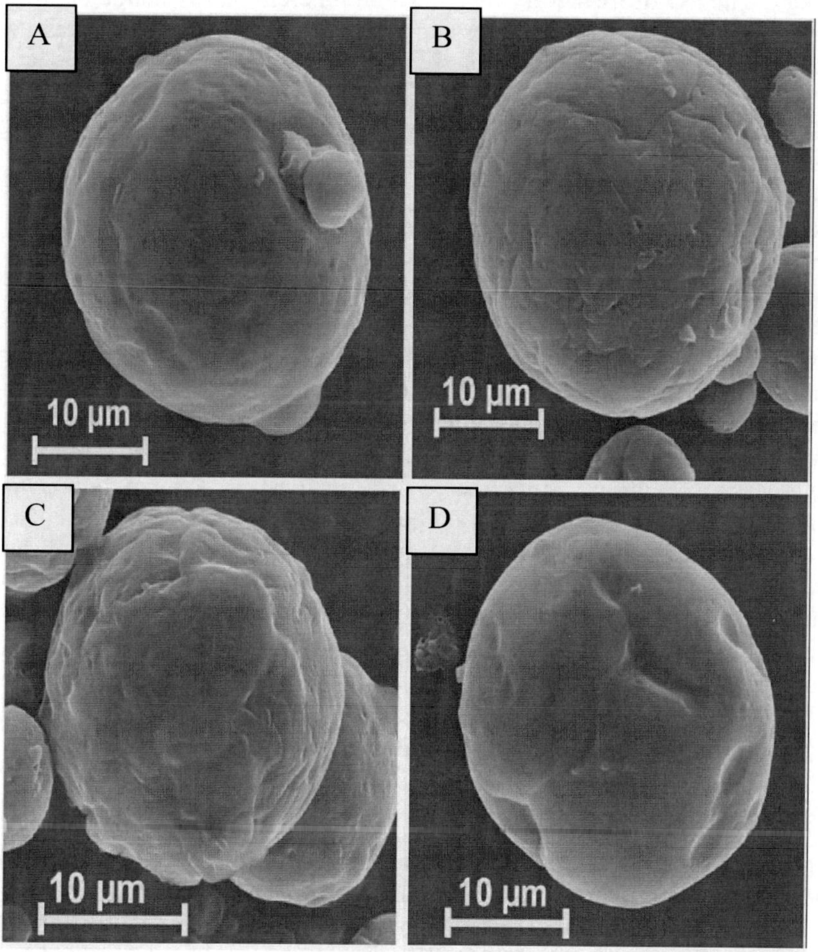

Figure 2.3 Scanning electron micrographs of fish oil microcapsules which were microencapsulated with different concentrations of sugar beet pectin (A and D: 1.1%; B and C: 2.2%) and different oil loads (A and C: 50%; B and D: 20%). Reproduced from Drusch (2007) with permission.

All these microcapsules were found to be roughly spherical in shape and of similar size (14–18 µm) (Drusch, 2007). A mixture of chitosan, maltodextrin and whey protein isolate was also found to be also a good wall material for encapsulation of fish oil using an ultrasonic atomizer followed by freeze-drying of the microcapsules (Klaypradit and Huang,

2008). The use of freeze-drying avoids an inherent disadvantage of spray drying, *i.e.*, the elevated processing temperature, which accelerates the oxidation of polyunsaturated lipids.

Protein-coated emulsion droplets have a strong tendency to aggregate especially in the presence of calcium ions (Eliot and Dickinson, 2003; Sok Line *et al.*, 2005). This phenomenon can be exploited in the formation of emulsion gels (see section 4 in chapter six for more details) and gel-particle carriers (micro-beads in the size range from 3 μm to 2 mm) (Beaulieu *et al.*, 2002; Chen *et al.*, 2006a). In particular, such micro-beads have application as oral delivery vehicles for heat-sensitive and fat-soluble nutraceuticals having different target-release characteristics. For example, it has been demonstrated that a cold-set β-lactoglobulin emulsion gel could be used to protect α-tocopherol under simulated gastrointestinal conditions, and that its delivery (*i.e.* controlled release) could be regulated by gel biodegradation. The potential of whey protein micro-beads to protect components that are sensitive to oxidation (vitamins, fatty acids) was demonstrated experimentally (Chen *et al.*, 2006a). These authors showed, through degradation via enzymatic hydrolysis, that β-lactoglobulin micro-beads containing a fat-soluble bioactive molecule such as retinol can exhibit resistance to pepsin, whilst being completely degraded by pancreatin, so leading to the release of the bioactive compound in the intestine.

3.3. *Mixtures and Complexes of Proteins with Polysaccharides as Delivery Vehicles*

Mixed protein/polysaccharide micro-beads have also been found to be promising delivery vehicles for immobilized bifidobacteria (Guérin *et al.*, 2003). Such micro-beads were made by a transacylation reaction involving the formation of amide bonds between protein and alginate (Levy and Edwards-Levy, 1996). This produces a membrane on the bead surface, protecting the immobilized bifidobacteria against both the very acidic conditions (pH 1–2) and the pepsin activity in the stomach.

The electrostatic interaction between oppositely charged protein and polysaccharide can be utilized for encapsulation and delivery of hydrophobic nutraceuticals. As a result of this interaction, we may have either complex coacervation (and precipitation) or soluble complex formation, depending on various factors, such as the type of polysaccharide used (anionic/cationic), the solution pH, the ionic strength, and the ratio of polysaccharide to protein (see sections 2.1, 2.2 and 2.5 in chapter seven for more details) (Schmitt *et al.*, 1998; de Kruif *et al.*, 2004; Livney, 2008; McClements *et al.*, 2008, 2009). The phenomenon of complex coacervation involving hydrocolloids like gelatin and gum arabic has long had important practical applications in the area of microencapsulation. Zimet and Livney (2009) recently reported on the use of electrostatic nanocomplexes of β-lactoglobulin + low-methoxyl pectin to entrap docosahexaenoic acid (DHA). The nanocomplexes were found to possess outstanding colloidal stability (zeta potential more negative than –50 mV, mean particle size ~ 100 nm), resulting in transparent dispersions having good potential for the enrichment of non-fat acid drinks. In addition, this new technology was shown to confer protection against oxidation of DHA during an accelerated shelf-life stress test: only ~ 5–10 % was lost during 100 hours at 40 °C, as compared to ~ 80 % when the unprotected DHA was monitored (Zimet and Livney, 2009).

Electrostatic biopolymer complexation (sections 2.1, 2.2 and 2.5 of chapter seven) can be achieved by sequential deposition of biopolymers at the surface of solid or liquid droplets (McClements *et al.*, 2008, 2009). This can provide protective multilayer films around colloidal particles, like, for example, droplets of polyunsaturated lipids, thereby improving the stability of the encapsulated lipids to oxidation. Due to overcharging effects, each deposited layer not only fully compensates the charge of the previous templating layer, but it also imparts an uncompensated counter charge allowing for further layer-by-layer deposition. In principle, by adjusting pH or salt concentration, there is the possibility to alter the charge density, and consequently the thickness and permeability of the interfacial membrane. In practice, multilayer O/W emulsions can be made with a three-fold better stability to Fe^{2+}-catalysed lipid oxidation than the conventional single-layer O/W emulsions (McClements and Decker, 2000; Klinkesorn *et al.*, 2005; Guzey and McClements, 2006; McClements *et al.*, 2008, 2009). Moreover, it appears that the cationic nature of the outer chitosan layer can provide enhanced oxidative stability to food oils containing polyunsaturated fatty acids, probably due to electrostatic repulsion of pro-oxidative metal cations (Ogawa *et al.*, 2003; Klinkesorn *et al.*, 2005). Even more stable oil-filled carriers can be prepared by the interfacial deposition of polyelectrolyte multilayers on previously formed positively charged lecithin–chitosan bilayers. Thus, the anionic polysaccharide pectin has been used to form trilayered membranes on emulsion droplets, leading to a very high resistance against aggregation over a wide range of pH and salt concentrations (Ogawa *et al.*, 2004). In addition, it has been suggested that it might be possible to use this approach to encapsulate one or more charged functional components between the interfacial layers, so that they could be released in response to some specific environmental triggers (Guzey and McClements, 2006; McClements *et al.*, 2008, 2009). Nevertheless, these same authors point out that, in order to use this approach successfully, one must carefully control the system composition and the preparation conditions so as to form reliably stable multilayer colloidal particles. That is, it is important to ensure that there is sufficient polyelectrolyte present to cover all of the surfaces present so as to prevent bridging flocculation, but not so much as might lead to depletion flocculation or surfactant–protein cluster formation in bulk solution.

We believe that covalent protein–polysaccharide conjugates produced by the Maillard reaction also have potential for effective encapsulation and delivery of hydrophobic nutraceuticals (see section B in chapter seven for full details). The Maillard browning reaction occurs without any added harmful chemicals under controlled conditions of dry heating (60–80 °C at a relative humidity of 45–80 %) for any binary system of protein + polysaccharide, *e.g.*, BSA + dextran, β-lactoglobulin + dextran, legumin + dextran, WPI + dextran and WPI + maltodextrin (Dickinson and Galazka, 1991, 1992; Dickinson and Semenova, 1992; Dickinson, 2003, 2009; Akhtar and Dickinson, 2003, 2007). Emulsion-stabilizing properties of such covalent conjugates have been found to be much better than those of the equivalent simple mixtures of protein + polysaccharide or the protein alone (especially in the close vicinity of p*I*). The excellent emulsion stabilizing properties of covalent conjugates are associated with the steric stabilizing ability of the bulky hydrophilic polysaccharide residue dangling from oil–water interface into the aqueous medium.

Electrostatic and non-electrostatic biopolymer complexes can also be used as effective steric stabilizers of double (multiple) emulsions. In this type of emulsion, the droplets of one liquid are dispersed within larger droplets of a second immiscible liquid (the dispersion medium for the smaller droplets of the first liquid). In practice, it is found that the so-called 'direct' water-in-oil-in-water (W/O/W) double emulsions are more common than 'inverse' oil-in-water-in-oil (O/W/O) emulsions (Grigoriev and Miller, 2009). In a specific example, some W/O/W double emulsions with polyglycerol polyricinoleate (PGPR) as the primary emulsifier and WPI–polysaccharide complexes as the secondary emulsifying agent were found to be efficient storage carriers for sustained release of entrapped vitamin B_1 (Benichou *et al.*, 2002).

3.4. *Food-grade Self-Assembled Biopolymer Particles*

The milk protein casein is a well-known self-assembling food protein (see section 1.1 in chapter six for further details). Semo and coworkers (2007) have demonstrated that casein-based nanoparticles reassembled from sodium caseinate and ionic calcium can provide partial protection against UV-light-induced degradation of vitamin D_2 that has been encapsulated within them. Moreover, it was revealed that the incorporation of the vitamin D_2 produced only a relatively small effect on the morphology of the casein nanoparticles. These authors proposed that re-assembled casein micelles could be used in the nano-encapsulation of many hydrophobic nutraceutical substances for the potential enrichment of low-fat or non-fat food products. However, it would appear that further experiments are needed to identify the actual binding zones for such nutraceuticals within the re-assembled micelle structure. In an extra refinement, casein nanogel particles stabilized by covalent cross-linking with the enzyme transglutaminase appear also to have some potential in encapsulation and protection technologies (Huppertz and de Kruif, 2008).

Studies are currently underway in Moscow on the suitability of using sodium caseinate nanoparticles as carriers for phosphatidylcholine (lecithin) containing > 80% unsaturated fatty acids (oleic, linoleic, linolenic). In particular, it has been established that phosphatidylcholine oxidation can be reduced, or effectively eliminated altogether, in dispersed systems containing complexes with the protein (see Figure 2.4).

A powerful technique for investigating protein self-assembly in dilute aqueous solutions is static light scattering. This technique has been used recently (Istarova *et al.*, 2005; Semenova *et al.*, 2008) to assess structural attributes of complexes formed by a cooperative binding mechanism involving various kinds of protein–phospholipid interactions (electrostatic, hydrogen bonding, and hydrophobic) favourable to protecting polyunsaturated lipids against oxidation. These inferred favourable nanostructural features are:

(i) a greater size of the protein particle relative to the phospholipid liposome, so that the protein can completely cover the lipid self-assembled structure;
(ii) a large extent of protein association in the complexes;
(iii) a high density for the complex particles;
(iv) a compact architecture of the complex particles; and
(v) an optimal total negative charge of the complexes, in order to inhibit unfavourable electrostatic attraction of pro-oxidative metal cations.

On considering the foaming capacity of these systems, we have found a synergistic effect for complexes of sodium caseinate with phosphatidylcholine, *i.e.*, a four-fold increase in the half-life the foam as compared to the pure protein foam in the range of experimental conditions studied (pH 5.5–7.0; ionic strength 0.001–0.01 M). We note also here that pure phosphatidylcholine did not give fine stable foams at all under these same experimental conditions. Thus, it is evident that food-grade sodium caseinate nanoparticles can potentially possess dual functionality in food systems, namely to act as a promising carrier for bioactive compounds, protecting them against oxidation, and also to function as a traditional structure-forming and stabilizing agent.

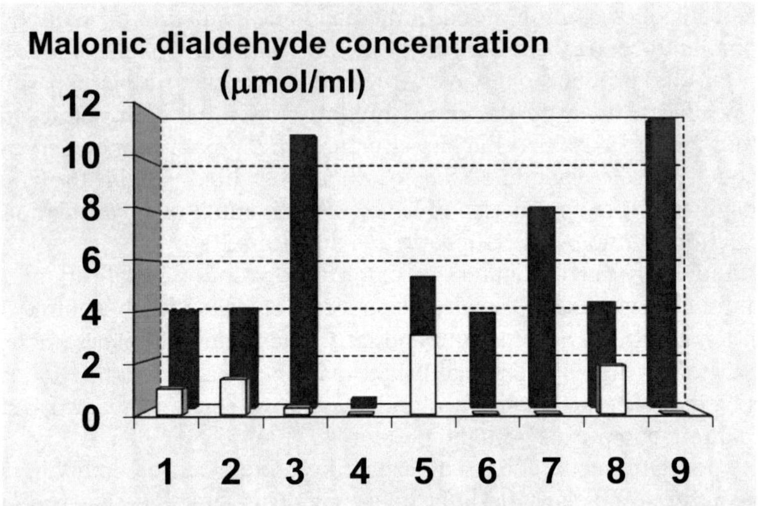

Figure 2.4 Effect on lipid oxidation of complex formation between phosphatidylcholine (10^{-3} M) and sodium caseinate nanoparticles (1 wt/vol%) at different pH and ionic strength (I). The concentration of malonic dialdehyde per gram of phosphatidylcholine produced as a result of lipid peroxidation of the tested samples under conditions of the accelerated oxidation (10^{-5} M $CuSO_4$; 3 hours at 70 °C) is plotted for the different samples. Black columns refer to free phosphatidylcholine and white columns refer to the complexes. Key: **1**: pH = 7.0, I = 0.1 M; **2**: pH = 7.0, I = 0.01 M; **3**: pH = 7.0, I = 0.001 M; **4**: pH = 6.0, I = 0.1 M; **5**: pH = 6.0, I = 0.01 M; **6**: pH = 6.0, I = 0.001 M; **7**: pH = 5.5, I = 0.1 M; **8**: pH = 5.5, I = 0.01 M; **9**: pH = 5.5, I = 0.001 M.

Another interesting type of food-grade nanostructured protein particle which could be promising for encapsulation is the α-lactalbumin nanotube (see section 1.2 in chapter six for full details). This structure is formed when the globular protein self-assembles into micrometre-long tubes having a hollow core with a diameter of only 16 nm (as shown in Figure 1.1 in chapter one). Because of the hollow core, these α-lactalbumin nanotubes could serve as delivery vehicles for drugs or other encapsulated molecules, such as vitamins or enzymes, or act as structures to protect or mask encapsulated compounds (Graveland-Bikker and de Kruif, 2006). According to these authors, besides their nano-delivery function, there may be additional benefits for health from using nanocapsules made from α-lactalbumin:

(i) the protein is relatively rich in tryptophan (four residues per molecule), and plasma tryptophan is known to have a positive effect on satiety and mood (alertness and brain-sustained attention);
(ii) it possesses bacterial or anti-tumour activity;
(iii) it can be converted *in vitro* to an apoptosis-inducing folding variant with oleic fatty acid; and
(iv) it can be used as the basis for design of anti-tumour agents, acting through the disorganization of chromatin structure, due to electrostatic interaction between α-lactalbumin and histone proteins.

Furthermore, dispersions of α-lactalbumin nanotubes could have promising industrial application as viscosity enhancing agents or gelling agents.

Chitosan and its derivatives have received much attention in recent years. Self-assembled N-alkyl chitosan derivatives can be made to form polymer micelles in aqueous medium (Zhang *et al.*, 2006). The polymer micelles have potential for controlled delivery and encapsulation. In addition, chitosan nanoparticles, prepared by ionic gelation of chitosan with tripolyphosphate anions, have been loaded with quercetin and demonstrated to show good bioavailability (Zhang *et al.*, 2008). Advantages of using chitosan derivatives as nanocarriers are possible positive effects of chitosan on human health due to its hypolipidemic, antigenotoxic and anticarcinogenic activities. However, despite the plausible claims made, it seems that additional research is required to validate the practicability of utilizing these chitosan-derived nano-materials in controlled release and targeted delivery of bioactive food compounds.

Concluding Remark

The successful application of food-grade biopolymers in the formulation of the next generation of smart delivery systems requires sound insight into the various intermolecular and colloidal interactions involved in the food matrix, along with some knowledge of the bioavailability *in vivo*. Furthermore, the impact of incorporated nutraceuticals on *all* the properties of a formulated functional food — appearance, physical/chemical stability, texture, mouthfeel, taste, flavour, bioavailability, and health impact — need to be simultaneously considered and addressed in order to achieve a balanced and acceptable solution for consumers.

BIBLIOGRAPHY

Acosta, E. (2009). Bioavailability of nanoparticles in nutrient and nutraceutical delivery. *Current Opinion in Colloid and Interface Science*, **14**, 3–5.

Akhtar, M., Dickinson, E. (2003). Emulsifying properties of whey protein–dextran conjugates at low pH and different salt concentrations. *Colloids and Surfaces B: Biointerfaces*, **31**, 125–132.

Akhtar, M., Dickinson, E. (2007). Whey protein–maltodextrin conjugates as emulsifying agents: an alternative to gum arabic. *Food Hydrocolloids*, **21**, 607–616.

Albanes, D. (1999). Beta-carotene and lung cancer: a case study. *American Journal of Clinical Nutrition*, **69**, 1345S–1350S.

Amar, I., Aserin, A., Garti, N. (2003). Solubilization patterns of lutein and lutein esters in food grade nonionic microemulsions. *Journal of Agricultural and Food Chemistry*, **51**, 4775–4781.

Amidon, G.L., Lennernas, H., Shah, V.P., Crison, J.R. (1995). A theoretical basis for a biopharmaceutic drug classification: the correlation of *in vitro* drug product dissolution and *in vivo* bioavailability. *Pharmaceutical Research*, **12**, 413–420.

Augustin, M.A., Hemar, Y. (2009). Nano- and micro-structured assemblies for encapsulation of food ingredients. *Chemical Society Reviews*, **38**, 902–912.

Balk, E.M., Lichtenstein, A.H., Chung, M., Kupelnick, B., Chew, P., Lau, J. (2006). Effects of omega-3 fatty acids on serum markers of cardiovascular disease risk: a systematic review. *Atherosclerosis*, **189**, 19–30.

Bangalore, D., McGlynn, W., Scott, D. (2005). Effect of beta-cyclodextrin in improving the correlation between lycopene concentration and ORAC values. *Journal of Agricultural and Food Chemistry*, **53**, 1878–1883.

Basu, A., Imrhan, V. (2007). Tomatoes *versus* lycopene in oxidative stress and carcinogenesis: conclusions from clinical trials. *European Journal of Clinical Nutrition*, **61**, 295–303.

Beaulieu, L., Savoie, L., Paquin, P., Subirade, M. (2002). Elaboration and characterization of whey protein beads by an emulsification/cold gelation process: application for the protection of retinol. *Biomacromolecules*, **3**, 239–248.

Benichou, A., Aserin, A., Garti, N. (2002). Double emulsions stabilized by new molecular recognition hybrids of natural polymers. *Polymers for Advanced Technologies*, **13**, 1019–1031.

Benichou, A., Aserin, A., Garti, N. (2004). Double emulsions stabilized with hybrids of natural polymers for entrapment and slow release of active matters. *Advances in Colloid and Interface Science*, **108–109**, 29–41.

Benichou, A., Aserin, A., Garti, N. (2007). O/W/O double emulsions stabilized with WPI–polysaccharide conjugates. *Colloids and Surfaces A: Physicochemical and Engineering Aspects*, **297**, 211–220.

Benita, S. (Ed.) (1998). *Submicron Emulsions in Drug Targeting and Delivery*, Boca Raton, FL: CRC Press.

Bhandari, B.R., Dumoulin, H.M.J., Richard, H.M.J. (1992). Flavor encapsulation of spray drying: application to citral and linalyl acetate. *Journal of Food Science*, **51**, 1301–1306.

Bhardwaj, V., Hariharan, S., Bala, I., Lamprecht, A., Kumar, N., Panchagnula, R., Kumar, M.N.V.R. (2005). Pharmaceutical aspects of polymeric nanoparticles for oral delivery. *Journal of Biomedical Nanotechnology*, **1**, 235–258.

Boileau, A.C., Merchen, N.R., Wasson, K., Atkinson, C.A., John, J., Erdman, W. (1999). *Cis*-lycopene is more bioavailable than *trans*-lycopene *in vitro* and *in vivo* in lymph-cannulated ferrets. *Journal of Nutrition*, **129**, 1176–1181.

Boileau, T.W., Boileau, A.C., Erdman, J.J.W. (2002). Bioavailability of all-*trans* and *cis*-isomers of lycopene. *Experimental Biology and Medicine*, **227**, 914–919.

Brown, M.J., Ferruzzi, M.G., Nguyen, M.L., Cooper, D.A., Eldridge, A.L., Schwartz, S.J., White, W.S. (2004). Carotenoid bioavailability is higher from salads ingested with full-fat than with fat-reduced salad dressings as measured with electrochemical detection. *American Journal of Clinical Nutrition*, **2004**, 396–403.

Cercaci, L., Rodriguez-Estrada, M.T., Lercker, G., Decker, E.A. (2007). Phytosterol oxidation in oil-in-water emulsions and bulk oil. *Food Chemistry*, **102**, 161–167.

Chaiyasit, W., Silvestre, M.P.C., McClements, D.J., Decker, E.A. (2000). Ability of surfactant hydrophobic tail group size to alter lipid oxidation in oil-in-water emulsions. *Journal of Agricultural and Food Chemistry*, **48**, 3077–3080.

Charych, D., Cheng, Q., Reichert, A., Kuziemko, G., Stroh, N., Nagy, J., Spevak, W., Stevens, R. (1996). A "litmus test" for molecular recognition using artificial membranes. *Chemistry and Biology*, **3**, 113–120.

Chen, C.C., Wagner, G. (2004). Vitamin E nanoparticle for beverage applications. *Chemical Engineering Research and Design*, **82**, 1432–1437.

Chen, L., Remondetto, G.E., Subirade, M. (2006a). Food protein-based materials as nutraceutical delivery systems. *Trends in Food Science and Technology*, **17**, 272–283.

Chen, H., Weiss, J., Shahidi, F. (2006b). Nanotechnology in nutraceuticals and functional foods. *Food Technology*, **60**(3), 30–36.

Chu, B.S., Ichikawa, S., Kanafusa, S., Nakajima, M. (2007). Preparation of protein-stabilized β-carotene nanodispersions by emulsification–evaporation method. *Journal of the American Oil Chemists' Society*, **84**, 1053–1062.

Cleland, L.G., French, J.K., Betts, W.H., Murphy, G.A., Elliott, M.J. (1988). Clinical and biochemical effects of dietary fish oil supplements in rheumatoid-arthritis. *Journal of Rheumatology*, **15**, 1471–1475.

Damodaran, S. (2005). Protein stabilization of emulsions and foams. *Journal of Food Science*, **70**, 54–66.

de Kruif, C.G., Weinbreck, F., de Vries, R. (2004). Complex coacervation of proteins and anionic polysaccharides. *Current Opinion in Colloid and Interface Science*, **9**, 340–349.

Delie, F. (1998). Evaluation of nano- and microparticle uptake by the gastrointestinal tract. *Advanced Drug Delivery Reviews*, **34**, 221–233.

des Rieux, A., Fievez, V., Garinot, M., Schneider, Y.J., Preat, V. (2006). Nanoparticles as potential oral delivery systems of proteins and vaccines: a mechanistic approach. *Journal of Controlled Release*, **116**, 1–27.

Desai, K.G., Park, H.J. (2005). Encapsulation of vitamin C in tripolyphosphate cross-linked chitosan microspheres by spray drying. *Journal of Microencapsulation*, **2**, 179–192.

Dickinson, E. (2001). Milk protein adsorbed layers and the relationship to emulsion stability and rheology. *Studies in Surface Science and Catalysis*, **132**, 973–978.

Dickinson, E. (2003). Hydrocolloids at interfaces and the influence on the properties of dispersed systems. *Food Hydrocolloids*, **17**, 25–39.

Dickinson, E. (2004). Food colloids: the practical application of protein nanoscience in extreme environments. Editorial overview. *Current Opinion in Colloid and Interface Science*, **9**, 295–297.

Dickinson, E. (2006). Structure formation in casein-based gels, foams, and emulsions. *Colloids and Surfaces A: Physicochemical and Engineering Aspects*, **288**, 3–11.

Dickinson, E. (2008). Interfacial structure and stability of food emulsions as affected by protein–polysaccharide interactions. *Soft Matter*, **4**, 932–942.

Dickinson, E. (2009). Hydrocolloids as emulsifiers and emulsion stabilizers. *Food Hydrocolloids*, **23**, 1473–1482.

Dickinson, E., Galazka, V.B. (1991). Emulsion stabilization by ionic and covalent complexes of β-lactoglobulin with polysaccharides. *Food Hydrocolloids*, **5**, 281–296.

Dickinson, E., Galazka, V.B. (1992). Emulsion stabilization by protein/polysaccharide complexes. In Phillips, G.O., Wedlock, D.J., Williams, P.A. (Eds). *Gums and Stabilisers for the Food Industry 6*, Oxford: IRL Press, pp. 351–362.

Dickinson, E., Semenova, M.G. (1992). Emulsifying properties of covalent protein–dextran hybrids. *Colloids and Surfaces*, **64**, 299–310.

Drusch, S. (2007). Sugar beet pectin: a novel emulsifying wall component for microencapsulation of lipophilic food ingredients by spray-drying. *Food Hydrocolloids*, **21**, 1223–1228.

Dziubla, T.D., Karim, A., Muzykantov, V.R. (2005). Polymer nanocarriers protecting active enzyme cargo against proteolysis. *Journal of Controlled Release*, **102**, 427–439.

Edge, R., McGarvey, D.J., Truscott, T.G. (1997). The carotenoids as antioxidants — a review. *Journal of Photochemistry and Photobiology B: Biology*, **41**, 189–200.

Eliot, C., Dickinson, E. (2003). Thermoreversible gelation of caseinate-stabilized emulsions at around body temperature. *International Dairy Journal*, **13**, 679–684.

Erhardt, J.G., Meisner, C., Bode, J.C., Bode, C. (2003). Lycopene, beta-carotene and colorectal adenomas. *American Journal of Clinical Nutrition*, **78**, 1219–1224.

Faulks, R.M., Southon, S. (2008). Assessing the bioavailability of nutraceuticals. In Garti, N. (Ed.). *Delivery and Controlled Release of Bioactives in Foods and Nutraceuticals*. Cambridge, UK: Woodhead, pp. 3–25.

Fernstrom, J.D. (2005). Branched-chain amino acids and brain function. *Journal of Nutrition*, **135**, 1539S–1546S.

Ficarra, R., Tommasini, S., Raneri, D., Calabro, M.L., Bella, M.R.D., Rustichelli, C., Ferrieres, J. (2004). The French paradox: lessons for other countries. *Heart*, **90**, 107–111.

Gamberini, M.C., Ficarra, P. (2002). Study of flavonoids/β-cyclodextrins inclusion complexes by NMR, FT-IR, DSC, X-ray investigation. *Journal of Pharmaceutical and Biomedical Analysis*, **29**, 1005–1014.

Geusens, P., Wouters, C., Nijs, J., Jiang, Y.B., Dequeker, J. (1994). Long-term effect of omega-3-fatty-acid supplementation Zn active rheumatoid-arthritis — a 12-month, double-blind, controlled study. *Arthritis and Rheumatism*, **37**, 824–829.

Gharsallaoui, A., Roudaut, G., Chambin, O., Voilley, A., Saurel, R. (2007). Applications of spray drying in microencapsulation of food ingredients: an overview. *Food Research International*, **40**, 1107–1121.

Gibbs, B.F., Kermasha, S., Alli, I., Mulligan, C.N. (1999). Encapsulation in the food industry. *International Journal of Food Sciences and Nutrition*, **50**, 213–224.

Gouin, S. (2004). Microencapsulation: industrial appraisal of existing technologies and trends. *Trends in Food Science and Technology*, **15**, 330–347.

Graveland-Bikker, J.F., de Kruif, C.G. (2006). Unique milk protein-based nanotubes: food and nanotechnology meet. *Trends in Food Science and Technology*, **17**, 196–203.

Grigoriev, D.O., Miller R. (2009). Mono- and multilayer covered drops as carriers. *Current Opinion in Colloid and Interface Science*, **14**, 48–59.

Guérin, D., Vuillemard, J.C., Subirade, M. (2003). Protection of bifidobacteria encapsulated in polysaccharide–protein gel beads against gastric juice and bile. *Journal of Food Protection*, **66**, 2076–2084.

Guzey, D., McClements, D.J. (2006). Formation, stability and properties of multilayer emulsions for application in the food industry. *Advances in Colloid and Interface Science*, **128–130**, 227–248.

Hallikainen, M.A., Sarkkinen, E.S., Uusitupa, M.I.J. (2000). Plant stanol eaters affect serum cholesterol concentrations of hypercholesterolemic men and women in a dose-dependent manner. *Journal of Nutrition*, **130**, 767–776.

Haynes, L.C., Levine, H., Finley, J.W. (1992). Method and liposome composition for the stabilization of oxidizable substances. US Patent No. 5,139, 803.

Hecq, J., Deleers, M., Fanara, D., Vranckx, H., Amighi, K. (2005). Preparation and characterization of nanocrystals for solubility and dissolution rate enhancement of nifedipine. *International Journal of Pharmaceutics*, **299**, 167–177.

Hibbeln, J.R., Nieminen, L.R.G., Blasbalg, T.L., Riggs, J.A., Lands, W.E.M. (2006). Healthy intakes of n-3 and n-6 fatty acids: estimations considering worldwide diversity. *American Journal of Clinical Nutrition*, **83**, 1483S–1493S.

Hogan, S.A., McNamee, B.F., O'Riordan, E.D., O'Sullivan, M. (2001). Microencapsulating properties of sodium caseinate. *Journal of Agricultural and Food Chemistry*, **49**, 1934–1938.

Horn, D., Rieger, J. (2001). Organic nanoparticles in the aqueous phase — theory, experiment, and use. *Angewandte Chemie, International Edition*, **40**, 4330–4361.

Horter, D., Dressman, J.B. (2001). Influence of physicochemical properties on dissolution of drugs in the gastrointestinal tract. *Advanced Drug Delivery Reviews*, **46**, 75–87.

Huppertz, T., de Kruif, C. G. (2008). Structure and stability of nanogel particles prepared by internal cross-linking of casein micelles. *International Dairy Journal*, **18**, 556–565.

Hussain, N., Jaitley, V., Florence, A.T. (2001). Recent advances in the understanding of uptake of microparticulates across the gastrointestinal lymphatics. *Advanced Drug Delivery Reviews*, **50**, 107–142.

Istarova, T.A., Semenova, M.G., Sorokoumova, G.M., Selishcheva, A.A., Belyakova, L.E., Polikarpov, Yu.N. (2005). Effect of pH on caseinate interactions with soy phospholipids in relation to surface activity of their mixtures. *Food Hydrocolloids*, **19**, 429–440.

Jafari, S.M., He, Y., Bhandari, B. (2007). Effectiveness of encapsulating biopolymers to produce sub-micron emulsions by high energy emulsification techniques. *Food Research International*, **40**, 862–873.

Jani, P., Halbert, G.W., Langridge, J., Florence, A.T. (1990) Nanoparticle uptake by the rat gastrointestinal mucosa: quantitation and particle size dependency. *Journal of Pharmacy and Pharmacology*, **42**, 821–826.

Kagami, Y., Sugimura, S., Fujishima, N., Matsuda, K., Kometani, T., Matsumura, Y. (2003). Oxidative stability, structure, and physical characteristics of microcapsules formed by spray drying of fish oil with protein and dextrin wall materials. *Journal of Food Science*, **68**, 2248–2255.

Kataoka, K., Kwon, G.S., Yokoyama, M., Okano, T., Sakurai, Y. (1993). Block copolymer micelles as vehicles for drug delivery. *Journal of Controlled Release*, **24**, 119–132.

Kaur, I.P., Kapila, M., Agrawal, R. (2007). Role of novel delivery systems in developing topical antioxidants as therapeutics to combat photoageing. *Ageing Research Reviews*, **6**, 271–288.

Keogh, M.K., O'Kennedy, B.T., Kelly, J., Auty, M.A., Kelly, P.M., Fureby, A., Haahr, A.-M. (2001). Stability to oxidation of spray-dried fish oil powder microencapsulated using milk ingredients. *Journal of Food Science*, **66**, 217–224.

King, A.H. (1983). Brown seaweed extracts (alginates). In Glicksman, M. (Ed.). *Food Hydrocolloids*, Boca Raton, FL: CRC Press, vol. 2, pp. 115–188.

Kirby, C.J. (1991). Microencapsulation and controlled delivery of food ingredients. *Food Science and Technology International*, **5**, 74–78.

Klaypradit, W., Huang, Y.-W. (2008). Fish oil encapsulation with chitosan using ultrasonic atomizer. *LWT — Food Science and Technology*, **41**, 1133–1139.

Klinkesorn, U., Sophanodora, P., Chinachoti, P., McClements, D.J., Decker, E.A. (2005). Increasing the oxidative stability of liquid and dried tuna oil-in-water emulsions with electrostatic layer-by-layer deposition technology. *Journal of Agricultural and Food Chemistry*, **53**, 4561–4566.

Kommuru, T.R., Gurley, B., Khan, M.A., Reddy, I.K. (2001). Self-emulsifying drug delivery systems (SEDDS) of coenzyme Q10: formulation development and bioavailability assessment. *International Journal of Pharmaceutics*, **212**, 233–246.

Kremer, J.M., Michalek, A.V., Lininger, L., Huyck, C., Bigauoette, J., Timchalk, M.A., Rynes, R.I., Zieminski, J., Bartholomew, L.E. (1985). Effects of manipulation of dietary fatty-acids on clinical manifestations of rheumatoid-arthritis. *Lancet*, **325**, 184–187.

Kreuter, J. (2001). Nanoparticulate systems for brain delivery of drugs. *Advanced Drug Delivery Reviews*, **47**, 65–81.

Kumar, M.N.V.R., Muzzarelli, R.A.A., Muzzarelli, C., Sashiwa, H., Domb, A.J. (2004). Chitosan chemistry and pharmaceutical perspectives. *Chemical Reviews*, **104**, 6017–6084.

Kwon, S.S., Nam, Y.S., Lee, J.S., Ku, B.S., Han, S.H., Lee, J.Y., Chang, I.S. (2002). Preparation and characterization of coenzyme Q10-loaded PMMA nanoparticles by a new emulsification process based on microfluidization. *Colloids and Surfaces A: Physicochemical and Engineering Aspects*, **210**, 95–104.

Lai, S.K., O'Hanlon, D.E., Harrold, S., Man, S.T., Wang, Y.-Y., Cone, R., Hanes, J. (2007). Rapid transport of large polymeric nanoparticles in fresh undiluted human mucus. *Proceeding of the National Academy of Sciences USA*, **104**, 1482–487.

Lamprecht, A., Koenig, P., Ubrich, N., Maincent, P., Neumann, D. (2006). Low molecular weight heparin nanoparticles: mucoadhesion and behaviour in Caco-2 cells. *Nanotechnology*, **17**, 3673–3680.

Langer, R., Peppas, N.A. (2003). Advances in biomaterials, drug delivery, and bionanotechnology. *AIChE Journal*, **49**, 2990–3006.

Lefèvre, T., Subirade, M. (2000). Molecular differences in the formation and structure of fine-stranded and particulate β-lactoglobulin gels. *Biopolymers*, **54**, 578–586.

Lemay, D.G., Dillard, C.J., German, J.B. (2007). Food structure for nutrition. In Dickinson, E., Leser, M.E. (Eds). *Food Colloids: Self-Assembly and Material Science*, Cambridge, UK: Royal Society of Chemistry, pp. 1–15.

Lesser, S., Cermak, R., Wolffram, S. (2004). Bioavailability of quercetin in pigs is influenced by the dietary fat content. *Journal of Nutrition*, **134**, 1508–1511.

Levine, M., Conry-Cantilena, C., Wang, Y., Welch, R.W., Washko, P.W., Dhariwal, K.R., Park, J.B., Lazarev, A., Graumlich, J.F., Kings, J., Cantilena, L.R. (1996). Vitamin C pharmacokinetics in healthy volunteers: evidence for a recommended dietary allowance. *Proceedings of the National Academy of Sciences USA*, **93**, 3704–3709.

Levy, M.C., Edwards-Levy, F. (1996). Coating alginate beads with cross-linked biopolymers: a novel method based on a transacylation reaction. *Journal of Microencapsulation*, **13**, 169–183.

Lipinski, C.A., Lombardo, F., Dominy, B.W., Feeney, P.J. (2001). Experimental and computational approaches to estimate solubility and permeability in drug discovery and development settings. *Advanced Drug Delivery Reviews*, **46**, 3–26.

Livney, Y.D. (2008). Complexes and conjugates of biopolymers for delivery of bioactive ingredients via food. In Garti, N. (Ed.). *Delivery and Controlled Release of Bioactives in Foods and Nutraceuticals*, Cambridge, UK: Woodhead, pp. 234–250.

Macleod, G.S., Collett, J.H., Fell, J.T. (1999). The potential use of mixed films of pectin, chitosan and HPMC for bimodal drug release. *Journal of Controlled Release*, **58**, 303–310.

Maltais, A., Remondetto, G.E., Gonzales, R., Subirade, M. (2005). Formation of soy protein isolate cold-set gels: protein and salt effects. *Journal of Food Science*, **70**, 67–73.

McClements, D.J. (2004). Protein-stabilized emulsions. *Current Opinion in Colloid and Interface Science,* **9**, 305–13.

McClements, D.J., Decker, E.A. (2000). Lipid oxidation in oil-in-water emulsions: impact of molecular environment on chemical reactions in heterogeneous food systems. *Journal of Food Science*, **65**, 1270–1282.

McClements, D.J., Decker, E.A., Park, Y., Weiss, J. (2008). Designing food structure to control stability, digestion, release and absorption of lipophilic food components. *Food Biophysics*, **3**, 219–228.

McClements, D.J., Decker, E.A., Park, Y., Weiss, J. (2009). Structural design principles for delivery of bioactive components in nutraceuticals and functional foods. *Critical Reviews in Food Science and Nutrition*, **49**, 577–606.

Medina, C., Santos-Martinez, M.J., Radomski, A., Corrigan, O.I., Radomski, M.W. (2007). Nanoparticles: pharmacological and toxicological significance. *British Journal of Pharmacology*, **150**, 552–558.

Memiolu, E., Bochot, A., Sen, M., Sharon, D., Duchêne, D., Hincal, A.A. (2002). Amphiphilic β-cyclodextrins modified on the primary face: synthesis, characterization and evaluation of their potential as novel excipients in the preparation of nanocapsules. *Journal of Pharmaceutical Sciences*, **91**, 1214–1224.

Merisko-Liversidge, E., Liversidge, G.G., Cooper, E.R. (2003). Nanosizing: a formulation approach for poorly water-soluble compounds. *European Journal of Pharmaceutical Sciences*, **18**, 113–120.

Minko, T., Stefanov, A., Pozharov, V. (2002). Lung hypoxia: antioxidant and antiapoptotic effects of liposomal α-tocopherol. *Journal of Applied Physiology*, **93**, 1550–1560.

Moghimi, S.M., Hunter, A.C., Murray, J.C. (2001). Long-circulating and target-specific nanoparticles: theory to practice. *Pharmacological Reviews*, **53**, 238–318.

Mun, S., Decker, E.A., McClements, D.J. (2007). Influence of emulsifier type on *in vitro* digestibility of lipid droplets by pancreatic lipase. *Food Research International*, **40**, 770–781.

Nishioka, Y., Yoshino, H. (2001). Limphatic targeting with a nanoparticulate system. *Advanced Drug Delivery Reviews*, **47**, 55–64.

Ogawa, S., Decker, E.A., McClements, D.J. (2003). Influence of environmental conditions on the stability of oil-in-water emulsions containing droplets stabilized by lecithin–chitosan membranes. *Journal of Agricultural and Food Chemistry*, **51**, 5522–5527.

Ogawa, S., Decker, E.A., McClements, D.J. (2004). Production and characterization of O/W emulsions containing droplets stabilized by lecithin–chitosan–pectin multilayered membranes. *Journal of Agricultural and Food Chemistry*, **52**, 3595–3600.

Omenn, G.S., Goodman, G.E., Thornquist, M.D. (1996). Effects of a combination of beta-carotene and vitamin A on lung cancer and cardiovascular disease. *New England Journal of Medicine*, **334**, 1150–1155.

Ostlund, R.E. (2004). Phytosterols and cholesterol metabolism. *Current Opinion in Lipidology*, **15**, 37–41.

Placzek, M., Gaube, S., Kerkmann, U., Gilbertz, K.P., Herzinger, T., Haen, E., Przybilla, B. (2005). Ultraviolet B-induced DNA damage in human epidermis is modified by the antioxidants ascorbic acid and D-α-tocopherol. *Journal of Investigative Dermatology*, **124**, 304–307.

Playne, M.J., Bennett, L.E., Smithers, G.W. (2003). Functional dairy foods and ingredients. *Australian Journal of Dairy Technology*, **58**, 242–264.

Qaqish, R.B., Amiji, M.M. (1999). Synthesis of a fluorescent chitosan derivative and its application for the study of chitosan–mucin interactions. *Carbohydrate Polymers*, **38**, 99–107.

Ransley, J.K., Donnelly, J.K., Read, N.W. (Eds). (2001). *Food and Nutritional Supplements: Their Role in Health and Disease*. Berlin: Springer-Verlag.

Ratnam, D.V., Ankola, D.D., Bhardwaj, V., Sahana, D.K., Kumar, M.N.V.R. (2006). Role of antioxidants in prophylaxis and therapy: a pharmaceutical perspective. *Journal of Controlled Release*, **113**, 189–207.

Redgwell, R.J., Fischer, M. (2005). Dietary fibre as a versatile food component: an industrial perspective. *Molecular Nutrition and Food Research*, **49**, 521–535.

Remondetto, G.E., Subirade, M. (2003). Molecular mechanisms of Fe^{2+}-induced β-lactoglobulin cold gelation: an interactions story. *Biopolymers*, **69**, 461–469.

Remondetto, G.E., Beyssac, E., Subirade, M. (2004). Influence of the microstructure of biodegradable whey protein hydrogels on iron release: an *in vitro* study. *Journal of Agricultural and Food Chemistry*, **52**, 8137–8143.

Remondetto, G.E., Paquin, P., Subirade, M. (2002). Cold gelation of β-lactoglobulin in the presence of iron. *Journal of Food Science*, **67**, 586–595.

Renard, D., Robert, P., Lavenant, L., Melcion, D., Popineau, Y., Guéguen, J., Duclairoir, C., Nakache, E., Sanchez, C., Schmitt, C. (2002). Biopolymeric colloidal carriers for encapsulation or controlled release applications. *International Journal of Pharmaceutics*, **242**, 163–166.

Roff, C.F., Foegeding, E.A. (1996). Dicationic-induced gelation of pre-denatured whey protein isolate. *Food Hydrocolloids*, **10**, 193–198.

Ruxton, C.H.S., Calder, P.C., Reed, S.C., Simpson, M.J.A. (2005). The impact of long-chain ω-3 polyunsaturated fatty acids on human health. *Nutrition Research Reviews*, **18**, 113–129.

Ruxton, C.H.S., Reed, S.C., Simpson, M.J.A., Millington, K.J. (2004). The health benefits of omega-3 polyunsaturated fatty acids: a review of the evidence. *Journal of Human Nutrition and Dietetics*, **17**, 449–459.

Sakuma, S., Suzuki, N., Sudo, R., Hiwatari, K.-I., Kishida, A., Akashi, M. (2002). Optimized chemical structure of nanoparticles as carriers for oral delivery of salmon calcitonin. *International Journal of Pharmaceutics*, **239**, 185–195.

Sarker, D.K. (2005). Engineering of nanoemulsions for drug delivery. *Current Drug Delivery*, **2**, 297–310.

Schmitt, C., Sanchez, C., Sobry-Banon, S., Hardy, J. (1998). Structure and technofunctional properties of protein–polysaccharide complexes. *Critical Reviews in Food Science and Nutrition*, **38**, 689–753.

Semenova, M.G., Belyakova, L.E., Polikarpov, Yu.N., Antipova, A.S., Anokhina, M.S. (2008). Utilization of sodium caseinate nanoparticles as molecular nanocontainers for delivery of bioactive lipids to food systems: relationship to the retention and controlled release of phospholipids in the simulated digestion conditions. In Williams, P.A., Phillips, G.O. (Eds). *Gums and Stabilisers for the Food Industry 14*, Cambridge, UK: Royal Society of Chemistry, pp. 326–333.

Semo, E., Kesselman, E., Danino, D., Livney, Y.D. (2007). Casein micelle as a natural nano-capsular vehicle for nutraceuticals. *Food Hydrocolloids*, **21**, 936–942.

Sergent, O., Morel, I., Cillard, J. (1999). Involvement of metal ions in lipid peroxidation: biological implications. In Sigel, A., Sigel, H. (Eds). *Metal Ions in Biological Systems, Vol. 36: Interrelation between Free Radicals and Metal Ions in Life Processes*, Boca Raton, FL: CRC Press, pp. 251–287.

Shahidi, F., Miraliakbari, H. (2005). Omega-3 fatty acids in health and disease. Part 2. Health effects of omega-3 fatty acids in autoimmune diseases, mental health, and gene expression. *Journal of Medicinal Food*, **8**, 133–148.

Shaw, L.A., McClements, D.J., Decker, E.A. (2007). Spray-dried multilayered emulsions as a delivery method for ω-3 fatty acids into food systems. *Journal of Agricultural and Food Chemistry*, **55**, 3112–3119.

Shea, T.B., Ortiz, D., Nicolosi, R.J., Kumar, R., Watterson, A.C. (2005). Nanosphere-mediated delivery of vitamin E increases its efficacy against oxidative stress resulting from exposure to amyloid beta. *Journal of Alzheimer's Disease*, **7**, 297–301.

Shima, M., Kobayashi, Y., Fujii, T., Tanaka, M., Kimura, Y., Adachi, S. (2004). Preparation of fine W/O/W emulsion through membrane filtration of coarse W/O/W emulsion and disappearance of the inclusion of outer phase solution. *Food Hydrocolloids*, **18**, 61–70.

Silvestre, M.P.C., Chaiyasit, W., Brannan, R.G., McClements, D.J., Decker, E.A. (2000). Ability of surfactant headgroup size to alter lipid and antioxidant oxidation in oil-in-water emulsions. *Journal of Agricultural and Food Chemistry*, **48**, 2057–2061.

Singh, H., Ye, A., Horne, D.S. (2009). Structuring food emulsions in the gastrointestinal tract to modify lipid digestion. *Progress in Lipid Research*, **48**, 92–100.

Sinha, J., Das, N., Basu, M.K. (2001). Liposomal antioxidants in combating ischemia-reperfusion injury in rat brain. *Biomedicine and Pharmacotherapy*, **55**, 264–271.

Skiba, M., Nemati, F., Puisieux, F., Duchêne, D., Wouessidjewe, D. (1996). Spontaneous formation of drug-containing amphiphilic β-cyclodextrin nanocapsules. *International Journal of Pharmaceutics*, **145**, 241–245.

Sok Line,V.L., Remondetto, G.E., Subirade, M. (2005). Cold gelation of β-lactoglobulin oil-in-water emulsions. *Food Hydrocolloids*, **19**, 269–278.

Spernath, A., Aserin, A. (2006). Microemulsions as carriers for drugs and nutraceuticals. *Advances in Colloid and Interface Science*, **128–130**, 47–64.

Spernath, A., Yaghmur, A., Aserin, A., Hoffman, R.E., Garti, N. (2002). Food grade microemulsions based on nonionic emulsifiers: media to enhance lycopene solubilization. *Journal of Agricultural and Food Chemistry*, **50**, 6917–6922.

Stringham, J.M., Hammond, B.R. (2005). Dietary lutein and zeaxanthin: possible effects on visual function. *Nutrition Reviews*, **63**, 59–64.

Takeuchi, H., Yamamoto, H., Kawashima, Y. (2001). Mucoadhesive nanoparticulate systems for peptide drug delivery. *Advanced Drug Reviews*, **47**, 39–54.

Tiede, K., Boxall, A.B.A., Tear, S.P., Lewis, J., David, H., Hassellov, M. (2008). Detection and characterization of engineered nanoparticles in food and the environment. *Food Additives and Contaminants*, **25**, 795–821.

Tur, J.A., Romaguera, D., Pons, A. (2005). Does the diet of the Balearic population, a Mediterranean-type diet, ensure compliance with nutritional objectives for the Spanish population? *Public Health Nutrition*, **8,** 275–283.

Ubbink, J., Krüger, J. (2006). Physical approaches for the delivery of active ingredients in foods. *Trends in Food Science and Technology*, **17**, 244–254.

Vandamme, Th.F., Lenourry, A., Charrueau, C., Chaumeil, J.-C. (2002). The use of polysaccharides to target drugs to the colon. *Carbohydrate Polymers*, **48**, 219–231.

Veerman, C., Baptist, H., Sagis, L.M.C., van der Linden, E. (2003). A new multistep Ca^{2+}-induced cold gelation process for β-lactoglobulin. *Journal of Agricultural and Food Chemistry*, **51**, 3880–3885.

Vega, C., Roos, Y.H. (2006). Spray-dried dairy and dairy-like emulsions — compositional considerations. *Journal of Dairy Science*, **89**, 383–401.

Velikov, K.P., Pelan, E. (2008). Colloidal delivery systems for micronutrients and nutraceuticals: tools and resources. *Soft Matter*, **4**, 1964–1980.

Versantvoort, C.H.M., van de Kamp, E., Rompelberg, C.J.M. (2004). Development of an *in vitro* digestion model to determine the bioaccessibility of contaminants from food. Report no. 320102002, Available from <http:/www.rivm.nl/en/>, National Institute for Public Health and the Environment, Bilthoven, the Netherlands.

Weber, C., Bysted, A., Holmer, G. (1997a). The coenzyme Q10 content of the average Danish diet. *International Journal of Vitamin and Nutrition Research*, **67**, 123–129.

Weber, C., Bysted, A., Holmer, G. (1997b). Intestinal absorption of coenzyme Q10 in a meal or as capsules to healthy subjects. *Nutrition Research*, **17**, 941–945.

Wong, N.C.W. (2001). The beneficial effects of plant sterols on serum cholesterol. *Canadian Journal of Cardiology*, **17**, 715–721.

Yokoyama, M., Miyauchi, M., Yamada, N., Okano, T., Sakurai, Y., Kataoka, K., Inoue, S. (1990). Characterization and anticancer activity of the micelle-forming polymeric anticancer drug adriamycin conjugated with poly(ethyleneglycol)–poly(aspartic acid) block copolymer. *Cancer Research*, **50**, 1693–1700.

Yoncheva, K., Gomez, S., Campanero, M.A., Gamazo, C., Irache, J.M. (2005). Bioadhesive properties of pegylated nanoparticles. *Expert Opinion on Drug Delivery*, **2**, 205–218.

Yuan, Y., Gao, Y., Zhao, J., Mao, L. (2008). Characterization and stability evaluation of β-carotene nanoemulsions prepared by high pressure homogenization under various emulsifying conditions. *Food Research International*, **41**, 61–68.

Zhang, C., Ding, Y., Ping, Q.E., Yu, L.L. (2006). Novel chitosan-derived nanomaterials and their micelle-forming properties. *Journal of Agricultural and Food Chemistry*, **54**, 8409–8416.

Zhang, Y., Yang, Y., Tang, K., Hu, X., Zou, G. (2008). Physicochemical characterization and antioxidant activity of quercetin-loaded chitosan nanoparticles. *Journal of Applied Polymer Science*, **107**, 891–897.

Zhu, Y.-Y., Zhang, G.-Y., Hong, X.-L., Dong, J.-F., Zhang, X.-G., Zeng, H. (2005). Synthesis of nanocapsules by complex coacervation in microemulsion. *Acta Chimica Sinica*, **63**, 1505–1509.

Zimet, P., Livney, Y.D. (2009). Beta-lactoglobulin and its nanocomplexes with pectin as vehicles for ω-3 polyunsaturated fatty acids. *Food Hydrocolloids*, **23**, 1120–1126.

CHAPTER THREE

THE THERMODYNAMIC APPROACH: ITS IMPORTANCE FOR UNDERSTANDING AND MANIPULATING THE MOLECULAR INTERACTIONS OF BIOPOLYMERS

Thermodynamics is based on the notion of the transformation of various kinds of energy through a set of exacting quantity relationships and laws. The universality and versatility of the thermodynamic laws and their consequences for arbitrary processes at equilibrium constitute the main strength and advantage of the thermodynamic approach. While equilibrium thermodynamics cannot say anything useful about kinetics of processes, it can nevertheless give us some insight into the possibility, in principle, of the occurrence of certain processes; and it can also estimate the extent of the approach to equilibrium (Prigogine and Defay, 1954; Edsall and Gutfreund, 1983).

In the field of food colloids, the use of molecular thermodynamics provides a set of qualitative and quantitative relationships describing fundamental phenomena occurring in the equilibrium state of systems for which the intermolecular interactions of biopolymers (proteins and polysaccharides) play a key role. The phenomena and processes amenable to discussion from the thermodynamic point of view are:

(i) biopolymer solubility and osmosis,
(ii) biopolymer aggregation and self-assembly,
(iii) biopolymer phase separation,
(iv) biopolymer adsorption at various types of interfaces, and
(v) interactions between the surfaces covered by adsorbed biopolymer layers (relating to the stability/instability of colloidal systems).

At its best, the thermodynamic approach can reveal both the role and the significance of the biopolymer interactions involved in these phenomena and processes, as well as allowing prediction or manipulation of the equilibrium under specific experimental conditions.

In general terms, the strength of the force F between a pair of biopolymer molecules or colloidal particles can be defined as the change in the universal potential of mean force $W(r)$ between the two molecules or particles as a function of their the centre-to-centre distance r (de Kruif, 1999; Prausnitz, 2003; de Kruif and Tuinier, 2005):

$$F = -\,\mathrm{d}W(r)/\mathrm{d}r\,. \qquad (3.1)$$

The quantity $W(r)$ is a free energy, which means that it includes both pure energy and entropic contributions. The expression 'mean force' indicates that, for any two particles *i* and *j* in the medium, the potential of mean force is the interaction free energy of the *i*–*j* pair, averaged with respect to all possible configurations of the particles in the medium. The 'medium' here includes solvent molecules, small ions, and all other species not explicitly considered to form part of the interacting particles. The potential of mean force $W(r)$ can be considered to be the sum of the free energies of attraction (A) and repulsion (R) between particles:

$$W(r) = W_{\mathrm{A}}(r) + W_{\mathrm{R}}(r)\,. \tag{3.2}$$

As a general principle, particles (or biopolymers) tend to become aggregated together at those separations where the attractive energy is greater in magnitude than the repulsive energy, and where $W(r)$ is most negative.

Theories or computer simulations used to calculate the potential of mean force $W(r)$ are typically based on numerous simplifying assumptions and approximations (de Kruif, 1999; Bratko *et al.*, 2002; Prausnitz, 2003; de Kruif and Tuinier, 2005; Horne *et al.*, 2007; Jönsson *et al.*, 2007). Therefore they can provide only a qualitative or, at best, semi-quantitative description of the potential of mean force. Such calculations are nevertheless useful because they can serve as a guide for trends in the factors determining the interactions of both biopolymers and colloidal particles. Thus, an increase in the absolute value of the calculated negative depth of $W(r)$ may be attributed to a predominant type of molecular feature favouring aggregation or self-association. To assist with such a theoretical analysis, expressions for some of the mean force potentials will be presented here in the discussion of specific kinds of interactions occurring between pairs of colloidal particles covered by biopolymers in food colloids.

To compile quantitatively reliable information, we need a source of experimental measurements. One way to determine the nature of intermolecular forces between biopolymer molecules in a solvent medium is to measure the so-called osmotic second virial coefficient A_2. Expressed in molar (biopolymer) terms, the quantity A_2 can be related to the two-body potential of mean force $W(r)$ by the following equation (Vrij, 1976; de Kruif, 1999; Prausnitz, 2003; de Kruif and Tuinier, 2005):

$$A_2 = -2\pi N_{\mathrm{A}} \int_0^{\infty} \left[\exp\left(\frac{-W(r)}{kT} \right) - 1 \right] r^2 \mathrm{d}r\,. \tag{3.3}$$

In equation (3.3), N_A is the Avogadro number, k is Boltzmann's constant, and T is the absolute temperature.

The osmotic second virial coefficient is traditionally measured by osmometry (Tombs and Peacocke, 1974), especially for rather small biopolymer molecules (number-average molar weight $M_n \leq 10^5$ Da). More commonly it is determined by laser light scattering for larger biopolymer molecules (weight-average molar weight in range 10^4 Da $\leq M_w \leq 10^9$ Da) (Evans, 1972, Kratochvil and Sudelof, 1986; Kaddur and Strazielle, 1986; Burchard, 1994; Semenova, 1996, 2007). Other useful techniques are equilibrium sedimentation (Wills *et al.*, 1996; Winzor *et al.*, 2001) and chromatography (Dumetz *et al.*, 2008) (see section 2 in chapter five for more details).

Regrettably, though, even the most careful and precise measurements of A_2 can give us only rough information about the form of the interaction potential. This is because $W(r)$ appears under an integral sign in equation (3.3), and so various alternative forms of $W(r)$ can satisfy the expression for A_2 (de Kruif, 1999; Prausnitz, 2003). Hence, experimental data for A_2 on any particular sample cannot give us a unique potential of mean force. In order interpret experimental data in terms of an exact expression for the potential of mean force at a fixed temperature, we must make some simplifying assumptions. Nevertheless, despite this limitation, precise thermodynamic criteria based on the value of the second virial coefficient can be used successfully to describe many kinds of properties, such as biopolymer solubility, phase separation, solvation, self-assembly, and osmosis in bulk biopolymer solutions, as well as the formal equilibrium conditions for reversible flocculation in colloidal systems stabilized by biopolymers (Vrij, 1976; Semenova *et al.*, 1990, 1999; Dickinson *et al.*, 1998a, 2001; de Kruif, 1999; Winzor *et al.*, 2001; Prausnitz, 2003; de Kruif and Tuinier, 2005; Semenova, 1996, 2007).

Before discussing all these biopolymer applications, we first take this opportunity to remind the reader that, in general, any thermodynamic variable can be expressed as the sum of two functions, one of which depends only on the temperature and pressure, and another which depends on the system composition (expressed as the mole fraction x_i of the i-component). Therefore, for example, the chemical potential μ_i of the i-component of the system at constant temperature T and pressure p (the general experimental conditions), *i.e.*, partial molar Gibbs free energy $(\partial G/\partial n_i)_{T,p}$ may be expressed as (Prigogine and Defay, 1954):

$$(\partial G/\partial n_i)_{T,p} = \mu_i = \mu_i^{0}(T,p) + \mu_i^{\text{mix}}(T,p,x_1,...,x_i) \quad . \qquad (3.4)$$

The first term in equation (3.4) is the standard chemical potential, and the second term is the mixing chemical potential, which is determined by the system composition. For the case of a thermodynamically ideal system, the quantity μ_i^{mix} is governed only by the mole fraction, *i.e.*:

$$\mu_i^{\text{mix}} = RT \ln x_i \quad , \tag{3.5}$$

where R is the gas constant. For the thermodynamically non-ideal (real) system, where the components interact, an activity term a_i is introduced to take into account the effect of the interactions between components, while keeping the mathematical form of the relationship unchanged:

$$\mu_i^{\text{mix}} = RT \ln a_i = RT \ln \gamma_i x_i \quad . \tag{3.6}$$

The quantity $\gamma_i = a_i/x_i$ is known as the activity coefficient (Prigogine and Defay, 1954).

1. *Two-Component Solutions: Biopolymer + Solvent*

We first consider the case of a two-component solution (biopolymer + solvent) over a moderately low range of biopolymer concentrations, *i.e.*, $C < 20$ % wt/wt. The quantities μ_i^{mix} in the equations for the chemical potentials of solvent and biopolymer may be expressed as a power series in the biopolymer concentration, with some restriction on the required number of terms, depending on the steepness of the series convergence and the desired accuracy of the calculations (Prigogine and Defay, 1954). This approach is based on simplified equations for the chemical potentials of both components as a virial series in biopolymer concentration, as developed by Ogston (1962) at the level of approximation of just pairwise molecular interactions:

$$\mu_1 = \mu_1^0 - (RT/m_1) \times (m_2 + \tfrac{1}{2} A_2^* m_2^2) \quad , \tag{3.7}$$

$$\mu_2 = \mu_2^0 + RT\,[\ln(m_2/m^0) + A_2^* m_2] \quad . \tag{3.8}$$

Here, μ_i^0 and m_i are the standard chemical potential and concentration (molal scale) of the i-component ($i = 1$ for solvent, $i = 2$ for biopolymer); A_2^* is the second virial coefficient (in molal scale units of cm^3/mol, *i.e.*, taking the polymer molar mass into account); and m^0 is the standard-state molality for the polymer.

Following on from equation (3.5), we note that it is the value of the second virial coefficient A_2 that determines the osmotic pressure of the biopolymer solution:

$$\Pi = (\mu_1{}^0 - \mu_1)/V_1 = RT(n + A_2C^2)\,. \tag{3.9}$$

Here, the quantities $\mu_1{}^0$ and μ_1 are, respectively, the chemical potentials of pure solvent and of the solvent at a certain biopolymer concentration; V_1 is the molar volume of the solvent; and n is the biopolymer number density, defined as $n = C/M$, where C is the biopolymer concentration (% wt/wt) and M is the number-averaged molar weight of the biopolymer. The second virial coefficient has (weight-scale) units of cm^3 mol g^{-2}. Hence, the more positive the second virial coefficient, the larger is the osmotic pressure in the bulk of the biopolymer solution. This has consequences for the fluctuations in the biopolymer concentration in solution, which affects the solubility of the biopolymer in the solvent, and also the stability of colloidal systems, as will be discussed later on in this chapter.

It is essential to take note here of the relationship between concentrations expressed in molal and weight units (Wells, 1984):

$$m_i = \frac{\dfrac{c_i\,\overline{v_1}}{M_i} \times 1000}{1 - c_i\,\overline{v_i}}\;. \tag{3.10}$$

Here, m_i is the molal concentration of the i-component in mol/1000 g water; c_i is the weight concentration of the i-component in g/ml, v_1 and v_i (with bars) are the specific volumes of water and biopolymer in ml/g; and M_i is the molar weight of the biopolymer in g/mol (Da). In turn, the relationship between the second virial coefficients expressed in the different units (molal, A_i^*; weight, A_i) is as follows (Wells, 1984):

$$A_i^* = \frac{2\,A_i\,M_i^2}{1000}\;. \tag{3.11}$$

The set of equations (3.7–3.9) shows that the sign and magnitude of the second virial coefficient provides information on how the behaviour of the macromolecular solution deviates from that of the thermodynamically ideal state, thus reflecting the nature and intensity of the intermolecular pair interactions (both biopolymer–biopolymer and biopolymer–solvent) (Prigogine and Defay, 1954; Tanford, 1961; Ogston 1962;

Nagasawa and Takahashi, 1972). More specifically, the value of the second virial coefficient determines the excess chemical potential, μ_i^{E} (also known as the excess partial molar Gibbs free energy), which characterizes the formation of biopolymer–solvent and biopolymer–biopolymer pair contacts:

$$\mu_1^{\mathrm{E}} = \mu_1^{\mathrm{mix}} - \mu_1^{\mathrm{mix,id}} = -\tfrac{1}{2}\, RT\,(A_2^{*} m_2^{2}/m_1) \quad , \tag{3.12}$$

$$\mu_2^{\mathrm{E}} = \mu_2^{\mathrm{mix}} - \mu_2^{\mathrm{mix,id}} = RT\, A_2^{*} m_2 \quad . \tag{3.13}$$

From the experimental temperature dependence of A_2^{*} (and the corresponding inferred temperature dependence of μ_i^{E}), the other basic excess thermodynamic functions can be determined using general thermodynamic relationships. This then provides a complete thermodynamic characterization of the system as a whole. Thus, for the determination of the excess molar enthalpy of the system at constant pressure, the following equation can be used (Prigogine and Defay, 1954):

$$h^{\mathrm{E}} = -T^{2} \frac{\partial(\mu^{\mathrm{E}}/T)}{\partial T} \quad . \tag{3.14}$$

(The excess molar enthalpy h^{E} is simply the heat of mixing at constant pressure related to 1 mole of solution.) And from the excess molar enthalpy and the excess chemical potential, we can obtain the excess molar entropy of the system from the following equation:

$$s^{\mathrm{E}} = \frac{h^{\mathrm{E}} - \mu^{\mathrm{E}}}{T} \quad . \tag{3.15}$$

The above listed excess thermodynamic functions all relate to 1 mole of solution. For the system as a whole, containing a total number of moles of all components, $n = \sum_i n_i$, we have:

$$G^{\mathrm{E}} = n\mu^{\mathrm{E}}; \quad H^{\mathrm{E}} = n h^{\mathrm{E}}; \quad S^{\mathrm{E}} = n s^{\mathrm{E}} \quad . \tag{3.16}$$

A knowledge of the magnitude of these quantities and their quantitative contributions to μ^{E} can give insight into the detailed character of the intermolecular interactions in a biopolymer solution, including the means by which their properties may be manipulated. The sign of the second virial coefficient provides a simple indicator of the type of interactions present in a biopolymer solution. Hence, a negative value of A_2 indicates thermodynamically favourable biopolymer–biopolymer interactions in a solution (decrease in the magnitude of μ_2^E) — in other words a mutual biopolymer attraction. A negative A_2 also indicates thermodynamically *un*favourable biopolymer–*solvent* interactions (increase in the magnitude of μ_1^E) — in other words a mutual repulsion. The exact opposite is the case for a positive value of the second virial coefficient.

The presence of thermodynamically favourable biopolymer–solvent interactions implies a good biopolymer solubility. Conversely, the presence of thermodynamically favourable biopolymer–biopolymer interactions ('poor solvent' conditions) is the driving force for the coil–helix transitions of polysaccharides, the folding of proteins, and many kinds of biopolymer associative self-assembly, leading to, for instance, the formation of soluble protein aggregates (de Kruif and Tuinier, 2001; Semenova *et al.*, 2005, 2007), or precipitates, or even protein crystals under optimum experimental conditions (George *et al.*, 1997; de Kruif, 1999; de Kruif and Tuinier, 2001; Prausnitz, 2003). In addition, on a macroscopic scale, the presence of attractive biopolymer–biopolymer interactions is associated with solutions of high viscosity, low diffusion rates, and gel-like structuring (de Kruif, 1999).

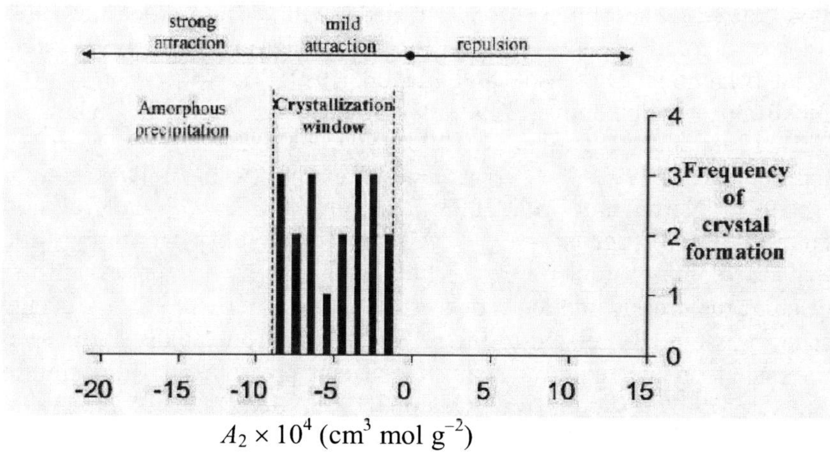

Figure 3.1 Dependence of protein crystallization on the thermodynamic conditions. The frequency of crystal formation is plotted against the osmotic second virial coefficient A_2. Data are taken from a broad range of proteins and crystallizing solvent conditions (George *et al.*, 1997). Reproduced from Prausnitz (2003) with permission.

An interesting example of the use of the experimental osmotic second virial coefficient is as a thermodynamic screening criterion for the conditions favourable for the crystallization of proteins (George *et al.*, 1997). Figure 3.1 shows that there is an accessible crystallization window, corresponding to a value of the second virial coefficient in the range from -1×10^{-4} to -8×10^{-4} cm^3 mol g^{-2}. In order for a crystal to form and grow, the dissolved protein molecules must be in the correct orientation. If forces of attraction are too strong ($A_2 < -8 \times 10^{-4}$ cm^3 mol g^{-2}), the individual protein molecules do not have sufficient time to 'get together' in just the right way, and so they agglomerate into an amorphous precipitate. On the other hand, if there are no significant forces of attraction ($A_2 \geq 0$), the solution is stable, and there is no formation of any precipitate, either amorphous or crystalline. Therefore, for the crystals to form, the forces of attraction should be rather weak; in other words, the value of the second virial coefficient should be negative, but not excessively so.

2. *Three-Component Solutions: Biopolymer*$_1$ + *Biopolymer*$_2$ + *Solvent*

The interactions in solution between two different biopolymers (protein + protein, polysaccharide + polysaccharide, protein + polysaccharide) are of primary importance in determining the microstructure and stability of food colloids. These interactions affect the bulk system properties (biopolymer incompatibility/compatibility and rheology) as well as the interfacial behaviour (Dickinson and Euston, 1991; Tsapkina *et al.*, 1992; Dickinson and Semenova, 1992; Pavlovskaya *et al.*, 1993; Dickinson, 1993, 1998, 2003; Tolstoguzov, 1996, 1997, 2000, 2002, 2003; Antipova and Semenova, 1997a,b; Dickinson and Pawlowsky, 1997; Dickinson *et al.*, 1998a; Norton and Frith, 2003; Semenova *et al.*, 1999; de Kruif and Tuinier, 2001; Turgeon *et al.*, 2003). The quantitative thermodynamic analysis of the character of such interactions is expressed through the value of the cross osmotic second virial coefficient A_{ij} (with $i \neq j$). This quantity is directly related to the set of chemical potentials, μ_i, of each component in the mixed system at constant pressure and temperature (Ogston, 1962; Edmond and Ogston, 1968; Semenova *et al.*, 1990, 1991a,b, 1999; Antipova and Semenova, 1995; Semenova and Savilova, 1998; Wasserman *et al.*, 1997; Semenova, 1996, 2007). In particular, for the ternary system, biopolymer$_1$ + biopolymer$_2$ + solvent, we have the following set of equations written at the level of approximation of the osmotic second virial coefficient (*i.e.*, taking into account only the pair interactions) (Ogston, 1962; Edmond and Ogston, 1968):

$$\mu_2 = \mu_2^0 + RT\,[\ln(m_2/m^0) + A_{22}^* m_2 + A_{24}^* m_4]\ , \tag{3.17}$$

$$\mu_4 = \mu_4^0 + RT\,[\ln(m_4/m^0) + A_{44}^* m_4 + A_{24}^* m_4]\ , \tag{3.18}$$

$$\mu_1 = \mu_1^0 - (RT/m_1)[m_2 + m_4 + ½(A_{22}^* m_2^2 + A_{44}^* m_4^2) + A_{24}^* m_2 m_4]\,. \tag{3.19}$$

Here, the solvent is component number 1, the biopolymer$_1$ is component number 2, and the biopolymer$_2$ is component number 4;[1] μ_i^0 and m^0 are the standard chemical potential and concentration (molal scale) of the *i*-component (i = 1, 2, 4), respectively; m_i is the concentration of the *i*-component in the system; A_{22}^* and A_{44}^* are the second virial coefficients (molal scale) characterizing the like pair interactions of types biopolymer$_1$–biopolymer$_1$ and biopolymer$_2$–biopolymer$_2$, respectively; and A_{24}^* is the cross second virial coefficient (molal scale) for the unlike biopolymer$_1$–biopolymer$_2$ pair interaction. We can see from equations (3.17) and (3.18) that a positive value of the cross second virial coefficient A_{24}^*, corresponding to a thermodynamically unfavourable (repulsive) interaction between unlike biopolymers, leads to an increase in the magnitude of their chemical potentials, *i.e.* to an increase in their thermodynamic activities in the mixed solution. The opposite is the case for a negative value of A_{24}^*.

As a consequence of thermodynamically unfavourable interactions, the phenomenon of phase separation can occur in mixed biopolymer solutions through the segregative mechanism ('simple coacervation') at moderately high biopolymer concentrations (Albertsson, 1971; Grinberg and Tolstoguzov, 1997; Schmitt *et al.*, 1998; Doublier *et al.*, 2000; de Kruif and Tuinier, 2001; Tolstoguzov, 2000, 2002, 2003; Benichou *et al.*, 2002; Turgeon *et al.*, 2003). This leads to the formation of water-in-water (W/W) emulsions with each of the two aqueous phases separately enriched by one of the polymer components. A phase may exist as the dispersed phase or the dispersion medium, depending on the system composition relative to the rectilinear diameter in the phase diagram (Albertsson, 1971). Two peculiarities of such emulsions are the co-solubility of the biopolymers in the coexisting phases and the very low interfacial tension (10^{-5}–10^{-6} N m^{-1}) (Guido *et al.*, 2002; Van Puyvelde *et al.*, 2002). This type of thermodynamic incompatibility commonly occurs in systems where one or both of the biopolymers are uncharged, or where both bio-

[1] In identifying the species in this biopolymer solution, we are following the convention of allocating odd numbers to low-molecular-weight components (solvent, salts) and even numbers to the polymeric components. So, for this system of biopolymer$_1$ + biopolymer$_2$ + solvent, there is no component labelled number 3.

polymers have similar signs of net electrical charge. Effects of excluded volume interactions and electrostatic repulsion between polymers of similar charge sign leads to segregative interactions between the incompatible biopolymers (Semenova, 1996; Tolstoguzov, 1997; Schaink and Smit, 1997; Semenova and Savilova, 1998; Tuinier and de Kruif, 1999; Tuinier *et al.*, 2000; Doublier *et al.*, 2000). The unwanted effect of phase separation in a food product can be suppressed by maintaining biopolymer concentrations within the stable homogeneous region. Conversely, water-in-water emulsions can be used as a positive structural feature in innovative food formulations (Norton and Frith, 2001; Tolstoguzov, 2003). For example, we can envisage the use of different types of multiple emulsions (oil-in-water$_1$-in-water$_2$, *etc.*) (see Figure 3.2) (Kim *et al.*, 2006) in the production of reduced fat foods with desirable flavour profiles (Malone *et al.*, 2003) and in the controlled release of lipophilic components from oil droplets (Malone and Appelqvist, 2003). Moreover, an understanding of the character of the interactions between biopolymers may allow the manipulation of food sensory properties by adjusting the interactions in a desirable way. For instance, novel food structures could be (and are being) produced by letting a mixed system phase separate partially. By 'freezing' the transition through cold-set gelation of gelatin or by the thermal denaturation and aggregation of heat-sensitive globular proteins, the desired phase-separated microstructures can be made to be effectively permanent (de Kruif and Tuinier, 2001).

When interactions are thermodynamically favourable ($A_{24}^* < 0$) there are three possible states of the system: biopolymer compatibility, soluble complex formation, or phase separation by 'complex coacervation'. In the case of complex coacervation, one of the phases contains predominantly the macromolecular complex and the other phase contains mainly the solvent (Tolstoguzov, 1997; Dickinson *et al.*, 1998a; Schmitt *et al.*, 1998; Semenova *et al.*, 1991a,b, 1999; Maroziene and de Kruif, 2000; Doublier *et al.*, 2000; de Kruif and Tuinier, 2001; Turgeon *et al.*, 2003; Dickinson, 2003; de Kruif *et al.*, 2004). Typically we find associative interactions when there are oppositely charged ionic groups or hydrogen bonding between biopolymers. Some authors have interpreted electrostatic complexation between the proteins and polysaccharides as being mainly enthalpy driven, causing a decrease in electrostatic free energy of the system (de Kruif and Tuinier, 2001; Girard *et al.*, 2003), whereas others have indicated that the processes of biopolymer compatibility and complexation are mainly entropy driven, due to the liberation of counterions and water molecules (Semenova *et al.*, 1991b; Appelqvist and Debet, 1997; Cai and Arntfield, 1997; Dautzenberg, 2001; Ball *et al.*, 2002). The generation of attractive electrostatic interactions at oil–water interfaces can be used for the creation of food emulsions with improved stability towards environmental stresses or with novel encapsulation–release characteristics (McClements, 2006).

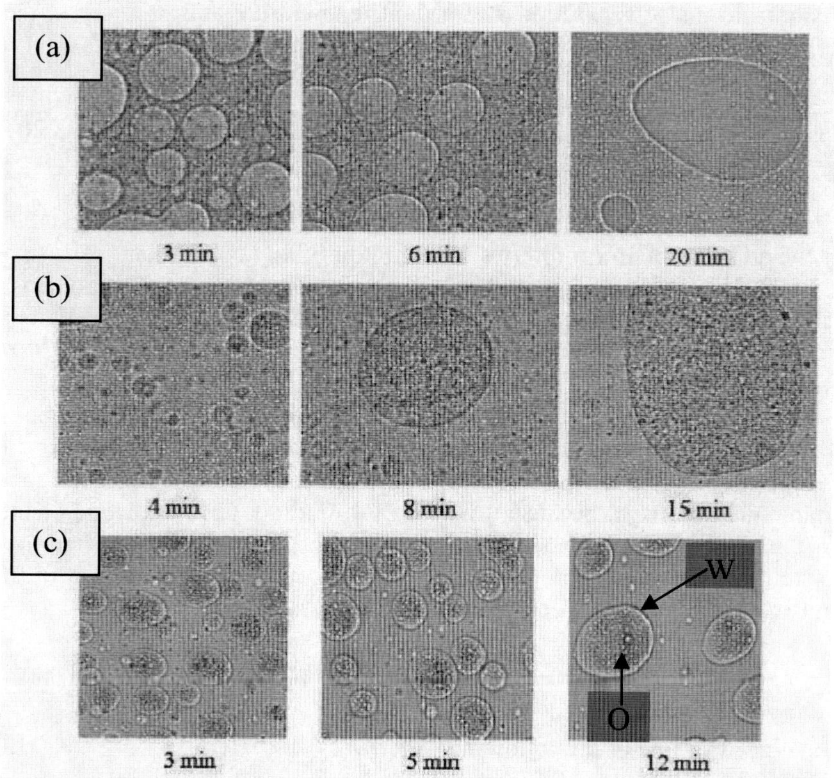

Figure 3.2 Evolution of the microstructure of phase-separated biopolymer emulsion system containing pectin and 0.5 wt% heat-denatured (HD) whey protein isolate (WPI) stabilized oil droplets. (a) Composition 1U:3L (one-to-three mass ratio of upper and lower phases). The large circles are the water droplets (W), while the small circles are the oil droplets (O). This system forms a W_2/W_1–O/W_1 emulsion, where O is oil, W_1 is HD-WPI-rich and W_2 is pectin-rich. (b) Composition 2U:2L. This system forms an $O/W_1/W_2$ emulsion, where O is oil, W_1 is HD-WPI-rich and W_2 is pectin-rich. (c) Composition 3U:1L. This system forms an $O/W_1/W_2$ emulsion, where O is oil, W_1 is HD-WPI-rich and W_2 is pectin-rich. Reproduced from Kim *et al.* (2006) with permission.

Knowledge of the expressions for the chemical potentials of each of the components allows theoretical prediction of the critical concentration boundaries of the phase diagram for ternary solutions of biopolymer$_1$ + biopolymer$_2$ + solvent. According to Prigogine and Defay (1954), a sufficient condition for material stability of this multicomponent system in relation to phase separation at constant temperature and pressure is the following set of inequalities for all the components of the system:

$$\mu_{ii} = \frac{\partial(\mu_i)_{T,p}}{\partial n_i} > 0 \ , \quad \mu_{ij} = \frac{\partial(\mu_i)_{T,p}}{\partial n_j} < 0 \quad (i \neq j) \ . \tag{3.20}$$

Equation (3.20) implies that the system will be thermodynamically stable if the addition of an infinitely small amount of any component leads to a decrease in chemical potentials of all the other constituent components. The fulfilment of the second inequality in equation (3.20) is a sufficient condition for the stability of the multicomponent system with respect to mutual diffusion.

There is, however, another statement of the necessary and sufficient condition of thermodynamic stability of the multicomponent system in relation to mutual diffusion and phase separation that is less stringent than equation (3.20) because it may be fulfilled not for every component of the multicomponent system. For example, in the case of the ternary system biopolymer$_1$ + biopolymer$_2$ + solvent, it appears enough to fulfil only two of the inequalities (Prigogine and Defay, 1954)

$$\mu_{11} > 0 , \ \mu_{22} > 0 , \ \mu_{44} > 0 , \tag{3.21}$$

together with one of the following:

$$\begin{aligned} \mu_{11}\mu_{22} - \mu_{12}^2 \geq 0 \quad , \\ \mu_{11}\mu_{44} - \mu_{14}^2 \geq 0 \quad , \\ \mu_{22}\mu_{44} - \mu_{24}^2 \geq 0 \quad . \end{aligned} \tag{3.22}$$

So, for stability, an arbitrary set of any three of these inequalities could be satisfied, *e.g.*,

$$\mu_{22} > 0 , \ \mu_{44} > 0 , \ \mu_{22}\mu_{44} - \mu_{24}^2 \geq 0 \ . \tag{3.23}$$

The fulfilment of the inequalities in equation (3.20) implies fulfilment of the inequalities in equation (3.23). But the reverse is not true: from the fulfilment of the inequalities in equation (3.23), it follows that equation (3.20) is satisfied only if we have

$$|\mu_{24}| \leq \sqrt{\mu_{22}\mu_{44}} \quad . \tag{3.24}$$

Moreover, the conditions in equation (3.20) are always fulfilled for the thermodynamically ideal system; so the thermodynamically ideal system is always stable with respect to fluctuations in the system composition. (For a full understanding of the development of the preceding equations, the interested reader should refer to the seminal work of Prigogine and Defay (1954).)

According to the above criteria, the analysis of the dependence of the chemical potentials of the components of the system on its composition can allow us to estimate whether the system will be thermodynamically stable with respect to compositional fluctuations and hence to phase separation. Thus, in order to determine the boundary conditions separating the region of absolute instability from both the metastable region and the stable region, one should analyse the set of inequalities in equation (3.23) and determine which of them fails first on passing from the region of stability to the region of instability. First and foremost, it is evident that, if the condition

$$\mu_{22}\mu_{44} - \mu_{24}^2 = 0 \tag{3.25}$$

coexists with either $\mu_{22} = 0$ or $\mu_{44} = 0$, then such a composition must lie inside the unstable region. Hence, the boundary separating the region of absolute instability from the metastable and stable regions is described by the Gibbs equation of the spinodal curve, which is in fact equation (3.25) (Prigogine and Defay, 1954).

On the basis of these relationships, using expressions for the chemical potentials of the components at the level of approximation of the second virial coefficients (equations 3.17 and 3.18), the spinodal curve (equation 3.25) can be expressed mathematically in the following form (Edmond and Ogston, 1968):

$$(\frac{1}{m_2} + A_{22}^*)(\frac{1}{m_4} + A_{44}^*) - A_{24}^{*2} = 0 \quad . \tag{3.26}$$

In practice, from a knowledge of measured values of the osmotic second virial coefficients it is rather easy to calculate the spinodal curve. It is worthy of note here to observe that reciprocal values of m_i for biopolymers of rather high molecular weight ($\geq 10^4$ g/mol) are often comparable with the magnitude of A_{24}^*. This requires that, as well as values of the osmotic second virial coefficients, the molecular weight should also be taken into account in the prediction of the boundary conditions relating to phase separation.

The binodal (or coexistence) curve, on which the compositions of the immiscible solutions (phases) lie at equilibrium, can be described by a set of equations involving equilibrium between the chemical potentials of the components in the coexisting phases (Prigogine and Defay, 1954):

$$\mu_1' = \mu_1'' \ , \quad \mu_2' = \mu_2'' \ , \quad \mu_3' = \mu_3'' \ . \tag{3.27}$$

On the basis of these relationships, using the expressions of the chemical potentials of the components at the level of approximation of the second virial coefficients, the binodal curve can be expressed by the following set of equations (Edmond and Ogston, 1968):

$$\frac{Q}{A_{24}^*} + (m_4' - m_4'') + \frac{A_{22}^* Q^2}{2A_{24}^{*2}}\left(\frac{e^P+1}{e^P-1}\right) + \frac{A_{44}^*}{2}(m_4'^2 - m_4''^2) + \frac{Q}{e^P-1}(m_4' e^P - m_4'') = 0 \ ,$$

$$P = \frac{A_{22}^*}{A_{24}^*}\ln\left(\frac{m_4'}{m_4''}\right) + \frac{A_{22}^* A_{44}^* - A_{24}^{*2}}{A_{24}^*}(m_4' - m_4'') \ ,$$

$$Q = -\left[\ln\left(\frac{m_4'}{m_4''}\right) + A_{44}^*(m_4' - m_4'')\right] \ ,$$

$$m_2' = m_2'' e^P = \frac{Q e^P}{A_{24}^*(e^P - 1)} \ . \tag{3.28}$$

In practice, it turns out that it is an extremely time-consuming procedure to calculate the binodal curve.

As well as the spinodal and binodal curves, the phase diagram of the system is characterized by the coordinates of the critical point. This is the single common point of intersection of the spinodal and binodal curves, where compositions of coexisting phases are identical. The critical point is defined by (Prigogine and Defay, 1954):

$$\left(\frac{\partial \mu_i}{\partial n_i}\right)_{T,p,n_j,\mu_j} = 0 \ , \quad \left(\frac{\partial^2 \mu_i}{\partial n_i^2}\right)_{T,p,n_j,\mu_j} = 0 \ . \tag{3.29}$$

On the basis of the relationships in equation (3.29), as considered at the level of approximation of the second virial coefficients, Edmond and Ogston (1968) derived the following set of equations for describing the coordinates of the critical point:

$$\frac{1}{m_2^c} + A_{22}^* = A_{24}^* \left(\frac{m_4^*}{m_2^*}\right)^{\frac{2}{3}} \ ,$$

$$\frac{1}{m_4^c} + A_{44}^* = A_{24}^* \left(\frac{m_2^*}{m_4^*}\right)^{\frac{2}{3}} \ . \tag{3.30}$$

Numerous research papers and reviews have demonstrated the usefulness and significance of the above approach for calculating the phase diagram from the experimentally determined osmotic second virial coefficients (Edmond and Ogston, 1968; Semenova *et al.*, 1990, 1991a,b; Semenova, 1996, 2007; Semenova and Savilova, 1998). By way of illustration, Figure 3.3 presents the phase diagram of the ternary system composed of glycinin + pectinate + water (Semenova *et al.*, 1990). We note that the calculated spinodal line and critical point are in good agreement with the experimentally obtained critical point and binodal curve.

Another important application of experimentally determined values of the osmotic second virial coefficient is in the estimation of the corresponding values of the Flory–Huggins interaction parameters χ_{12}, χ_{14} and χ_{24}. In practice, these parameters are commonly used within the framework of the Flory–Huggins lattice model approach to the thermodynamic description of solutions of polymer + solvent or polymer$_1$ + polymer$_2$ + solvent (Flory, 1942; Huggins, 1942; Tanford, 1961; Zeman and Patterson, 1972; Hsu and Prausnitz, 1974; Johansson *et al.*, 2000):

$$\chi_{1i} = \frac{1}{2} - \frac{\bar{v}_1 M_1}{\bar{v}_i^2} A_{1i} \ , \qquad (i = 2,4) \tag{3.31}$$

$$\chi_{24} = \frac{2\bar{v}_1 M_1}{\bar{v}_2 \bar{v}_4} A_{24} - 1 + \chi_{12} + \chi_{14} \quad . \tag{3.32}$$

Here $\bar{v}_i$ and $\bar{v}_1$ are the partial specific volumes of the polymer (i = 2,4) and the solvent, respectively; M_1 is the molar weight of the solvent; and χ_{1i} and χ_{24} are the Flory–Huggins interaction parameters, quantifying the energy of interaction between unlike lattice-based polymer segments (χ_{24}) or between polymer segments and solvent molecules (χ_{1i}).

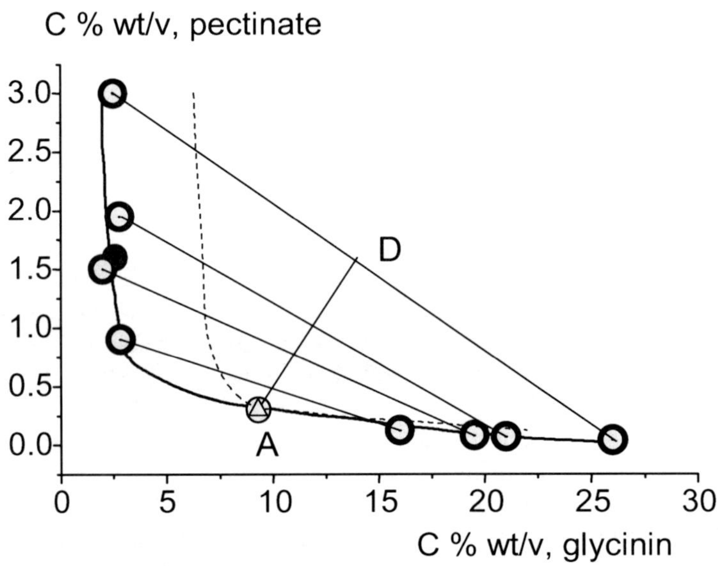

Figure 3.3 Illustration of the calculation of the phase diagram of a mixed biopolymer solution from the experimentally determined osmotic second virial coefficients. The phase diagram of the ternary system glycinin + pectinate + water (pH = 8.0, 0.3 mol/dm^3 NaCl, 0.01 mol/dm^3 mercaptoethanol, 25 °C): —, experimental binodal curve; ----, calculated spinodal curve; ○, experimental critical point; △, calculated critical point; **O—O**, binodal tie-lines; AD, rectilinear diameter; ●, the threshold of phase separation (defined as the point on the binodal curve corresponding to minimal total concentration of biopolymer components). Reproduced from Semenova *et al.* (1990) with permission.

The Flory–Huggins theory has been found to be useful for the qualitative theoretical analysis of the phase behaviour of several uncharged polymer-based ternary systems. But it has been found to fail in cases where the molecular weight polydispersity is extensive (Clark, 2000). The Flory–Huggins approach conveniently includes qualitative effects of the signs and magnitudes of the energies of interaction amongst all the components on the basis of a 'mean field' approximation. That is, the composition of nearest neighbours around every lattice site is assumed to be the same as the average composition of the phase. A disadvantage of the theory, however, is that the macromolecules are treated as uncharged 'random walk' unbranched chains, in which each segment of every molecule is free to sample any part of the lattice. Hence, this theory would seem in practice to be of limited usefulness for most charged biopolymers, including many food proteins having pronounced secondary and tertiary structure and food polysaccharides having rigid or semi-rigid conformations. That having been said, the Flory–Huggins approach has proven useful for qualitatively describing the general features of protein distributions in two-phase aqueous solutions, especially at ionic strengths that are sufficiently high to screen the electrostatic interactions (Walter *et al.,* 1991; Johansson *et al.*, 1998, 2000). On balance, though, it is considered by the authors of this book that the approach using osmotic second virial coefficients offers a greater universal applicability for the thermodynamic description of mixed biopolymer solutions.

An additional experimentally accessible (and useful) thermodynamic parameter for quantitatively characterizing the biopolymer–biopolymer interactions in solution is the enthalpy change measured directly from isothermal titration/mixing calorimetry (h^E), *i.e.*, the heat of mixing of the biopolymers in solution (Semenova *et al.*, 1991b; Cesàro *et al.*, 1999; Girard *et al.*, 2003). From the experimentally determined temperature dependence of h^E, the temperature dependence of μ_i^E and consequently s^E can be inferred using general thermodynamic relationships, thereby providing a complete thermodynamic characterization of the system as a whole (see equations 3.14 and 3.15). Moreover, the nature of the different kinds of biopolymer–biopolymer interactions (electrostatic, hydrophobic, hydrogen bonding, *etc.*) can be inferred from the sign of the measured enthalpy change (see chapter four for more details). Thus, the knowledge of the sign and absolute value of measured enthalpy changes allows one to make use of heating and cooling procedures in the tailor-made manipulation of biopolymer interactions for the establishment of desirable macroscopic properties of food colloids.

3. *Food Colloids Stabilized by Biopolymers*

3.1. *Surface Behaviour of Biopolymers*

Information on the chemical potentials of components in a solution of biopolymers can serve as a guide to trends in surface activity of the biopolymers at fluid interfaces (air–water, oil–water). In the thermodynamic context we need look no further than the Gibbs adsorption equation,

$$d\gamma = -\sum_i \Gamma_i d\mu_i \, , \tag{3.33}$$

where γ is the surface (interfacial) tension, Γ_i is the surface concentration of the *i*-component of the solution, and μ_i is the chemical potential of the *i*-component at the interface. At equilibrium the chemical potentials of the biopolymer in the bulk and at the interface should be equal. Hence, if the chemical potential of the biopolymer in the bulk increases, then that at the interface will increase also.

Let us first consider the case of a thermodynamically unfavourable interaction ($A_{24}^* > 0$) between protein and polysaccharide. Mixing causes enhanced values of the chemical potentials (thermodynamic activities) in the bulk of the aqueous phase (see equations (3.17) and (3.18)). This has important consequences for both the surface activity of the adsorbing protein and the colloid stability of corresponding oil-in-water emulsions (Tsapkina *et al.*, 1992; Dickinson and Semenova, 1992; Pavlovskaya *et al.*, 1993; Antipova and Semenova, 1997a; Dickinson *et al.*, 1998a; Semenova *et al.*, 1999; Semenova, 2007). A consequence of biopolymer incompatibility is a drastic increase in protein surface activity at the planar oil–water interface. This is indicated by an increase in both the rate of protein adsorption and the steady-state value of the surface (interfacial) pressure[2] (π) of the protein adsorbed layer. There is also a significant increase in protein loading (Γ) on the emulsion droplets (see Table 3.1). It is important to be aware as to whether the biopolymer mixture is close to or within the compositional area corresponding to liquid–liquid phase separation, since, if this is the case, thermodynamically unfavourable interactions will tend to induce multilayer formation at the interface by the adsorbing biopolymer. Table 3.1 illustrates this situation for the system containing legumin + dextran (Tsapkina *et al.*, 1992; Dickinson and Semenova, 1992).

[2] Surface pressure π is defined by $\pi = \gamma_0 - \gamma$, where γ_0 and γ are the values of the surface tensions for the solvent and mixed solution, respectively.

Table 3.1 The influence of the character of the biopolymer–biopolymer interactions on the protein loading on emulsion oil droplets.

System	$A_{24}^{*} \times 10^{-5}$ (cm^3/mol)	$\Delta\Gamma = \Gamma_{pr}^{mixture} - \Gamma_{pr}^{alone}$ (mg/m^2)	Reference
legumin + dextran (in region above binodal)	0.3	4.5	Tsapkina *et al.*, 1992
β-casein + pectinate (in region below binodal)	26.5	3.7	Semenova *et al.*, 1999
α_{s1}-casein + pectinate	– 334	0.2	Semenova *et al.*, 1999
sodium caseinate + pectinate	– 155	0.1	Dickinson *et al.*, 1998a

As a result of the increase in the biopolymer surface activity, there is commonly a decrease in the mean size of emulsion droplets produced, *i.e.*, an increase in the total surface area of the fresh emulsions stabilized by the mixture of the thermodynamically unfavourable biopolymers (Dickinson and Semenova, 1992; Tsapkina *et al.*, 1992; Semenova *et al.*, 1999). A moderate increase in protein thermodynamic activity can therefore lead to improved emulsifying capacity, but the opposite may be the case for strong incompatibility. This is illustrated by the comparison in Figure 3.4 of the sizes of emulsion droplets stabilized by legumin in the presence of dextran having two alternative molar masses of 48 kDa and 500 kDa (Dickinson and Semenova, 1992; Semenova and Savilova, 1998). It is assumed that incipient phase separation during emulsification in the mixed aqueous phase containing legumin and the high-molecular-weight dextran makes it more difficult for the protein to get rapidly to the surface of newly formed droplets in order to stabilize them properly against recoalescence.

Let us turn now to consider systems with thermodynamically favourable interaction ($A_{24}^{*} < 0$) (*i.e.*, mutual attraction) between protein and polysaccharide. Here there is little measurable effect on the protein loading (see Table 3.1) (Semenova *et al.*, 1999). However, an important consequence of the protein–polysaccharide attraction is a substantial increase in the surface shear viscosity of the protein adsorbed layer (Dickinson and Euston, 1991; Dickinson and Galazka, 1992; Dickinson *et al.*, 1998a; Semenova *et al.*, 1999; Dickinson, 2003; Semenova, 2007).

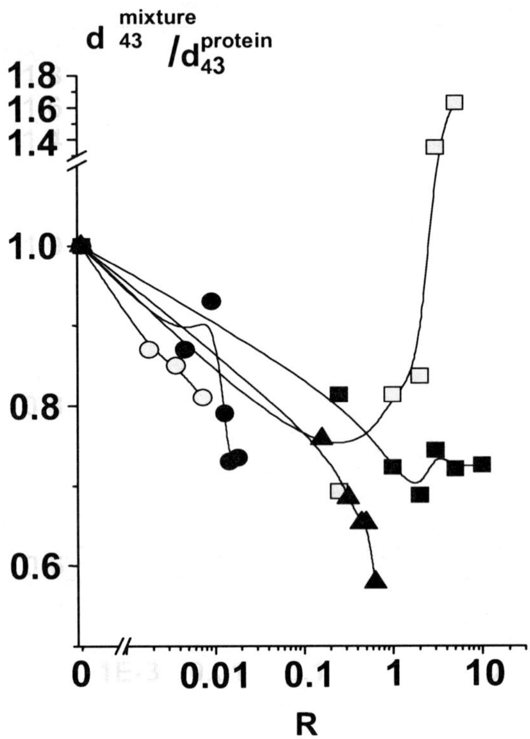

Figure 3.4 Effect of polysaccharide on protein-stabilized emulsions. The diameter ratio, $d_{43}^{\text{mixture}} / d_{43}^{\text{protein}}$, is plotted against the molar ratio R (moles polysaccharide / moles protein). Here d_{43}^{mixture} is average droplet diameter in fresh emulsion prepared with protein + polysaccharide, and d_{43}^{protein} is average diameter in emulsion stabilized by protein alone. Key: ■, □, legumin + dextran (48 kDa) or legumin + dextran (500 kDa), respectively (0.5 w/v % protein, 10 vol% oil, pH = 8.0, I = 0.1 M) (Dickinson and Semenova, 1992); ○, ●, α_{s1}-casein + pectinate and β-casein + pectinate at pH = 7.0, I = 0.01 M (2.0 w/v % protein, 40 vol% oil), respectively; ▲, β-casein + pectinate at pH = 5.5, I = 0.01 M (2.0 w/v % protein, 40 vol% oil) (Semenova *et al.*, 1999). Reproduced from Semenova (2007) with permission.

For an adsorbed layer of α_{s1}-casein at pH = 5.5 and ionic strength = 0.01 M ($A_{24}^* = -334 \times 10^5$ cm^3/mol), the interfacial shear viscosity measured after 24 hours was found to increase fivefold in the presence of pectinate (Semenova *et al.*, 1999). The reduction in measured droplet size recorded for this same system could be attributable to the associative adsorption of polysaccharide onto the surface of droplets covered by the protein, giving a thicker and more highly charged layer, which could then confer a greater degree of electrostatic and steric stabilization on the emulsion droplets during and after the emulsification (Dickinson *et al.*, 1998a; Semenova *et al.*, 1999).

3.2. *Interactions between Colloidal Particles Covered by Biopolymer*

The interactions between the dispersed particles covered by biopolymer adsorbed layers determine the stability properties of many food colloidal systems. As an example, let first us mention some findings from model experiments on the interactions between emulsion droplets stabilized by adsorbed layers of α_{s1}-casein or β-casein shown in Figure 3.5. The data were from laser light scattering (Semenova *et al.*, 1999) and from the shear-driven formation of particle doublets at a wall using a specially built colloidal particle scattering apparatus (Whittle *et al.*, 2000; Semenova *et al.*, 2001). In explaining the data, it was postulated that the irreversible sticking (capture) of the casein-coated particles was determined, firstly, by the strength of the protein–protein attraction (absolute value of the negative A_2) and, secondly, by the thickness/extension of the protein adsorbed layers (as indicated, by the values of M_w and R_G measured in solution). The irreversible sticking of the colloidal particles was found to correlate in qualitative terms with the measured viscoelasticity of the corresponding concentrated protein-stabilized emulsions (Dickinson *et al.*, 1998b).

In general terms, the interactions between the colloidal particles with surfaces covered by adsorbed biopolymer layers can be described qualitatively and quantitatively using the appropriate expression for the potential of mean force $W(r)$. Extending the formalism of equation (3.2), at least four separate contributions to $W(r)$ can contribute to the total free energy of interaction between a pair of colloidal particles in the aqueous dispersion medium (a biopolymer solution) (Snowden *et al.*, 1991; Dickinson, 1992; Israelachvili, 1992; Vincent, 1999; de Kruif, 1999; Prausnitz, 2003):

$$W_{\mathrm{tot}}(r) = W_{\mathrm{A}}(r) + W_{\mathrm{E}}(r) + W_{\mathrm{dep}}(r) + W(r)_{\mathrm{steric}}. \qquad (3.34)$$

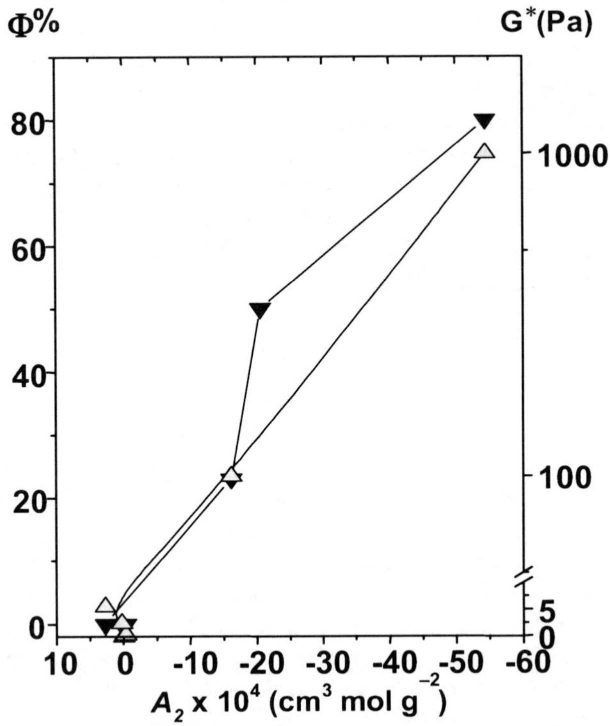

Figure 3.5 Demonstration of correlation between the stickiness of protein-coated droplet pair encounters in shear flow (left ordinate axis) and visco-elasticity of concentrated emulsions (right ordinate axis) with the strength of protein–protein attraction as indicated by the second virial coefficient A_2 determined from static light scattering: ▼, percentage capture efficiency (Φ%); △, complex shear modulus (G*) for emulsions stabilized by α_{s1}-casein or β-casein (pH = 5.5, ionic strength in the range 0.01–0.2 M).

The four terms in equation (3.34) are defined as follows.

(i) W_A is the attractive van der Waals potential. It has electromagnetic character. It arises from the ubiquitous London dispersion forces that act between the fluctuating dipoles of the polarizable molecules of the interacting particles and their adsorbed biopolymer molecules.

(ii) W_E is the electrostatic repulsive potential arising from overlap of electrical double-layers around the charged particles.

(iii) W_{dep} is the attractive potential induced by the depletion of non-adsorbed biopolymer from the particle surfaces.
(iv) W_{steric} is the steric repulsive potential arising from the interaction between biopolymer adsorbed layers. Two entropic contributions to W_{steric} are known as the mixing (or osmotic) term and the elastic (or volume restriction) term. These two terms are normally taken as additive: $W_{steric} = W_{s,mix} + W_{s,el}$.

For the case of a pair of spherical particles, the van der Waals attractive potential is given by

$$W_A = -\frac{B}{6}\left(\frac{2}{S^2-4} + \frac{2}{S^2} + \ln\frac{S^2-4}{S^2}\right) , \tag{3.35}$$

where B is an effective Hamaker (van der Waals) constant, and S is the ratio of the separation distance between the centres of the two particles and the particle radius a. The composite Hamaker constant B depends on the vales of the individual Hamaker constants of the internal colloidal particle material (B_1), the solvent (B_2), and the adsorbed biopolymer (B_3), through the following relationship:

$$B = \left\{B_1^{0.5} - \left[\phi_2^a B_3^{0.5} + \left(1 - \phi_2^a\right)B_2^{0.5}\right]\right\}^2 . \tag{3.36}$$

The quantity ϕ_2^a is the average volume fraction of biopolymer in the adsorbed layer.

The electrostatic repulsive potential is given by

$$W_E = 2\varepsilon_0\varepsilon_r a\pi\psi_0^2 \ln\{1 + \exp[-\tau(S-2)]\} , \tag{3.37}$$

where ψ_0 is the theoretical surface potential, which is conventionally equated to the measurable zeta potential (ζ); ε_0 and ε_r are the permittivity in vacuum and the relative dielectric constant, respectively; and the parameter τ is the ratio of the particle radius a to the double-layer thickness κ^{-1} (Dickinson, 1992). The Debye–Hückel screening length is defined by

$$1/\kappa = \left(\frac{\varepsilon_r\varepsilon_0}{4\pi e^2 \sum_i C_i Z_i^2}\right)^{0.5} , \tag{3.38}$$

where e is the elementary charge, C_i is the concentration of ions of type i, and Z_i is the ion valency. The summation in the denominator of equation (3.38) is the same as that contained in the definition of ionic strength:

$$I = 0.5 \sum_i C_i Z_i^2 . \qquad (3.39)$$

As the ionic strength increases, the Debye screening length decreases, resulting in interparticle electrostatic interactions of shorter range. The theoretical surface potential can be further related to the net protein charge Q (Horne *et al.*, 2007):

$$\psi_0 = \frac{Q}{a \varepsilon_0 \varepsilon_r (1 + \kappa a)} . \qquad (3.40)$$

The depletion potential W_{dep} can be calculated using the following equation (Vincent *et al.*, 1986; Vincent, 1999):

$$W_{\text{dep}} = 2\pi a \left(\frac{\mu_1 - \mu_1^0}{V_1^0}\right)\left(\Delta + \delta - p - \frac{H}{2}\right)^2 \times \left(1 + \frac{2(\Delta + \delta - p)}{3a} + \frac{H}{6a}\right) . \qquad (3.41)$$

Here, Δ is the depletion layer thickness (assumed equal to the radius of gyration of the polymer, R_G), $H = r - 2a$ is the surface-to-surface particle separation, V_1^0 is the molar volume of the solvent, and μ_1 and μ_1^0 are the solvent chemical potentials for the polymer solution and the pure solvent. It appears that the assumption $\Delta = R_G$ is generally acceptable providing that the polymer solution is in the dilute concentration regime. At higher polymer concentrations, however, the value of Δ is reduced according to the relationship (Vincent, 1990):

$$\Delta = R_G \left(\frac{\phi_{\text{b}}}{\phi_{\text{overlap}}}\right)^m , \qquad (3.42)$$

where ϕ_{b} is the bulk polymer concentration, and the index m has values of – 0.75 and – 1.0 in athermal and theta solvents, respectively. Values of the critical overlap concentration can be determined approximately from the relationship $\phi_{\text{overlap}} = 4/[\eta]$, where $[\eta]$ is the intrinsic viscosity of the polymer in bulk solution, as defined by Tanford (1961).

The first term in brackets in equation (3.41) is equal to $-\Pi$, where Π is the osmotic pressure of the polymer solution. The quantities δ and p

are special parameters associated with the adsorbed polymer layer: δ is the thickness of the layer, and p is the distance of interpenetration that can take place between the adsorbed layer and free polymer before any depletion force is experienced. The value of p is obviously dependent on the nature of the adsorbed layer; in principle, it can be calculated from theoretical equations for the adsorbed layer profile. For example, for the linear adsorbed layer profile, we have (Vincent *et al.*, 1986):

$$p = -\left[\frac{\left(\mu_1 - \mu_1^0\right)\delta\Delta}{RT\,(1-2\chi)\phi_2^a\phi_2^b}\right]^{1/2} . \tag{3.43}$$

In the above equation, χ is the Flory–Huggins interaction parameter, R is the universal gas constant, ϕ_2^a is the average volume fraction of polymer in the adsorbed layer, and ϕ_2^b is the bulk polymer concentration.

For particles covered by an adsorbed polymer layer, it is also necessary to take into account the effect on the steric repulsive forces of the structure of the adsorbed layer. For example, for the case of separations corresponding to $\delta < H < 2\delta$, the values of $W_{s,mix}$ associated with linear adsorbed layer profiles may be obtained from the equation (Vincent *et al.*, 1986):

$$W_{s,mix} = \frac{16\pi aRT}{3V_1^0}\frac{\left(\phi_2^a\right)^2}{\delta^2}\left(\frac{1}{2}-\chi\right)\left(\delta-\frac{H}{2}\right)^4 . \tag{3.44}$$

For $0 < H < \delta$, the equation to be used for $W_{s,mix}$ is the same as that for a uniform adsorbed layer profile:

$$W_{s,mix} = \frac{4\kappa a\delta^2 RT}{V_1^0}\left(\phi_2^a\right)^2\left(\frac{1}{2}-\chi\right)\left(\frac{H}{2\delta}-\frac{1}{4}-\ln\frac{H}{\delta}\right) . \tag{3.45}$$

The elastic contribution to the steric potential is given by

$$W_{s,el} = 2\pi aRT\Gamma_2 S_{el} , \tag{3.46}$$

where Γ_2 is the adsorbed amount of polymer (number of chains per unit area), and S_{el} is a geometric function that depends on the form of the segment concentration profile $\rho(z)$ in the adsorbed layer normal to the interface (Napper, 1977, 1983). Although $\rho(z)$ has been determined experimentally for individual particles using small-angle neutron scattering

(Fleer *et al.*, 1993), it is necessary to know how $\rho(z)$ varies with particle separation in order to calculate W_{steric}. For some specific systems, W_{steric} may be measured directly using the surface force apparatus (Israelachvili and Adams, 1978) or atomic force microscopy (Ducker *et al.*, 1991). It is important to be aware here that the steric repulsive force is of extremely short range, *i.e.*, of the order of ~1 nm (Leong *et al.*, 1995). Often, the contribution from $W_{steric}(r)$ to $W_{tot}(r)$ is predominant, so that particles are prevented from approaching each other close enough to experience the effect of W_E arising from charges on the bare particle surface (Snowden *et al.*, 1991). A more reliable way to calculate interparticle potentials between adsorbed polymer layers is using self-consistent-field theory, which can allow for the effects of charges on the polymer chains explicitly; the cases of adsorbed layers of α_{s1}-casein and β-casein have been considered in detail (Dickinson *et al.*, 1997).

For certain idealized cases, the pair interaction between colloidal particles can be represented by a potential energy diagram where the total energy of interaction W_{tot} is plotted as a function of particle separation H (Williams and Smith, 1995; Dickinson, 1992, 1997). The resultant potential energy curve often displays both a primary minimum and a secondary minimum. When the height of the primary maximum (W_{max}) is of the order of 15–20 kT, then particle aggregation in the primary minimum is prevented since the Brownian motion is not able to drive the particles over the kinetic barrier within the normal experimental timescale. For significantly lower values of W_{max}, strong irreversible aggregation is the inevitable result. If the depth of the secondary minimum (W_{min}) is greater than a few kT, then weak reversible aggregation will take place. The deeper the secondary minimum, the stronger is the flocculated structure that may be produced. When $W_{dep}(r)$ becomes more important, this term gives rise to an increase in the depth of the secondary minimum in the total potential energy curve, which is associated with a stronger, albeit still reversible, state of flocculation. It is readily inferred from equation (3.41) that the depletion interaction strength increases with increasing osmotic pressure of the polymer solution. Also the parameter Δ has a significant effect on W_{min}, since the depletion potential is proportional to the square of the depletion layer thickness. For model dispersed systems, namely polystyrene latex particles in the presence of polymers such as hydroxyethylcellulose or sodium carboxymethylcellulose, it has been calculated (Williams and Smith, 1995) that the value of W_{min} changes from around –1 kT to –25 kT as Δ increases from 14 nm to 100 nm; and also that it changes from –1 kT to –22 kT as the solution osmotic pressure increases from zero to 250 N/m^2.

In practice, for any real food colloid, the precise form of $W_{tot}(r)$ is never properly known. But what can sometimes be inferred is the likely mechanism of stabilization and the approximate magnitude of some limiting energy barrier or attractive well depth. The latter quantities are typically estimated from experimental observations of coagulation kinetics, phase transitions, or rheology. The link is achieved using established theory, which relates a measured quantity (turbidity, viscosity, *etc.*) to the interparticle interactions. But as the validity of any theory normally relies on assumptions that are of questionable validity when applied to complex food systems, the potential parameters inferred in this way can be regarded as only a rough indication of relative interaction strengths, and not numbers of absolute significance (Dickinson, 1997). On a more positive note, though, it is recognized that transitions between dispersed and flocculated colloidal states are controlled primarily by thermodynamic factors. The following example for a milk protein-based system provides evidence of the success of this thermodynamic approach.

The set of structural parameters (M_w, R_G, R_h, $\rho = R_G/R_h$) and interaction parameters (A_2) of sodium caseinate nanoparticles were determined from light-scattering experiments (Dickinson *et al.*, 2001). These quantities were then used to calculate the interaction potential between caseinate-stabilized emulsion droplets in the presence of unadsorbed protein. In particular, equation (3.41) was used to predict a pronounced minimum in the plot of the Gibbs free energy of the depletion interactions (ΔG_{dep}) between emulsion droplets for a specific range of calcium ion contents. Good colloidal stability of the emulsion was found experimentally over a narrow range of calcium ion concentrations, namely from 5 to 8 mM. As shown in Figure 3.6, this was dramatically different from the emulsion depletion flocculation observed at both low (< 5 mM) and high (< 8 mM) Ca^{2+} contents (Dickinson and Golding, 1998; Dickinson and Davies, 1999). The (in)stability behaviour of the emulsion system could be predicted theoretically. A minimum in the negative value of ΔG_{dep} was accounted for by a combination of two effects: the sharp decrease in Π (Figure 3.6) and a change in the self-assembled structure of the sodium caseinate nanoparticles, associated with calcium ion cross-linking in the interior of the protein nanoparticles (Dickinson *et al.*, 2001). In turn, the decrease in Π calculated from equation (3.41) could be attributed to two factors: (i) a decrease in the number density of caseinate nanoparticles, $n = C/M_w$, as controlled by the extent of protein self-assembly M_w at constant protein concentration C, and (ii) contributions to A_2 from electrostatic repulsion and excluded volume effects (Tanford, 1961, Edmond and Ogston, 1968; Nagasawa and Takahashi, 1972) due to the reduction in the net charge and the size of the caseinate nanoparticles, respectively (Dickinson *et al.*, 2001).

In principle, the expressions for pair potentials, osmotic pressure and second virial coefficients could be used as input parameters in computer simulations. The objective of performing such simulations is to clarify physical mechanisms and to provide a deeper insight into phenomena of interest, especially under those conditions where structural or thermodynamic parameters of the studied system cannot be accessed easily by experiment. The nature of the intermolecular forces responsible for protein self-assembly and phase behaviour under variation of solution conditions, including temperature, pH and ionic strength, has been explored using this kind of modelling approach (Dickinson and Krishna, 2001; Rosch and Errington, 2007; Blanch *et al.*, 2002).

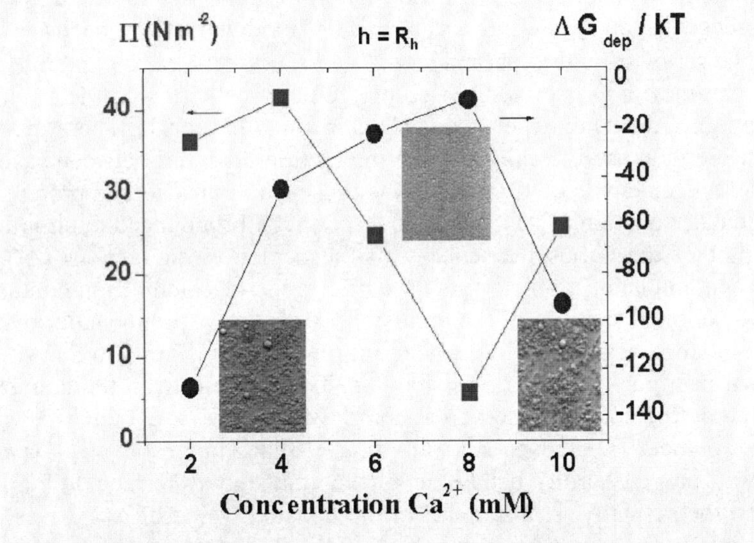

Figure 3.6 Effect of Ca^{2+} content on predicted values of osmotic pressure (Π, ■, left axis) of caseinate nanoparticles in emulsion continuous phase and the free energy of the depletion interaction (ΔG_{dep}, ●, right axis) between a pair of emulsion droplets (a = 250 nm) covered by sodium caseinate. The interdroplet separation h is equal the thickness of the depletion layer R_{h} (pH = 7.0, ionic strength = 0.05 M). The three inserts are light micrographs (magnification × 400 times) for emulsion samples of low, medium and high calcium contents. Reproduced from Semenova (2007) with permission.

Knowledge of the nanoscale structure of mixed biopolymer phases and protein–polysaccharide complexes is still very scarce. Under such circumstances, numerical simulations can show how the supramolecular structural complexity changes depending on ionic strength and pH, as well as the chain length, flexibility and charge density of the polyelectrolytes (Grymonpré *et al.*, 2001; Akinchina and Linse, 2002; Hayashi *et al.*, 2003). In the case of food colloids, the computer simulation and theoretical modelling of systems with well-defined interactions between the surfaces of the particles can allow us to gain deeper insight into the effect of the nature and strength of biopolymer interactions on the structural features of adsorbed layers under various environmental conditions (Whittle *et al.*, 2000; Ettelaie *et al.*, 2008).

Concluding Remark

In this chapter we have outlined how the use of a universal thermodynamic approach can provide valuable insight into the consequences of specific kinds of biopolymer–biopolymer interactions. The advantage of the approach is that it leads to clear quantitative analysis and predictions. It allows connections to be made between the molecular scale and the macroscopic scale, explaining the contributions of the biopolymer interactions to the mechanisms of microstructure formation, as well as to the appearance of novel functionality arising from the manipulation of food colloid formulations. Of course, we must remind ourselves that, taken by itself, the thermodynamic approach cannot specify the molecular or colloidal structures in any detail, nor can it give us information about the rates of the underlying kinetic processes.

BIBLIOGRAPHY

Akinchina, A., Linse, P. (2002). Monte Carlo simulations of polyion–macroion complexes. 1. Equal absolute polyion and macroion charges. *Macromolecules*, **35**, 5183–5193.

Albertsson, P.-Å. (1971). *Partition of Cell Particles and Macromolecules*, 2nd edn, Stockholm: Almqvist & Wiksell.

Antipova, A.S., Semenova, M.G. (1995). Effect of sucrose on the thermodynamic incompatibility of different biopolymers. *Carbohydrate Polymers*, **28**, 359–365.

Antipova, A.S., Semenova, M.G. (1997a). Effect of neutral carbohydrate structure in the set glucose / sucrose / maltodextrin / dextran on protein surface activity at the air–water interface. *Food Hydrocolloids*, **11**, 71–77.

Antipova, A.S., Semenova, M.G. (1997b). Influence of sucrose on the thermodynamic properties of the 11S globulin of *Vicia faba*–dextran–aqueous solvent system. *Food Hydrocolloids*, **11**, 415–421.

Appelqvist, I., Debet, M. (1997). Starch–biopolymer interactions. *Food Reviews International*, 13, 163–224.

Ball, V., Winterhalter, M., Schwinte, P., Lavalle, P., Voegel, J.C., Schaaf, P. (2002). Complexation mechanism of bovine serum albumin and poly(allylamine hydrochloride). *Journal of Physical Chemistry*, **106**, 2357–2364.

Benichou, A., Aserin, A., Garti, N. (2002). Protein–polysaccharide interactions for stabilization of food emulsions. *Journal of Dispersion Science and Technology*, **23**, 93–123.

Blanch, H.W., Prausnitz, J.M., Curtis, R.A., Bratko, D. (2002). Molecular thermodynamics and bioprocessing: from intracellular events to bioseparations. *Fluid Phase Equilibria*, **194–197**, 31–41.

Bratko, D., Striolo, A., Wu, J.Z., Blanch, J.M., Prausnitz, J.M. (2002). Orientation-averaged pair potentials between dipolar proteins or colloids. *Journal of Physical Chemistry B*, **106**, 2714–2720.

Burchard, W. (1994). Light scattering. In Ross-Murphy, S.B. (Ed.). *Physical Techniques for the Study of Food Biopolymers*, Glasgow: Blackie, pp. 151–214.

Cai, R., Arntfield, S. (1997). Thermal gelation in relation to binding of bovine serum albumin–polysaccharides systems. *Journal of Food Science*, **62**, 1129–1134.

Cesàro, A., Cuppo, F., Fabri, D., Sussich, F. (1999). Thermodynamic behaviour of mixed biopolymers in solution and in gel phase. *Thermochimica Acta*, **328**, 143–153.

Clark, A.H. (2000). Direct analysis of experimental tie line data (two polymer–one solvent systems) using Flory–Huggins theory. *Carbohydrate Polymers*, **42**, 337–351.

Dautzenberg, H. (2001). Polyelectrolyte complex formation of highly aggregating systems: methodical aspects and general tendencies. In Radeva, T. (Ed.). *Physical Chemistry of Polyelectrolytes*, Surfactant Science Series Vol. 99, New York: Marcel Dekker, pp. 743–792.

de Kruif, C.G. (1999). Attractive interactions and aggregation in food dispersions. In Dickinson, E., Rodriguez Patino, J.M. (Eds). *Food Emulsions and Foams: Interfaces, Interactions and Stability*, Cambridge, UK: Royal Society of Chemistry, pp. 29–44.

de Kruif, C.G., Tuinier, R. (2001). Polysaccharide–protein interactions. *Food Hydrocolloids*, **15**, 555–563.

de Kruif, C.G., Weinbreck, F., de Vries, R. (2004). Complex coacervation of proteins and anionic polysaccharides. *Current Opinion in Colloid and Interface Science*, **9**, 340–349.

de Kruif, C.G., Tuinier, R. (2005). Stabilization of food colloids by polymers. In Dickinson, E. (Ed.). *Food Colloids: Interactions, Microstructure and Processing*, Cambridge, UK: Royal Society of Chemistry, pp. 61–73.

Dickinson, E. (1992). *An Introduction to Food Colloids*, Oxford: Oxford University Press, chap. 1.

Dickinson, E. (1993). Protein–polysaccharide interactions. In Dickinson, E., Walstra, P. (Eds). *Food Colloids and Polymers: Stability and Mechanical Properties*, Cambridge, UK: Royal Society of Chemistry, pp. 77–93.

Dickinson, E. (1997). Aggregation processes, particle interactions and colloidal structure. In Dickinson, E., Bergenståhl, B. (Eds). *Food Colloids: Proteins, Lipids and Polysaccharides*, Cambridge, UK: Royal Society of Chemistry, pp. 107–126.

Dickinson, E. (1998). Stability and rheological implications of electrostatic milk protein–polysaccharide interactions. *Trends in Food Science and Technology*, **9**, 347–354.

Dickinson, E. (2003). Hydrocolloids at interfaces and the influence on the properties of dispersed systems. *Food Hydrocolloids*, **17**, 25–39.

Dickinson, E., Davies, E. (1999). Influence of ionic calcium on stability of sodium caseinate emulsions. *Colloids and Surfaces B: Biointerfaces*, **12**, 203–212.

Dickinson, E., Euston, S.R. (1991). Stability of food emulsions containing both protein and polysaccharide. In Dickinson E. (Ed.). *Food Polymers, Gels and Colloids*, Cambridge, UK: Royal Society of Chemistry, pp.132–146.

Dickinson, E., Galazka, V.B. (1992). Emulsion stabilization by protein–polysaccharide complexes. In Phillips, G.O., Wedlock, D.J., Williams, P.A. (Eds). *Gums and Stabilisers for the Food Industry 6*, Oxford: IRL Press, pp. 351–362.

Dickinson, E., Golding, M. (1998). Influence of calcium ions on creaming and rheology of emulsions containing sodium caseinate. *Colloids and Surfaces A: Physicochemical and Engineering Aspects*, **144**, 167–177.

Dickinson, E., Krishna, S. (2001). Aggregation in a concentrated model protein system: a mesoscopic simulation of β-casein self-assembly. *Food Hydrocolloids*, **15**, 107–115.

Dickinson, E., Pawlowsky, K. (1997) Effect of ι-carrageenan on flocculation, creaming, and rheology of a protein-stabilized emulsion. *Journal of Agricultural and Food Chemistry*, **45**, 3799–3806.

Dickinson, E., Semenova, M.G. (1992). Emulsifying behaviour of protein in the presence of polysaccharide under conditions of thermodynamic incompatibility. *Journal of the Chemical Society, Faraday Transactions*, **88**, 849–854.

Dickinson, E. Pinfield, V.J., Horne, D.S., Leermakers, F.A.M. (1997). Self-consistent-field modelling of adsorbed casein interaction between two protein-coated surfaces. *Journal of the Chemical Society, Faraday Transactions*, **93**, 1785–1790.

Dickinson, E., Semenova, M.G., Antipova, A.S., Pelan, E. (1998a). Effect of high-methoxy pectin on properties of casein-stabilized emulsions. *Food Hydrocolloids*, **12**, 425–432.

Dickinson, E., Semenova, M.G., Antipova, A.S. (1998b). Salt stability of casein emulsions. *Food Hydrocolloids*, **12**, 227–235.

Dickinson, E., Semenova, M., Belyakova, L., Antipova, A., Il'in, M., Tsapkina, E., Ritzoulis, C. (2001). Analysis of light scattering data on the calcium ion sensitivity of caseinate solution thermodynamics: relationship to emulsion flocculation. *Journal of Colloid and Interface Science*, **239**, 87–97.

Doublier, J.-L., Garnier, C., Renard, D., Sanchez, C. (2000). Protein–polysaccharide interactions. *Current Opinion in Colloid and Interface Science*, **5**, 202–214.

Ducker, W.A., Senden, T.J., Pashley, R.M. (1991). Direct measurement of colloidal forces using an atomic force microscope. *Nature*, **353**, 239–241.

Dumetz, A.C., Chockla, A.M., Kaler, E.W., Lenhoff, A.M. (2008). Effects of pH on protein–protein interactions and implications for protein phase behaviour. *Biochimica et Biophysica Acta*, **1784**, 600–610.

Edmond, E., Ogston, A. (1968). An approach to the study of phase separation in ternary aqueous systems. *Biochemistry Journal*, **109**, 569–576.

Edsall, J.T., Gutfreund, H. (1983). *Biothermodynamics: The Study of Biochemical Processes at Equilibrium*, Chichester: Wiley.

Ettelaie, R., Akinshina, A., Dickinson, E. (2008). Mixed protein–polysaccharide interfacial layers: a self consistent field calculation study. *Faraday Discussions*, **139**, 161–178.

Evans, J. M. (1972). Manipulation of light scattering data. In Huglin, M.B. (Ed.). *Light Scattering from Polymer Solutions*, London: Academic Press, chap. 5, pp. 89–164.

Fleer, G.J., Cohen Stuart, M.A., Scheutjens, J.M.H.M., Cosgrove, T., Vincent, B. (1993). *Polymers at Interfaces*, London: Chapman & Hall, chap. 11.

Flory, P.J. (1942). Thermodynamics of high polymer solutions. *Journal of Physical Chemistry*, **10**, 51–61.

George, A., Chiang, Y., Guo, B., Arabshahi, A., Cai, Z., Wilson, W.W. (1997). Second virial coefficient as predictor in protein crystal growth. *Methods in Enzymology*, **276**, 100–110.

Girard, M., Turgeon, S.L., Gauthier, S.F. (2003). Thermodynamic parameters of β-lactoglobulin–pectin complexes as assessed by isothermal titration calorimetry. *Journal of Agricultural and Food Chemistry*, **51**, 4450–4455.

Grinberg, V.Y., Tolstoguzov, V.B. (1997). Thermodynamic incompatibility of proteins and polysaccharides in solutions. *Food Hydrocolloids*, **11**, 145–158.

Grymonpré, K.R., Staggemeier, B.A., Dubin, P.L., Mattison, K.W. (2001). Identification by integral computer modelling and light scattering of an electrostatic serum albumin–hyaluronic acid binding site. *Biomacromolecules*, **2**, 422–429.

Guido, S., Simeone, M., Alfani, A. (2002). Interfacial tension of aqueous mixtures of Na-caseinate and Na-alginate by drop deformation in shear flow. *Carbohydrate Polymers*, **48**, 143–152.

Hayashi, Y., Ullner, M., Linse, P. (2003). Complex formation in solutions of oppositely charged polyelectrolytes at different polyion compositions and salt content. *Journal of Physical Chemistry B*, **107**, 8198–8207.

Horne, D.C., Lucey, J.A., Choi, J.-W. (2007). Casein interactions: does the chemistry really matter? In Dickinson, E., Leser, M.E. (Eds). *Food Colloids: Self-Assembly and Material Science*, Cambridge, UK: Royal Society of Chemistry, pp. 155–166.

Hsu, C.C., Prausnitz, J.M. (1974). Thermodynamics of polymer compatibility in ternary systems. *Macromolecules*, **7**, 320–324.

Huggins, M.L. (1942). Some properties of solutions of long-chain compounds. *Journal of Physical Chemistry*, **46**, 151–158.

Israelachvili, J.N. (1992). *Intermolecular and Surface Forces*, 2nd edn, London: Academic Press.

Israelachvili, J.N., Adams, G.E. (1978). Measurement of forces between two mica surfaces in aqueous electrolyte solutions in the range 0–100 nm. *Journal of the Chemical Society, Faraday Transactions 1*, **74**, 975–1001.

Johansson, H., Karlstrom, G., Tjerneld, F., Haynes, C. (1998). Driving forces for phase separation and partitioning in aqueous two-phase systems. *Journal of Chromatography B*, **711**, 3–17.

Johansson, H.-O., Brooks, D.E., Haynes, C.A. (2000). Macromolecular crowding and its consequences. *International Review of Cytology*, **192**, 155–170.

Jönsson, B., Lund, M., da Silva, F.L.B. (2007). Electrostatics in macromolecular solutions. In Dickinson, E., Leser, M.E. (Eds). *Food Colloids: Self-Assembly and Material Science*, Cambridge, UK: Royal Society of Chemistry, pp. 129–154.

Kaddur, L.O., Strazielle, C. (1986). Experimental investigation of light scattering by a solution of two polymers. *Polymer*, **28**, 459–468.

Kim, H.-J., Decker, E.A., McClements, D.J. (2006). Preparation of multiple emulsions based on thermodynamic incompatibility of heat-denatured whey protein and pectin solutions. *Food Hydrocolloids*, **20**, 586–595.

Kratochvil, P., Sudelof, L.O. (1986). Interactions in polymer solutions. *Acta Pharmaceutica Suecica*, **23**, 31–46.

Leong, Y.K., Scales, P.J., Healy, T.W., Boger, D.V. (1995). Interparticle forces arising from adsorbed polyelectrolytes in colloidal suspensions. *Colloids and Surfaces A: Physicochemical and Engineering Aspects*, **95**, 43–52.

Malone, M.E., Appelqvist, I.A.M. (2003). Gelled emulsion particles for the controlled release of lipophilic volatiles during eating. *Journal of Controlled Release*, **90**, 227–241.

Malone, M.E., Appelqvist, I.A.M., Norton, I.T. (2003). Oral behaviour of food hydrocolloids and emulsions. Part 2. Taste and aroma release. *Food Hydrocolloids*, **17**, 775–784.

Maroziene, A., de Kruif, C.G. (2000). Interaction of pectin and casein micelles. *Food Hydrocolloids*, **14**, 391–394.

McClements, D.J. (2006). Non-covalent interactions between proteins and polysaccharides. *Biotechnology Advances*, **24**, 621–625.

Nagasawa, M., Takahashi, A. (1972). Light scattering from polyelectrolyte solutions. In Huglin, M.B. (Ed.). *Light Scattering From Polymer Solutions*, London: Academic Press, pp. 671–723.

Napper, D.H. (1977). Steric stabilization. *Journal of Colloid and Interface Science*, **58**, 390–407.

Napper, D.H. (1983). *Polymeric Stabilization of Colloidal Dispersions*, London: Academic Press.

Norton, I.T., Frith, W. J. (2001). Microstructure design in mixed biopolymer composites. *Food Hydrocolloids*, **15**, 543–553.

Norton I.T., Frith W.J. (2003). Phase separation in mixed biopolymer systems. In Dickinson, E., van Vliet, T. (Eds). *Food Colloids, Biopolymers and Materials*, Cambridge, UK: Royal Society of Chemistry, pp. 282–297.

Ogston, A.G. (1962). Some thermodynamic relationships in ternary systems, with special reference to the properties of systems containing hyaluronic acid and protein. *Archives of Biochemistry and Biophysics,* Supplement **1**, 39–51.

Pavlovskaya, G., Semenova, M., Tsapkina, E., Tolstoguzov V. (1993). The influence of dextran on the interfacial pressure of adsorbing layers of 11S globulin *Vicia faba* at the planar *n*-decane/aqueous solution interface. *Food Hydrocolloids*, **7**, 1–10.

Prausnitz, J.M. (2003). Molecular thermodynamics for some applications in biotechnology. *Journal of Chemical Thermodynamics*, **35**, 21–39.

Preston, B.N., Davies, M., Ogston, A.G. (1965). The composition and physicochemical properties of hyaluronic acids prepared from ox synovial fluid and from a case of mesothelioma. *Biochemical Journal*, **96**, 449–471.

Prigogine, I., Defay, R. (1954). *Chemical Thermodynamics*, London: Longmans Green.

Rosch, T.W., Errington, J.R. (2007). Investigation of the phase behaviour of an embedded charge protein model through molecular simulation. *Journal of Physical Chemistry*, **111**, 12591–12598.

Schaink, H.M., Smit, J.A.M. (1997). Mean field calculation of polymer segment depletion and depletion-induced demixing in ternary systems of globular proteins and flexible polymers in a common solvent. *Journal of Physical Chemistry*, **107**, 1004–1015.

Schmitt, C., Sanchez, C., Desobry-Banon, S., Hardy, J. (1998). Structure and techno-functional properties of protein–polysaccharide complexes. *Critical Reviews in Food Science and Nutrition*, **38**, 689–753.

Semenova, M.G. (1996). Factors determining the character of biopolymer–biopolymer interactions in multicomponent aqueous solutions modelling food systems. In Parris, N., Kato, A., Creamer, L.K., Pearce, J. (Eds). *Macromolecular Interactions in Food Technology*, ACS Symposium Series No. 650, Washington, D.C.: American Chemical Society, pp. 37–49.

Semenova, M.G. (2007). Thermodynamic analysis of the impact of molecular interactions on the functionality of food biopolymers in solution and in colloidal systems. *Food Hydrocolloids*, **21**, 23–45.

Semenova, M., Savilova, L. (1998). The role of biopolymer structure in interactions between unlike biopolymers in aqueous medium. *Food Hydrocolloids*, **12**, 65–75.

Semenova, M.G., Bolotina, V.S., Grinberg, V.Ya., Tolstoguzov, V.B. (1990). Thermodynamic incompatibility of the 11S fraction of soybean globulin and pectinate in aqueous medium. *Food Hydrocolloids*, **3**, 447–456.

Semenova, M., Pavlovskaya, G., Tolstoguzov, V. (1991a) Light scattering and thermodynamic phase behaviour of the system 11S globulin – κ-carrageenan – water. *Food Hydrocoll*oids, **4**, 469–479

Semenova, M.G., Bolotina, V.S., Dmitrochenko, A.P., Leontiev, A.L., Polyakov, V.I., Braudo, E.E., Tolstoguzov, V.B. (1991b). The factors affecting compatibility of serum albumin and pectinate in aqueous medium. *Carbohydrate Polymers*, **15**, 367–385.

Semenova, M., Antipova, A., Belyakova, L., Dickinson, E., Brown, R., Pelan, E., Norton, I. (1999). Effect of pectinate on properties of oil-in-water emulsions stabilized by α_{s1}-casein and β-casein. In Dickinson, E., Rodriguez Patino, J.M. (Eds). *Food Emulsions and Foams: Interfaces, Interactions and Stability*, Cambridge, UK: Royal Society of Chemistry, pp. 163–175.

Semenova, M.G., Chen, J., Dickinson, E., Murray, B.S., Whittle, M. (2001). Sticking of protein-coated particles in a shear field. *Colloids and Surfaces B: Biointerfaces*, **22**, 237–244.

Semenova, M.G., Belyakova, L.E., Dickinson, E., Eliot, C., Polikarpov, Yu.N. (2005). Caseinate interactions in solution and in emulsions: effect of temperature, pH and calcium ions. In Dickinson, E. (Ed.). *Food Colloids: Interactions, Microstructure and Processing*, Cambridge, UK: Royal Society of Chemistry, pp. 209–217.

Semenova, M.G., Belyakova, L.E., Polikarpov, Yu.N., Stankovic, I., Antipova, A.S., Anokhina, M.S. (2007). Analysis of light scattering data on the sodium caseinate assembly as a response to the interactions with likely charged anionic surfactant. *Food Hydrocolloids*, **21**, 704–715.

Snowden, M.J., Clegg, S.M., Williams, P.A., Robb I.D. (1991). Flocculation of silica particles by adsorbing and non-adsorbing polymers. *Journal of the Chemical Society, Faraday Transactions*, **87**, 2201–2207.

Tanford, C. (1961). *Physical Chemistry of Macromolecules*, New York: Wiley.

Tolstoguzov, V. (1996). Structure–property relationships in foods. In Parris, N., Kato, A., Creamer, L.K., Pearce, J. (Eds). *Macromolecular Interactions in Food Technology*, ACS Symposium Series No. 650, Washington, D.C.: American Chemical Society, pp. 2–14.

Tolstoguzov, V.B. (1997). Protein–polysaccharide interactions. In Damodaran, S., Paraf, A. (Eds). *Food Proteins and their Applications*, New York: Marcel Dekker, pp. 171–198.

Tolstoguzov, V.B. (2000) Compositions and phase diagrams for aqueous systems based on proteins and polysaccharides. *International Review of Cytology*, **192**, 3–31.

Tolstoguzov, V.B. (2002). Thermodynamic aspects of biopolymer functionality in biological systems, foods, and beverages. *Critical Reviews in Biotechnology*, **22**, 89–174.

Tolstoguzov, V.B. (2003). Some thermodynamic considerations in food formulation. *Food Hydrocolloids*, **17**, 1–23.

Tombs, M.P., Peacocke, A.R. (1974). *The Osmotic Pressure of Biological Macromolecules*, Oxford: Clarendon Press.

Tsapkina, E., Semenova, M., Pavlovskaya, G., Leontiev, A., Tolstoguzov, V. (1992). The influence of incompatibility on the formation of adsorbing layers and dispersion of *n*-decane emulsion droplets in aqueous solution containing a mixture of 11S globulin from *Vicia faba* and dextran. *Food Hydrocolloids*, **6**, 237–251.

Tuinier, R., de Kruif, C.G. (1999). Phase separation, creaming, and network formation of oil-in-water emulsions induced by an exocellular polysaccharide. *Journal of Colloid and Interface Science*, **218**, 201–210.

Tuinier, R., Dhont, J.K.G., de Kruif, C.G. (2000). Depletion-induced phase separation of aggregated whey protein colloids by an exocellular polysaccharide. *Langmuir*, **16**, 1497–1507.

Turgeon, S.L., Beaulieu, M., Schmitt, C., Sanchez, C. (2003). Protein–polysaccharide interactions: phase-ordering kinetics, thermodynamics and structural aspects. *Current Opinion in Colloid and Interface Science*, **8**, 401–414.

Van Puyvelde, P., Antonov, Y.A., Moldenaers, P. (2002). Rheo-optical measurement of the interfacial tension of aqueous biopolymer mixtures. *Food Hydrocolloids*, **16**, 395–402.

Vincent, B. (1990). The calculation of depletion layer thickness as a function of bulk polymer concentration. *Colloids and Surfaces*, **50**, 241–249.

Vincent, B. (1999). Dispersion stabilization and destabilization by polymers. In Dickinson, E., Rodriguez Patino, J.M. (Eds). *Food Emulsions and Foams: Interfaces, Interactions and Stability*, Cambridge, UK: Royal Society of Chemistry, pp. 19–28.

Vincent, B., Edwards, J., Emmett, S., Jones, A. (1986). Depletion flocculation in dispersions of sterically-stabilized particles ("soft spheres"). *Colloids and Surfaces*, **18**, 261–281.

Vrij, A. (1976). Polymers at interfaces and the interactions in colloidal dispersions. *Pure and Applied Chemistry*, **48**, 471–483.

Walter, H., Johansson, G., Brooks, D.E. (1991). Partitioning in aqueous two phase systems: recent results. *Analytical Biochemistry*, **155**, 215–242.

Wasserman, L., Semenova, M., Tsapkina, E. (1997). Thermodynamic properties of the 11S globulin of *Vicia faba*–ovalbumin–aqueous solvent system: phase behaviour and light scattering. *Food Hydrocolloids*, **11**, 327–337.

Wells, I.D. (1984). The transformation of virial equations for polymer solutions between different concentration scales. *Chemica Scripta*, **23**, 202–204.

Whittle, M., Murray, B.S., Chen, J., Dickinson, E. (2000). Simulation and experiments on colloidal particle capture in a shear field. *Langmuir*, **16**, 9784–9791.

Williams, P.A., Smith, N.J. (1995). Depletion flocculation. In Harding, S.E., Hill, S.E., Mitchell, J.R. (Eds). *Biopolymer Mixtures*, Leicestershire: Nottingham University Press, chap. 9, pp. 161–172.

Wills, P.R., Jacobsen, M.P., Winzor, D.J. (1996). Direct analysis of solute self-association by sedimentation equilibrium. *Biopolymers*, **38**, 119–130.

Winzor, D.J., Carrington, L.E., Harding, S.E. (2001). Analysis of thermodynamic non-ideality in terms of protein solvation. *Biophysical Chemistry*, **93**, 231–240.

Zeman, L., Patterson, D. (1972). Effect of the solvent on polymer incompatibility in solution. *Macromolecules*, **5**, 513–516.

PART TWO

BIOPOLYMER INTERACTIONS IN THE BULK AND AT THE INTERFACE

CHAPTER FOUR

ENERGY AND CHARACTER OF MAIN TYPES OF BIOPOLYMER INTERACTIONS

The wide diversity of the intramolecular and intermolecular interactions of food biopolymers is determined by the variety of different monomer units making up their polymeric chains. Table 4.1 reviews the chemical structures of the main monomer units comprising food proteins and polysaccharides. It is well known that proteins are built from a basic menu of 20 amino acids. In addition some food proteins contain extra amino-acid residues, *e.g.*, phosphoserine in casein and hydroxyproline in gelatin. Using their constituent carboxyl and amino groups, the amino acids are joined together by peptide bonds into a unique linear sequence (the polypeptide chain) which is the distinguishing characteristic of each individual protein. The composition and sequence of the amino-acid side chains, possessing non-polar (hydrophobic), polar, or charged functional groups, determines the overall amphiphilic nature of each protein molecule. In contrast, the polysaccharides are made from chains of monomeric sugar units. The carbohydrate groups are mainly polar or charged, which gives the polysaccharides a predominantly hydrophilic character (Cantor and Schimmel, 1980; Lehninger, 1982).

Owing to the diverse chemical nature of functional groups in proteins and polysaccharides, they are prone to a variety of types of molecular interactions, both in bulk aqueous media and at air–water or oil–water interfaces. To a first approximation one may consider an adsorbed layer of biopolymers at the interface as simply a special type of highly concentrated biopolymer solution. Thus, the same variety of interactions that are typically found for biopolymers in a bulk aqueous media also occur in biopolymer adsorbed layers at the interfaces in food colloids. Moreover, these same molecular interactions are also involved in the close encounters between pairs of colloidal particles covered by adsorbed biopolymer layers. In the rest of this chapter we shall briefly remind ourselves of the main basic types of intermolecular interactions; readers requiring more detailed background information are directed to other sources (Cantor and Schimmel, 1980; Lehninger, 1982; Israelachvili, 1992; Dickinson, 1998; Finkelstein and Ptitsyn, 2002; McClements, 2005, 2006; Min *et al.*, 2008).

Table 4.1. Structural formulae of the basic monomer units comprising the polymer chains of food proteins and food polysaccharides.

(a) Proteins

Amino-acid classification based on side-chain polarity (Lehninger, 1982)

Hydrophobic side chains:

Alanine

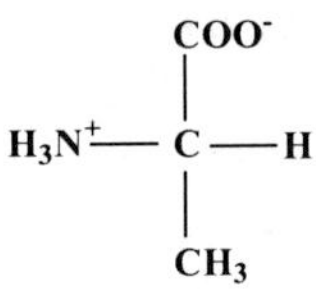

Valine

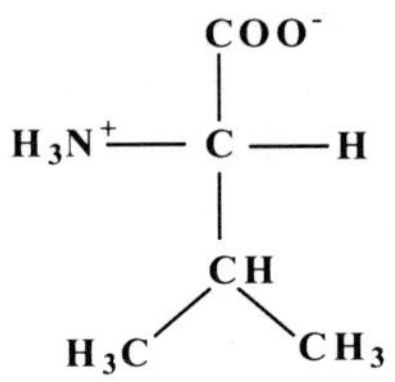

Leucine

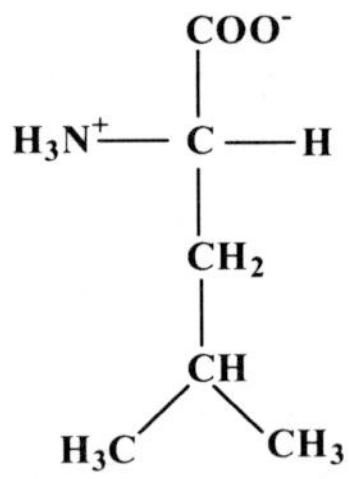

Isoleucine

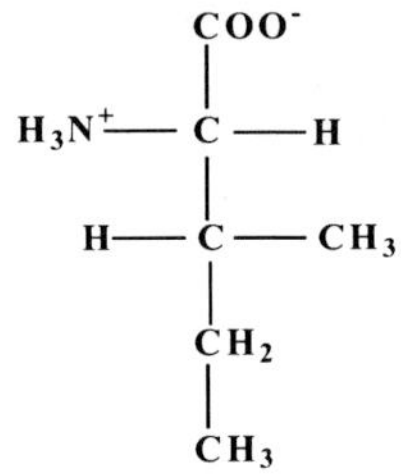

Proline

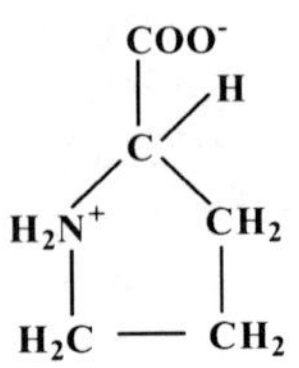

Methionine

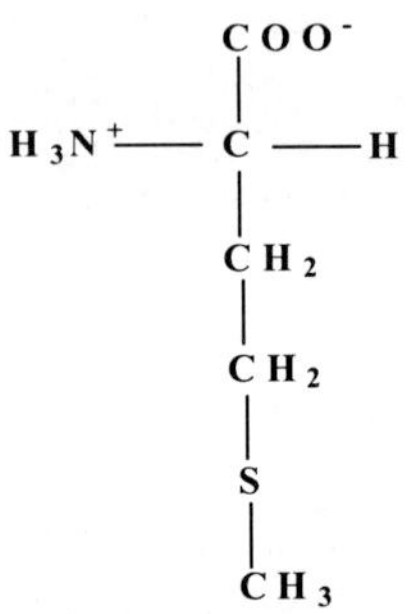

Phenylalanine

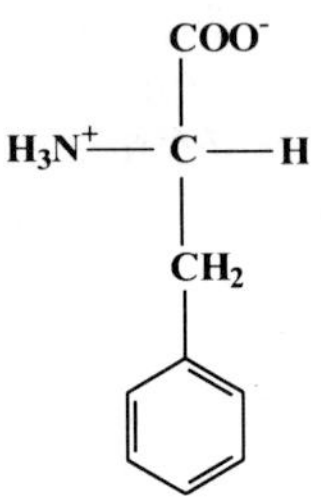

Tryptophan

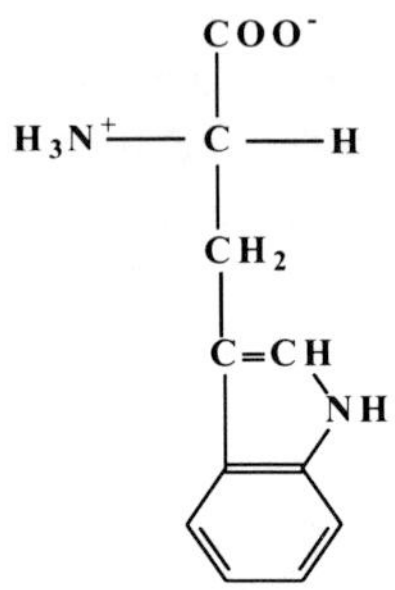

Polar side chains:

Glycine

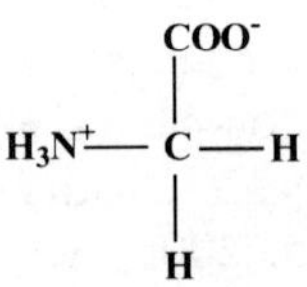

Serine

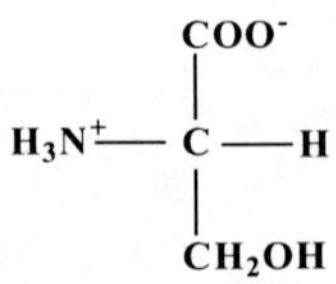

Threonine

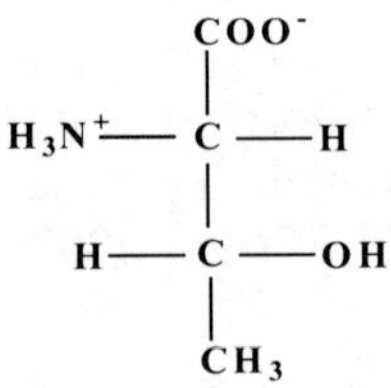

Cysteine

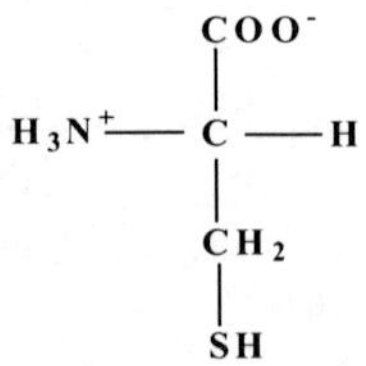

Table 4.1 *Continued*

Polar side chains (*continued*):

Tyrosine

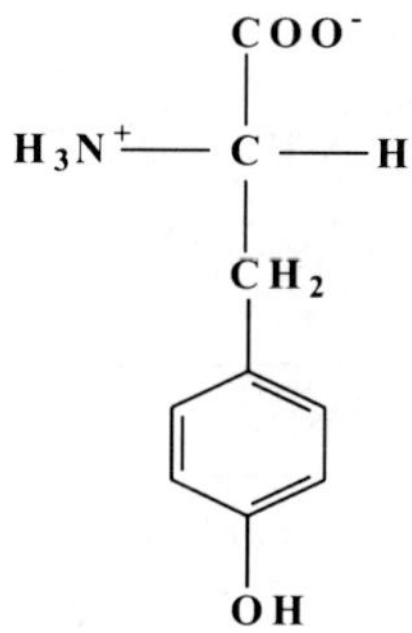

Asparagine

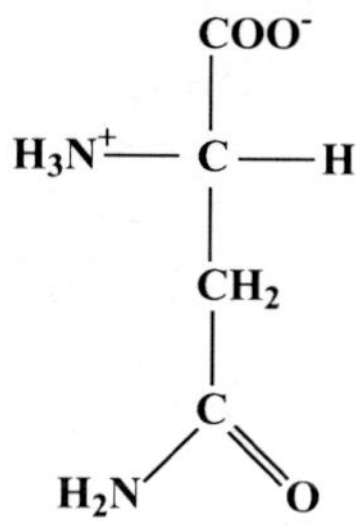

Glutamine

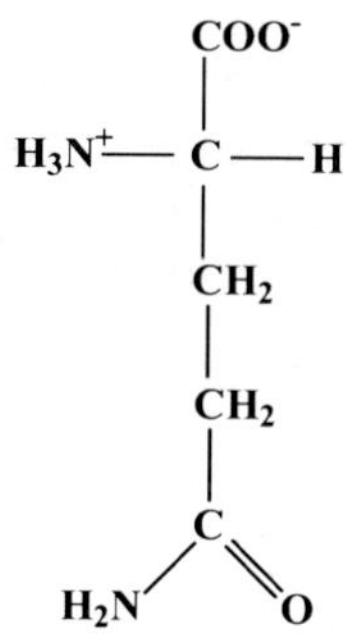

Hydroxyproline

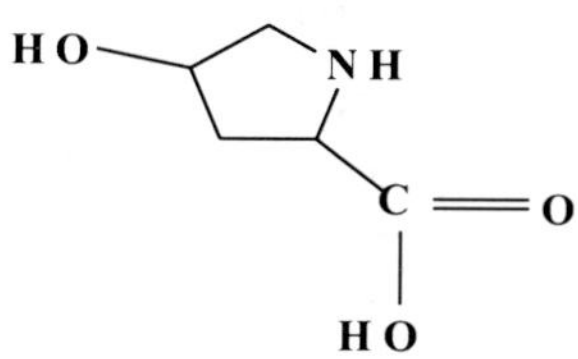

Potentially negatively charged side chains:

Aspartate

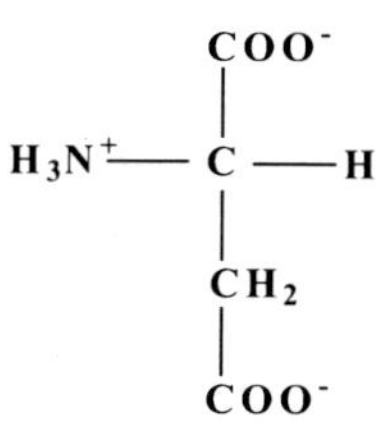

Glutamate

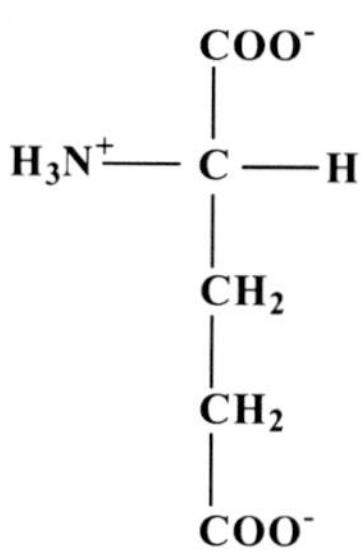

Table 4.1 *Continued*

Potentially negatively charged side chains (*continued*):

Phosphoserine

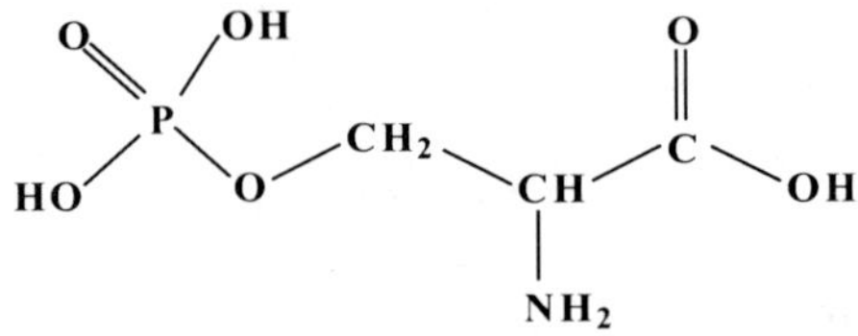

Potentially positively charged side chains:

Lysine

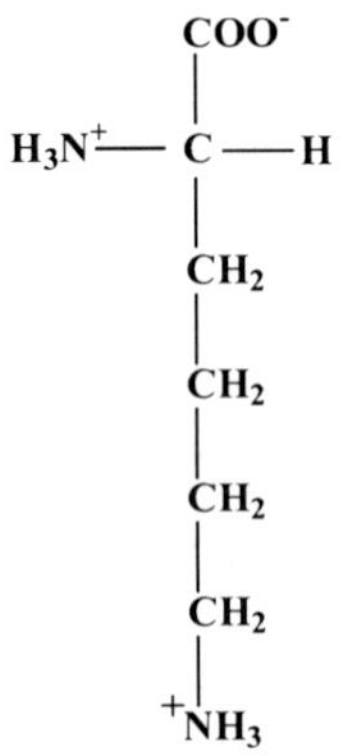

Arginine

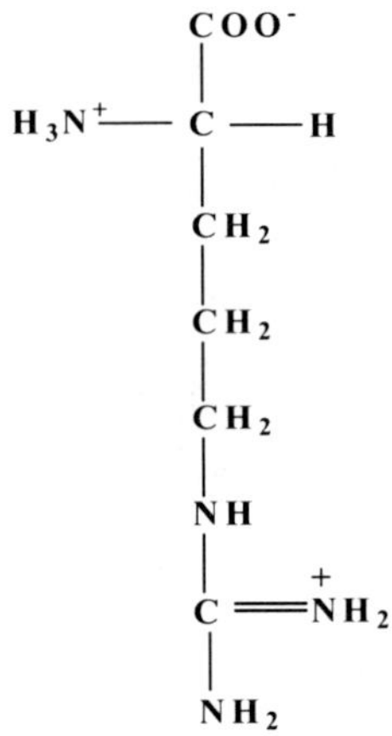

Histidine

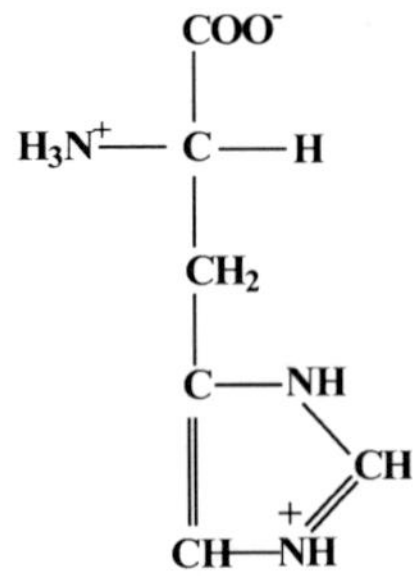

(b) Polysaccharides

Neutral polysaccharides

methylcellulose: (1-4)-linked glucose units in a linear chain

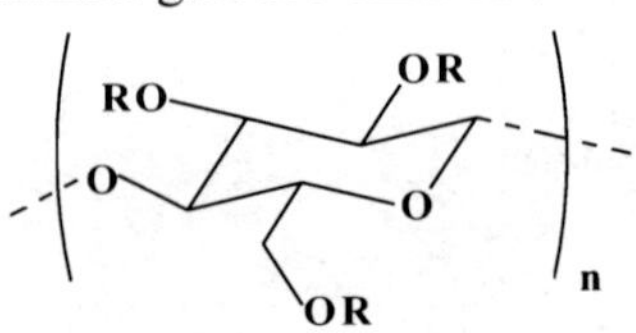

R = H or CH_3

Table 4.1 *Continued*

Neutral polysaccharides (*continued*)

dextran (glucan): α-(1-6) glycosidic linkages between glucose units; branching from α-(1-4) linkages (in some cases, α-(1-2) and α-(1-3))

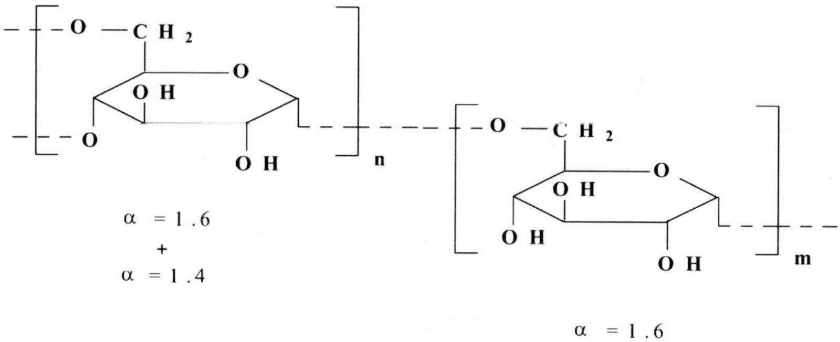

starch polysaccharides: amylose, amylopectin, maltodextrins (from acidic or enzymatic hydrolysis of starch)

amylose: α-(1-4) linearly linked glucose units

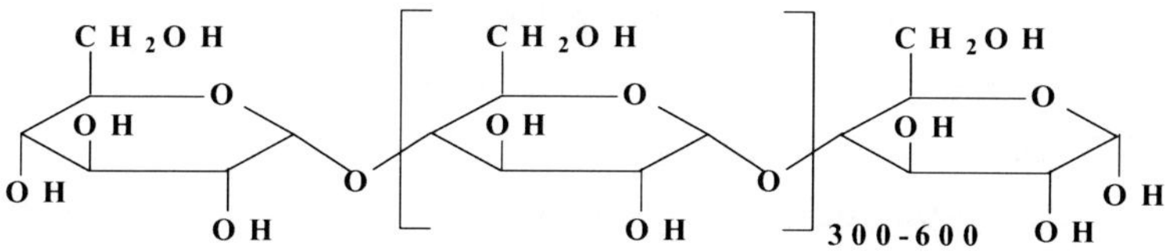

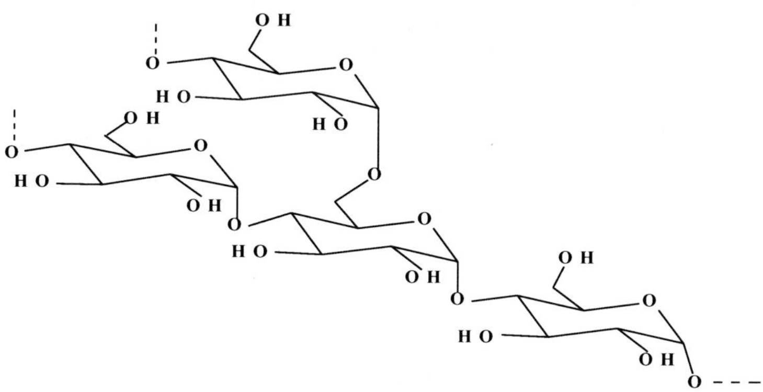

amylopectin: glucose units linked linearly via α(1-4) bonds; branching from α(1-6) bonds every 24 to 30 glucose units

Table 4.1 *Continued*

Neutral polysaccharides (*continued*)

galactomannans: (1-4)-linked β-D-mannopyranose backbone; branch points (1-6)-linked α-D-galactopyranose

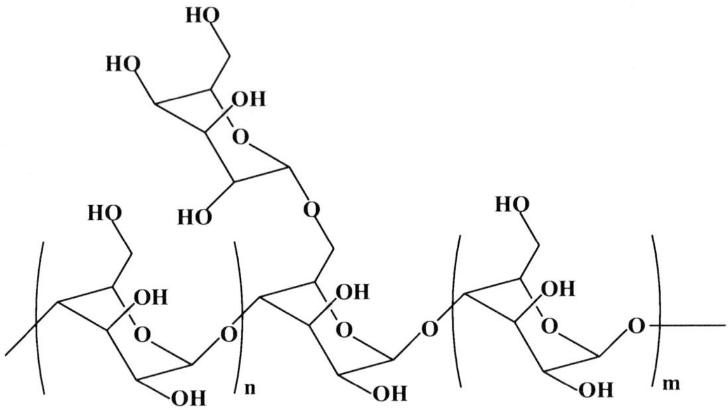

In order of increasing mannose-to-galactose ratio:

- **fenugreek gum**, mannose : galactose ~ 1:1
- **guar gum**, mannose : galactose ~ 2:1
- **tara gum**, mannose : galactose ~ 3:1
- **locust bean gum** or **carob gum**, mannose : galactose ~ 4:1

Anionic polysaccharides

carboxymethylcellulose: (1-4)-linked in a linear way glucose units

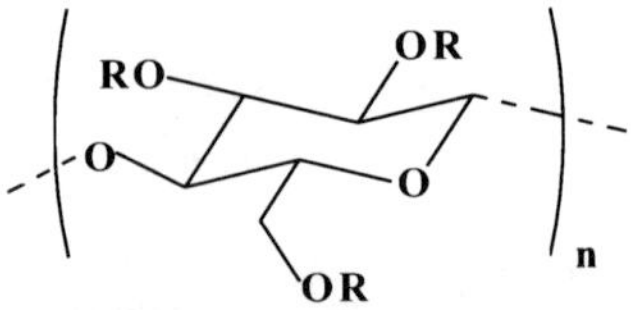

R = H or CH_2CO_2H

Table 4.1 *Continued*

Anionic polysaccharides (*continued*)

pectin: α-(1-4) linked D-galacturonic acid [m] with α-(1-2) linked L-rhamnose [n] (by convention, > 50 % of R = CH_3 for high-methoxyl pectin, and < 50 % for low-methoxyl pectin)

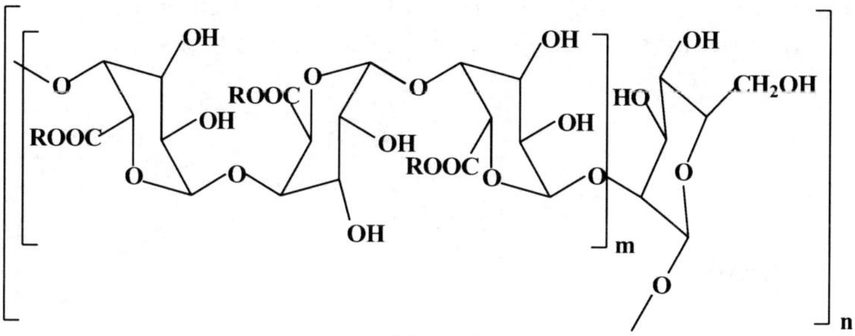

R = H or CH_3

alginate: (1-4)-linked β-D-mannuronate and its C-5 epimer α-L-guluronate

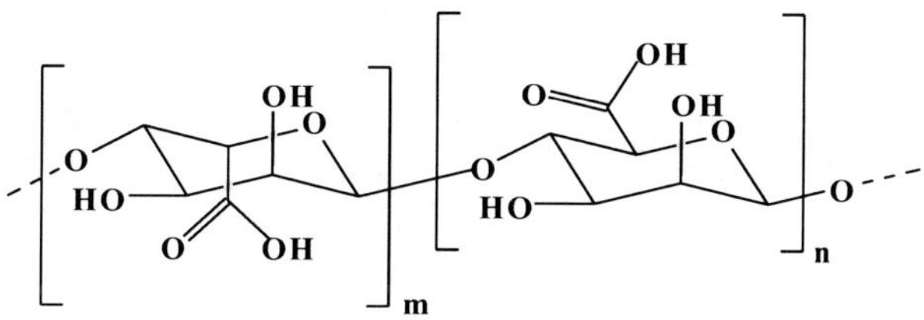

κ-carrageenan and **ι-carrageenan**: repeating disaccharide residues; Residue A is (1,3-linked) D-galactose-4-sulfate; Residue B is (1,4-linked) 3,6-anhydro-D-galactose for κ-carrageenan and (1,4-linked) 3,6-anhydro-D-galactose 2-sulfate for ι-carrageenan

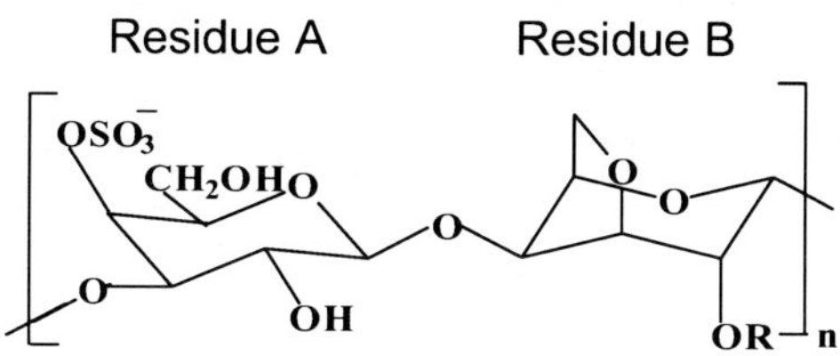

R = H or SO_3^-

Table 4.1 *Continued*

Anionic polysaccharides (*continued*)

xanthan: (1-4) linked β-D-glucose backbone; side chains of mannose and glucuronic acid (side chain linked to every other glucose of backbone at the 3 position; about half terminal mannose units with pyruvic acid group linked as a ketal to its 4 and 6 positions; other mannose unit with acetyl group at the 6 positions)

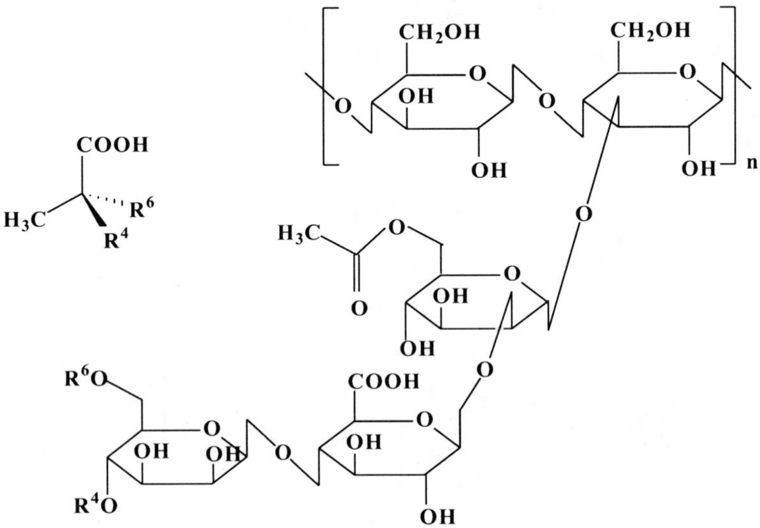

Cationic polysaccharide

chitosan: β-(1-4)-linked D-glucosamine (deacetylated unit) and N-acetyl-D-glucosamine (acetylated unit)

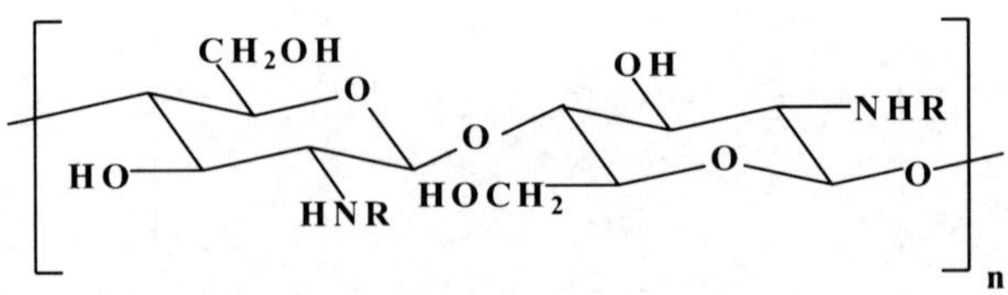

R = H or $-COCH_3$ (acetyl)

1. *Van der Waals Interactions*

This is a general term to describe the universal weak long-range attractive interactions (0.4–2 kJ/mol) of electromagnetic origin exhibited by all functional groups with permanent or induced dipoles. At very short range separations (2–3 nm), of course, such attractive interactions are transformed into strong repulsion (Finkelstein and Ptitsyn, 2002).

Equations for calculating van der Waals interaction forces/energies between macromolecules or colloidal particles are quite well established (Israelachvili, 1992; Dickinson and McClements, 1995; McClements, 2005). (For example, see equations (3.35) and (3.36) in chapter 3). The interactions between nanoparticles are potentially more complicated, however, because the nanoparticle size and interparticle separation are comparable in magnitude, precluding the use of the asymptotic forms of the equations; also nanoparticles are commonly anisotropic, and their dielectric properties are often not known (Min *et al.*, 2008).

Van der Waals forces act between all groups to some extent, but they rarely have a crucial or predominant influence on the net biopolymer–biopolymer interactions.

2. *Electrostatic Interactions*

Electrostatic interactions are Coulombic interactions between the charged functional groups of biopolymers. They may be either repulsive or attractive depending on whether the electrical charges are of the same or opposite sign. They are generally exothermic in character, *i.e.*, associated with evolution of heat. Compared with other (non-covalent) types of interaction, they are relatively strong (~ 440 kJ/mol at a distance of 0.3 nm between a pair of single charges) (Cantor and Schimmel, 1980; Dickinson and McClements, 1995; Finkelstein and Ptitsyn, 2002; McClements, 2005). The strength of the interactions decreases with increasing temperature.

Under the influence of electrostatic interactions in aqueous media, the interacting ions can either keep all their surrounding water molecules or lose part of them under the formation of ion pairs (Jenks, 1969). Moreover, the strength and range of these ionic interactions tend to decrease with increasing ionic strength (see equations (3.37), (3.38) and (3.39) in chapter 3). *Inter*molecular attractive electrostatic interactions form the main basis of biopolymer self-association and complex formation. *Intra*-molecular electrostatic interactions, both attractive and repulsive, control the level of folding/unfolding and expansion of macromolecules, as well as their macromolecular flexibility and their detailed biopolymer conformations — in particular, the secondary (helical or sheet-like), tertiary, and quaternary structures of proteins. All of these properties are sensitive to temperature and to the pH and ionic strength of the aqueous medium.

More generally, it is widely recognized that it is the establishment of repulsive electrostatic forces stronger than attractive van der Waals forces that is the main effective mechanism preventing aggregation of biopolymer molecules and colloidal particles in stable aqueous systems.

3. *Ion Bridging*

Ion bridging is a specific type of Coulombic interaction involving the simultaneous binding of polyvalent cations (*e.g.*, Ca^{2+}, Fe^{2+}, Cu^{2+}) to two different anionic functional groups on biopolymer molecules. This type of ionic interaction is commonly involved in associative self-assembly of biopolymers. As a consequence it is also an important contributory factor in the flocculation (via bridging or depletion) of colloidal particles or emulsion droplets in aqueous media containing adsorbed or non-adsorbed biopolymers (Dickinson and McClements, 1995).

4. *Hydrogen Bonding*

These are short-range attractive interactions with a specific orientational character. There are three main molecular types (Cantor and Schimmel, 1980; Lehninger, 1982; Dickinson and McClements, 1995; Finkelstein and Ptitsyn, 2002; McClements, 2005):

$$-\mathrm{O}-\mathrm{H}^{\delta+} \rightarrow {}^{\delta-}\mathrm{O}{<}^{\mathrm{H}}, \quad -\mathrm{N}-\mathrm{H}^{\delta+} \rightarrow {}^{\delta-}\mathrm{O}{<}^{\mathrm{H}}, \quad -\mathrm{N}-\mathrm{H}^{\delta+} \rightarrow {}^{\delta-}\mathrm{N}{<}^{\mathrm{H}} .$$

A hydrogen bonding interaction occurs when the hydrogen atom bound covalently to an electronegative atom approaches another electronegative atom. It is suggested that the hydrogen bond has dual character: that is, it is both electrostatic and covalent. Its character is determined by the interaction between the electronic clouds of atoms of the donor (hydrogen) and the acceptor (oxygen or nitrogen) accompanied by a shift of electron density from the donor to the acceptor (Jenks, 1969).

These interactions are usually exothermic in a character, and they are moderately strong (10–40 kJ/mol), with the strength depending on the bond orientation (Jenks, 1969). Thus, hydrogen bonding becomes less important with increasing temperature. *Inter*molecular hydrogen bonding is the predominant interaction responsible for biopolymer complex formation between a charged and a neutral biopolymer, or between a pair of neutral biopolymers. *Intra*molecular hydrogen bonding is commonly involved in macromolecular folding/unfolding, and in determining the detailed biopolymer conformations (secondary structure of proteins and helical structure of many polysaccharides), especially at ambient and physiological temperatures.

5. *Hydrophobic Interactions*

These are long-ranged attractive interactions between non-polar groups separated by water. The interactions are moderately strong (5–40 kJ/mol) and endothermic (up to around 60 °C).

Hydrophobic interactions are of entropic origin. That is to say, their formation is driven by the gain in the entropy of the system, especially involving the local structuring of the water molecules in the vicinity of the non-polar groups (Jenks, 1969; Cantor and Schimmel, 1980; Dickinson and McClements, 1995; McClements, 2005). A consequence of this entropic character is that the interactions become stronger with increasing temperature up to ~ 60 °C.

In an aqueous medium these *inter*molecular attractive interactions make a strong contribution to biopolymer self-association and inclusion complex formation, as well as to the flocculation of biopolymer-coated colloidal particles. *Intra*molecular hydrophobic interactions commonly influence the level of folding/unfolding of macromolecules as well as their detailed conformations.

6. *Excluded Volume ('Steric') Interactions*

These are repulsive interactions of short range (0.2–0.3 nm) arising from the highly unfavourable overlap of full electron clouds. They are sometimes called 'steric' interactions because they restrict the relative spatial arrangement of pairs of segments on the same or different macromolecules. The excluded volume expression arises from the fact that the volume occupied by one biopolymer molecule in solution is not available to other biopolymer molecules. Thus the size and shape of the biopolymer molecule/particle (as determined by the macromolecular conformation/flexibility or aggregate architecture) is of prime importance in relation to 'steric' interactions.

The excluded volume effect is associated with a reduction in the mixing entropy of the system. The resulting 'steric' interactions contribute predominantly to the thermodynamic phase separation (incompatibility) that is commonly observed in mixed biopolymer solutions. Excluded volume interactions also influence general biopolymer solubility issues and colloidal stability with respect to flocculation in aqueous media (Tanford, 1961; Dickinson and McClements, 1995).

7. *Solvation, Structural and Depletion Forces*

So-called solvation/structural forces, or (in water) hydration forces, arise in the gap between a pair of particles or surfaces when solvent (water) molecules become ordered by the proximity of the surfaces. When such ordering occurs, there is a breakdown in the classical continuum theories of the van der Waals and electrostatic double-layer forces, with the consequence that the monotonic forces they conventionally predict are replaced (or accompanied) by exponentially decaying oscillatory forces with a periodicity roughly equal to the size of the confined species (Min *et al.*, 2008). In practice, these confined species may be of widely variable structural and chemical types — ranging in size from small solvent molecules (like water) up to macromolecules and nanoparticles.

In the case of biopolymer molecules residing in the space between colloidal particles or droplets, the force associated with the deep energy minimum at contact is often referred to as the 'depletion force' because the intervening biopolymer species are depleted from the narrow gap between the pair of neighbouring particles. This attractive interparticle interaction underlies the phenomenon of reversible depletion flocculation in oil-in-water emulsions (see equation (3.41) in chapter 3).

8. *Covalent Bonding*

Covalent bonds are the very strong chemical linkages formed from the combination of unpaired electrons on atoms on different biopolymer molecules. *Intra*molecular covalent bonds are obviously ubiquitous. An example of *inter*molecular covalent bond formation is the reaction between a protein amino group and the reducing group of a polysaccharide, which occurs as a result of the Maillard reaction during dry-heating of a mixture of polysaccharide and protein. Such covalent bond formation can allow the preparation of biopolymer conjugates having enhanced functional properties (Akhtar and Dickinson, 2007; Dickinson, 2009).

Concluding Remarks

In addition to the considerations mentioned above, it is rather important to keep in mind that a common aspect in determining the overall thermodynamic behaviour of a biopolymer solution/dispersion is the necessity of taking into account *all* of the component interactions. This includes the interactions of the biopolymer(s) with the water molecules, including both hydration (attraction) and dehydration (release), as well as the interactions amongst the water molecules themselves.

The wide variety in the types of intermolecular interactions implies many different opportunities for their manipulation. So, for example, electrostatic interactions and ion bridging between biopolymers can be readily tuned by changing the pH and/or ionic strength of the aqueous medium. And the hydrogen bonding or hydrophobic interactions can be strengthened, respectively, by decreasing or increasing the temperature. In addition, there is the possibility of influencing the molecular interactions through application of external force fields — magnetic, electric, convective, hydrostatic and hydrodynamic (Min *et al.*, 2008). By these latter means, new processing technologies can provide possibilities for ordering and assembling all the various kinds of biopolymer-based building blocks, including individual macromolecules, soluble biopolymer aggregates, and colloidal particles/droplets covered by biopolymers.

BIBLIOGRAPHY

Akhtar, M., Dickinson, E. (2007). Whey protein–maltodextrin conjugates as emulsifying agents: an alternative to gum arabic. *Food Hydrocolloids*, **21**, 607–616.

Cantor, C.R., Schimmel, P.R. (1980). *Biophysical Chemistry*, Part I, San Francisco: Freeman.

Dickinson, E., McClements, D.J. (1995). *Advances in Food Colloids*, Glasgow: Blackie.

Dickinson E. (1998). Stability and rheological implications of electrostatic milk protein–polysaccharide interactions. *Trends in Food Science and Technology*, **9**, 347–354.

Dickinson, E. (2009). Hydrocolloids as emulsifiers and emulsion stabilizers. *Food Hydrocolloids*, **23**, 1473–1482.

Israelachvili, J.N. (1992). *Intermolecular and Surface Forces*, 2nd edn, London: Academic Press.

Finkelstein, A.V., Ptitsyn, O.B. (2002). *Protein Physics. A Course of Lectures (Soft Condensed Matter, Complex Fluids and Biomaterials)*, San Diego: Academic Press.

Jenks, W. P. (1969). *Catalysis and Chemistry in Enzymology*. New York: McGraw-Hill.

Lehninger, A.L. (1982). *Principles of Biochemistry*. New York: Worth Publishers, vol. 1.

McClements, D.J. (2005). *Food Emulsions: Principles, Practice, and Techniques*, 2nd edn, Boca Raton, FL: CRC Press.

McClements, D.J. (2006). Non-covalent interactions between proteins and polysaccharides. *Biotechnology Advances*, **24**, 621–625.

Min, Y., Akbulut, M., Kristiansen, K., Golan, Y., Israelachvili, J. (2008). The role of interparticle and external forces in nanoparticle assembly. *Nature Materials*, **7**, 527–538.

Tanford, C. (1961). *Physical Chemistry of Macromolecules*, New York: Wiley.

CHAPTER FIVE

PHYSICO-CHEMICAL CHARACTERIZATION OF BIOPOLYMERS IN SOLUTION IN TERMS OF THERMODYNAMIC PARAMETERS

1. *Basic Thermodynamic Functions and Parameters*

Any biopolymer solution or food colloid system can be considered as a closed thermodynamic system that is able to exchange energy (but not matter) with its environment under conditions of constant pressure and temperature (Prigogine and Defay, 1954; Edsall and Gutfreund, 1983). For such a system the total heat content Q is characterized by the value of the enthalpy H, as determined by the internal energy of the system, U, and the work pV that the system can do against the constant pressure (and therefore under the change of its volume V). In its integral form the enthalpy can therefore be defined as:

$$H = U + pV = Q \quad . \tag{5.1}$$

In turn, in its differential form the change in enthalpy is defined as:

$$\mathrm{d}H = \mathrm{d}U + p\mathrm{d}V + V\mathrm{d}p = \mathrm{d}Q \quad . \tag{5.2}$$

So, under the normal environmental conditions encountered by biopolymer solutions and food colloids, *i.e.*, constancy of pressure, the change in enthalpy is defined as:

$$\mathrm{d}H = \mathrm{d}U + p\mathrm{d}V \quad . \tag{5.3}$$

For liquid-like systems, where each molecule occupies only a very small volume under atmospheric pressure, the term $p\mathrm{d}V$ is generally negligibly small in comparison with the change in the internal energy of the system. Therefore one can ignore the $p\mathrm{d}V$ term in equation (5.3) and consider the change in enthalpy as equivalent to the change in internal energy of the system (Finkelstein and Ptitsyn, 2002). What this means in statistical thermodynamic terms is that the changes in biopolymer–biopolymer and biopolymer–solvent interactions contribute to the change in the enthalpy of the system, and consequently to its change in internal energy.

Let us denote the equilibrium state of the system existing before the occurrence of the interactions as A, and the equilibrium state reached as a result of the interactions as B. The change in enthalpy in passing from state A to state B, *i.e.*,

$$H_B - H_A = \Delta H = Q \,, \qquad (5.4)$$

is independent of the path taken because the internal energy and enthalpy are thermodynamic state functions. When a system takes up heat during any process, the values of ΔH and Q are positive and the process (interaction) is called an endothermic process (interaction). The opposite is the case for an exothermic process (interaction). Under the influence of any such process, the magnitude of the enthalpy change, ΔH, can be determined experimentally by direct calorimetric measurement (Semenova *et al.*, 1991; Blandamer *et al.*, 1998; Portnaya *et al.*, 2006).

In order to take into account the spontaneity and irreversibility of real processes (heat always goes from a hot substance to a cold one, but not the reverse), thermodynamics invokes the notion of the entropy S. In statistical terms entropy is defined as the probability of accessible states for each molecule in the system:

$$S = k \ln V \;. \qquad (5.5)$$

In equation (5.5), k is the Boltzmann constant (1.38066×10^{-23} J K^{-1}), which is related to the (thermal) energy of one molecule, and V is the number of all accessible states of the molecule. Generally, in physical chemistry, we consider not a single molecule (or particle), but rather one mole, *i.e.*, 6×10^{23} molecules (or particles). Thus equation (5.5) becomes

$$S = R \ln V \;, \qquad (5.6)$$

where $R = kN_A = 8.314$ J K^{-1} mol^{-1} is the universal gas constant, and N_A is Avogadro's number.

In classical thermodynamics the entropy S is formally related to the heat Q that a thermodynamically closed system can exchange with its surroundings. The second law of thermodynamics states that the entropy increases during all spontaneous chemical and physical processes, taking account of changes in both the system and its surroundings (Prigogine and Defay, 1954; Edsall and Gutfreund, 1983):

$$\Delta S = \Delta S_{\text{surroundings}} + \Delta S_{\text{system}} > 0 \;. \qquad (5.7)$$

In the limiting case of a reversible process, the change in entropy of the system, ΔS, is determined by the heat Q that is taken up by the system during the process divided by the (absolute) temperature T (Edsall and Gutfreund, 1983):

$$\Delta S = \frac{Q}{T} = \frac{\Delta H}{T} \quad . \tag{5.8}$$

The relationship between the statistical expression (equation (5.6)) and the classical expression (equation (5.8)) for determination of the entropy can be explained by the statement that, due to the additional heat taken up, the system acquires more available microstates (Edsall and Gutfreund, 1983). Equation (5.8) introduces a procedure for the direct calorimetric measurement of the entropy change for a specific process such as the reversible formation of a new set of biopolymer interactions.

Strictly speaking, equation (5.8) is only true for reversible processes. However, the entropy S, like the internal energy and the enthalpy, is a thermodynamic function of the state of the system. This means that the change in entropy is determined by the initial and final equilibrium states of the system, rather than the path between them. So, if there is a reversible way of going from state A to state B, then it can be assumed that, for another (irreversible) path, the change in entropy of the system is the same. But, if the path $A \rightarrow B$ is irreversible, then there is a corresponding increase in the entropy of the surroundings (*i.e.*, $\Delta S_{surroundings}$ in equation (5.7)) that does not occur under reversible conditions (Prigogine and Defay, 1954; Edsall and Gutfreund, 1983).

The heat taken up, dH, for an infinitesimally small increase in temperature, dT, at constant pressure p, defines another important calorimetrically measurable quantity — the heat capacity C_p. It is formally defined by the partial derivative

$$C_p = \left(\frac{\partial H}{\partial T}\right)_p \quad , \tag{5.9}$$

where the index 'p' indicates that the change occurs at constant pressure. Hence we have

$$dH = C_p dT \quad , \tag{5.10}$$

$$dS = C_p \frac{dT}{T} = C_p d(\ln T) \quad \cdot \tag{5.11}$$

By integrating over the specific temperature interval ($T_1 \rightarrow T_2$) one can determine the overall change in the entropy of the system due to this temperature change, *i.e.*,

$$\int_{T_1}^{T_2} C_p \, \mathrm{d}(\ln T) = \Delta S \quad . \tag{5.12}$$

Under the general conditions of constant pressure and temperature, the sum of the changes in the magnitude of the enthalpy and the entropy during any process determines the overall change in the Gibbs free energy of the system:

$$\Delta G = \Delta H - T\Delta S \quad . \tag{5.13}$$

The change in the free energy is, by definition, the amount of useful work/energy which has to be done/taken to transform the system from state A to state B, on the condition that the system constantly exchanges heat with the surroundings. Accordingly, the term 'free energy' is seen to be an extension of the conventional notion of energy under conditions of exchange of heat by the system with its local environment. For a chemical or physical process to proceed spontaneously, including formation of new biopolymer/colloidal particle interactions, the change in Gibbs free energy of the system must be negative. Thus, both an exothermic enthalpy change ($\Delta H < 0$) and an increase in the system's entropy ($\Delta S > 0$) will favour the spontaneity of the process. The balance of enthalpy and entropy terms in equation (5.13) therefore determines the thermodynamic nature of each process of interest, including those involving changes in interactions between biopolymer molecules and/or colloidal particles.

The quantities *G, H*, and *S* are called extensive thermodynamic functions because the magnitude of the quantity in each case depends on the amount of substance in the system. The change in Gibbs free energy under addition of unit concentration of component *i* at constant concentrations of the other components is called the partial Gibbs free energy of the *i*-component, *i.e.*, the chemical potential of the *i*-component in the system. The chemical potential is an intensive thermodynamic quantity, like temperature and concentrations. The formal definition is

$$(\partial G/\partial n_i)_{T,p,n_j} = \mu_i \quad , \tag{5.14}$$

where the index n_j indicates that the concentrations of all the other components ($i \neq j$) in the system are kept constant. Thus, the Gibbs energy of the system as a whole is

$$G = \sum_i n_i \mu_i \quad . \tag{5.15}$$

It is well recognized that the thermodynamic properties of a multicomponent system are uniquely determined if the chemical potentials of all the components of the system are known as a function of the independent variables T, n_1,... (Prigogine and Defay, 1954). In the case of a thermodynamically non-ideal system (where the components are said to be 'interacting'), the chemical potentials of solvent and biopolymer(s) can be expressed as a virial series in the biopolymer concentration(s) (see equations (3.7) and (3.8)). The osmotic second virial coefficient determines the excess chemical potential, μ_i^E, *i.e.*, the excess partial molar Gibbs free energy, characterizing the formation of various pair contacts, such as biopolymer–solvent and biopolymer–biopolymer (see equations (3.12) and (3.13)) (Wills *et al.*, 1993). From the experimentally determined temperature dependence of the second virial coefficient and the calculated temperature dependence of μ_i^E, we can use general thermodynamic relationships to determine other basic excess thermodynamic functions like molar excess enthalpy and molar excess entropy. This can provide a complete thermodynamic characterization of interactions in the system (see equations (3.14) and (3.15)) (Dhillon, 1979). Within the limits of this approach, the description of the thermodynamic behaviour of biopolymer solutions and colloidal systems is thereby quantitatively established. In particular, the osmotic second virial coefficient, which can be readily measured in various ways, is recognized as a useful quantity for describing and predicting thermodynamic behaviour of biopolymer solutions and biopolymer-based colloidal systems (Clark, 2000; Semenova, 2007; Su *et al.*, 2008).

2. *Experimental Techniques for Determining Thermodynamic Quantities of Biopolymer Interactions in Solution*

2.1. *Osmometry*

The membrane osmometer is a device for measuring the osmotic pressure, Π, of a biopolymer solution separated from the pure solvent by a semi-permeable membrane (Tanford, 1961; Edmond and Ogston, 1968; Tombs and Peacocke, 1974; Edsman and Sundelöf, 1987; Amur *et al.*,

1997). The difference between the chemical potential of the solvent in its pure state and that in the biopolymer solution determines the osmotic pressure of the biopolymer solution:

$$\Pi = \left(\frac{-\Delta\mu_1}{V_1}\right) = \frac{(\mu_1^0 - \mu_1)}{V_1} = RTC\left(M_n^{-1} + A_2 C + A_3 C^2 + A_4 C^3\right) \cdot \tag{5.16}$$

Here, $\mu_1{}^0$ and μ_1 are, respectively, the chemical potentials of pure solvent and solvent at a certain concentration of biopolymer; V_1 is the molar volume of the solvent; $M_n = \Sigma_i x_i M_i$ is the number-averaged molar mass of the biopolymer (sum of products of mole fractions, x_i, and molar masses, M_i, over all the polymer constituent chains (i) as determined by the polymer polydispersity) (Tanford, 1961); A_2, A_3 and A_4 are the second, third and fourth virial coefficients, respectively (in weight-scale units of cm^3 mol g^{-2}), characterizing the two-body, three-body and four-body interactions amongst the biopolymer molecules/particles, respectively; and C is the weight concentration (g ml^{-1}) of the biopolymer.

At small solute concentrations the second virial coefficient is the main contributor to the value of Π, and so in practice the general equation (5.16) is usually restricted to just the term containing the second virial coefficient. At this level of approximation, the osmotic pressure of a ternary solution (biopolymer$_i$ + biopolymer$_j$ + solvent) may be expressed in the following simple form using the molal scale (Edmond and Ogston, 1968):

$$\Pi = \left(\frac{\mu_1^0 - \mu_1}{V_1}\right) = \frac{RT}{V_1 m_1}\left(m_i + m_j + A_{ii}^* m_i^2/2 + A_{jj}^* m_j^2/2 + A_{ij}^* m_i m_j\right) \cdot \tag{5.17}$$

Here, as previously, the solvent is taken as component 1, one of the biopolymers is the i-component, while another is the j-component; m_1, m_i and m_j are the concentrations (moles per kg of water) of the components; A_{ii}^* and A_{jj}^* are the second virial coefficients (m^3/mol) characterizing the like pair interactions of the types biopolymer$_i$–biopolymer$_i$ and biopolymer$_j$–biopolymer$_j$, respectively; and A_{ij}^* is the cross second virial coefficient corresponding to the biopolymer$_i$–biopolymer$_j$ pair interaction.

In data analysis, it is conventional to plot Π/C *versus* C for the binary solution (biopolymer + solvent) or $\Pi/(m_i + m_j)$ *versus* $(m_i + m_j)$ for the ternary solution (biopolymer$_i$ + biopolymer$_j$ + solvent). The initial slope of the curve can then be used to determine the second virial coefficient

A_2 from equation (5.16) or the cross second virial coefficient A_{ij}^* from equation (5.17). In turn, this knowledge of the second virial coefficients and their temperature dependence allows calculation of the values of the chemical potentials of all components of the biopolymer solution or colloidal system, as well as enthalpic and entropic contributions to those chemical potentials. On the basis of this information, a full description and prediction of the thermodynamic behaviour can be realised (see chapter 3 and the first paragraph of this chapter for the details).

It is important for us to keep in mind that biopolymers are generally not monodisperse components. Proteins are typically paucidisperse — mixtures of monomers, dimers and multimers. And polysaccharides are polydisperse: their chain lengths and molar masses can be represented as a continuous distribution. For this reason the virial coefficients appearing in equations (5.16) and (5.17) should be interpreted as averages. So the inverse of the number-averaged molar mass of component i is given by

$$\frac{1}{\overline{M}_i} = \frac{\sum_{k=1}^{N} n_{i,k}}{\sum_{k=1}^{N} n_{i,k} M_{ik}} = \sum_{k=1}^{N} \frac{\widetilde{w}_{i,k}}{M_{i,k}} \quad , \qquad (5.18)$$

In equation (5.18), the quantity $n_{i,k}$ denotes the number of moles of fraction k of the i-component having a molar mass of $M_{i,k}$, and the quantity $\widetilde{w}_{i,k}$ represents the weight fraction of fraction k of the i-component.

The averaged second virial coefficients for a polydisperse system are given by (Schaink and Smit, 2007):

$$\overline{A}_{2,ij} = \sum_{kl} \widetilde{w}_{ik} \widetilde{w}_{jl} A_{ij,kl} \quad . \qquad (5.19)$$

The quantity $A_{ij,kl}$ represents the cross second virial coefficient characterizing the interaction between the fraction k of the i-component and the fraction l of the j-component.

Osmotic pressure is a colligative property: it depends on the number of solute molecules or particles in a given volume of solvent and not on their mass (or volume). This means that, in practice, the usefulness of osmometry is mainly limited to measurements involving rather small biopolymer molecules/particles of number-average molar weight $M_n \leq 10^5$ Da. For systems of larger macromolecules/particles (10^4 Da $\leq M_n \leq 10^9$ Da) measurements of laser light scattering are to be preferred (Evans, 1972, Kratochvil and Sudelof, 1986; Kaddur and Strazielle, 1986; Burchard, 1994; Semenova, 1996, 2007). That having been said, we wish to make the reader aware that Edsman and Sundelöf (1987) have shown that osmotic measurements can be generated with sufficiently good accuracy for the determination of the virial coefficients of some relatively high-molecular-weight polymer samples. This is because the osmotic pressure measured in some systems is relatively high at moderate concentrations due to the strongly contributing non-ideality terms.

2.2. *Laser Light Scattering*

Static laser light scattering probes structure on length scales of the order of ~ 1 μm and below. It is a particularly useful tool for exploring the different kinds of species and interactions in a dilute biopolymer solution, including the phenomenon of self-association (and hetero-association) of biopolymers on the nanometre scale. These properties of biopolymers in solution are of utmost interest in the basic biological sciences, in food science, and in biotechnology (Burchard, 1994; Murphy, 1997; Semenova, 1996, 2007; Semenova *et al.*, 2009). In practical terms, for studying biopolymer solutions, it is clear that light scattering has emerged as one of the key tools in the arsenal of experimental techniques that can be brought to bear on the problem.

The scattered light intensity from a polymer solution arises from the fluctuations in both the solvent density and the polymer concentration. These fluctuations are considered as stable during the timescale of the measurement in the static mode of light scattering (for more details, see Evans (1972)). The light scattered from just the polymer (in excess of the light scattered from the pure solvent) is given by (Burchard, 1994)

$$\Delta R_{\Theta} = K\langle \Delta c^2 \rangle \ , \qquad (5.20)$$

where $\Delta R_{\Theta} = R_{\Theta}^{\text{solution}} - R_{\Theta}^{\text{solvent}}$. The Rayleigh ratio (*i.e.*, the normalized scattering intensity) is defined by $R_{\Theta} = (i_{\Theta}/I_0)r^2$, where I_0 is the primary beam intensity, i_{Θ} is the corresponding scattering intensity at scattering angle Θ, and r is the distance of the detector from the centre of the scattering volume. The quantity K is an optical contrast constant; for vertically polarized laser light it is given by $K = [4\pi^2 n_0^2/(\lambda_0^4 N_A)](\mathrm{d}n/\mathrm{d}c)^2$, where λ_0 is the wavelength of the primary beam *in vacuo*, n_0 is the solvent refractive index, dn/dc is the polymer refractive index increment, and N_A is the Avogadro number.

The term $\langle \Delta c^2 \rangle$ in equation (5.20) denotes the mean-square polymer concentration fluctuation. It is related directly to the osmotic pressure $\Pi = -(\Delta\mu_1/V_1)$ in the polymer solution by the equation:

$$<\Delta c^2> = cRT(\mathrm{d}c/\mathrm{d}\Pi) \quad . \tag{5.21}$$

Therefore the scattering intensity defined in equation (5.20) is directly proportional to the osmotic compressibility:

$$\Delta R_\Theta = KcRT(\mathrm{d}c/\mathrm{d}\Pi) \quad . \tag{5.22}$$

This has a clear physical meaning: the large variation in the values of the osmotic pressure in neighbouring solution volumes hinders the development of polymer concentration fluctuations, and consequently it reduces the intensity of the light scattering.

An important measure of biopolymer size is the radius of gyration R_G. It is defined by $R_G^2 = (1/N)\,\Sigma_i < r_i^2>$ where r_i is the distance of monomer i from the centre of mass (sum taken from $i = 1$ to $i = N$) (Burchard, 1994; Tanford, 1961). For the case of a large biopolymer molecule/particle, we have $<R_G^2>^{1/2} > \lambda/25$; for example, at $\lambda = 633$ nm (the wavelength of the He–Ne laser), we have $<R_G^2>^{1/2} > 25$ nm. For such a biopolymer species, the phase difference between the light rays scattered from the different parts of the polymer molecule/particle leads to a decrease in the scattering intensity as a function of scattering angle Θ due to light interference (the Mie effect). On taking the Mie effect into account, one can transform equation (5.20) into a simple relationship which is often taken as the general scattering equation:

$$\Delta R_\Theta = KcRT(\mathrm{d}c/\mathrm{d}\Pi)P(q) \tag{5.23}$$

Here $P(q)$ is the particle scattering factor and $q = (4\pi/\lambda_0)\sin(\Theta/2)$ is the scattering vector. The value of P reflects the specific size and shape of the polymer particle. This parameter has been calculated and tabulated for many different kinds of idealized colloidal and macromolecular structures (Burchard, 1994; Evans, 1972; Tanford, 1961).

For analysis by means of the most general graphical treatment of light scattering measurements (the Zimm plot), the scattering equation (5.23) is rewritten in the following easy-to-use form (Burchard, 1994; Evans, 1972; Tanford, 1961):

$$(Kc)/\Delta R_\Theta = 1/(RT)(\mathrm{d}\Pi/\mathrm{d}c)P^{-1}(q) \quad . \tag{5.24}$$

Using the Gibbs–Duhem law ($m_1\mathrm{d}\mu_1 + m_2\mathrm{d}\mu_2 = 0$) (Prigogine and Defay, 1954), the equilibrium osmotic pressure in the biopolymer solution can be related to biopolymer concentration c in terms of virial coefficients by equation (5.16). In turn, the reciprocal particle scattering factor is given by (Burchard, 1994; Evans, 1972; Tanford, 1961):

$$P^{-1}(\Theta) = (1 + (q^2 \langle R_G^2 \rangle)/3 + \ldots) = 1 + (16\pi^2/3\lambda_0^2) \langle R_G^2 \rangle \sin^2(\Theta/2) + \ldots \quad (5.25)$$

Finally, at the level of pair interactions, after the differentiation of the osmotic pressure Π with respect to biopolymer concentration c, we get the following widely used form of the scattering equation, showing clearly the relationship between ΔR_Θ and the measured quantities M_w, R_G and A_2:

$$Kc/\Delta R_\Theta = 1/M_w + (16\pi^2/(3M_w\lambda_0^2)) \langle R_G^2 \rangle \sin^2(\Theta/2) + 2A_2c \ . \quad (5.26)$$

Here, M_w is the weight-average molar weight. In order to determine M_w, we make a plot of $Kc/\Delta R_\Theta$ *versus* $(q^2 + kc)$. The quantity k is an arbitrary constant that simply shifts the angle-dependent curves for the various concentrations along the q^2 axis so that the lines become well separated from each other. The intersection point on the ordinate is the reciprocal weight-average molar mass, $1/M_w$. The initial slope of the curve for $\Theta \rightarrow 0$ is equal to $2A_2$; and the initial slope for $c \rightarrow 0$, which is independent of the architecture of the biopolymer molecule/particle, is influenced by R_G (see Figure 5.1 for an example).

In those cases where the Zimm plot exhibits an upturn (non-linearity) or presents physically meaningless negative values (*e.g.*, with large particles ($M_w \geq 10^8$ Da) or branched macromolecules), the required extrapolations towards zero scattering angle ($q^2 = 0$) and zero concentration may be inaccurate or even impossible. As an alternative in such cases, we can make use of a Berry plot, *i.e.*, a plot of $(Kc/\Delta R_\Theta)^{1/2}$ against q^2. This type of plot can often produce satisfactory linearization, at least in the region of small q^2 (Burchard, 1994):

$$\left(\frac{Kc}{R_\Theta}\right)^{1/2} = \left(\frac{1}{M_w}\right)^{1/2}\left[1 + \frac{1}{6}(R_G q)^2\right] + A_2 M_w^{1/2} c + \ldots \ . \quad (5.27)$$

But for large globular particles, the angular dependence in the Berry plot can still show an upturn. In such a case the Guinier plot of $\ln(Kc/\Delta R_\Theta)$ against q^2 can be used to linearize the data (Burchard, 1994).

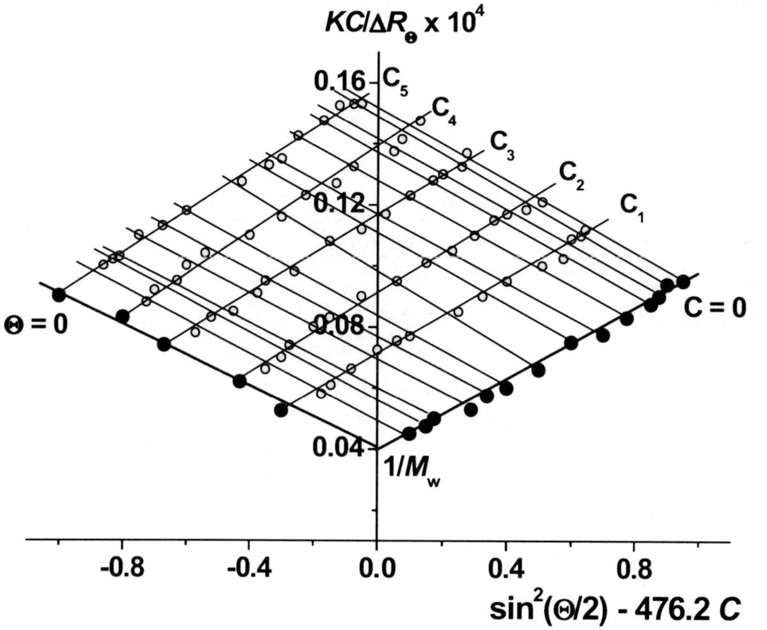

Figure 5.1 An experimental Zimm plot (equation (5.26)) for an aqueous solution of high-methoxy pectin (DE 58%) (pH = 8.0, 0.3 mol/dm^3 NaCl, 0.01 mol/dm^3 mercaptoethanol, 25 °C). The measurements were made at various pectin concentrations c_1, c_2, *etc.*, and at various scattering angles Θ. The intersection point on the ordinate for $c \to 0$ and $\Theta \to 0$ is the reciprocal weight-average molar mass, $1/M_w$. Replotted from Semenova *et al.* (1990).

The cross second virial coefficient, A_{ij}, can be determined from static light scattering measurements on a diluted ternary biopolymer solution (Kratochvil and Sundelöf, 1986; Kaddur and Strazielle, 1986). The conventional methodological scheme involves measuring the concentration dependence of the intensity of the scattered light from the set of mixed biopolymer$_1$ + biopolymer$_2$ solutions, whilst keeping constant the ratio of the concentrations of the two biopolymer components. That is, a solution of the mixed biopolymers of known composition is gradually diluted with pure solvent. We subtract the scattering contribution due to the pure solvent away from the scattering intensity of the solution, and then evaluate the results by means of a basic equation excluding terms containing the third and higher powers of the biopolymer concentrations, *i.e.*, restricted to pair interactions only. In its reciprocal form, the basic equation is written as (Kratochvil and Sundelöf, 1986):

$$\left(\frac{K'(C_i + C_j)}{[\Delta R_\Theta]_{\Theta=0}}\right) = \left(\frac{1}{\nu_i^2 M_{wi} w_i + \nu_j^2 M_{wj} w_j}\right) + \frac{2\left(\nu_i^2 M_{wi}^2 w_i^2 A_{ii} + 2\nu_i \nu_j M_{wi} M_{wj} w_i w_j A_{ij} + \nu_j^2 M_{wj}^2 w_j^2 A_{jj}\right)}{\left(\nu_i^2 M_{wi} w_i + \nu_j^2 M_{wj} w_j\right)^2}(C_i + C_j) \tag{5.28}$$

The quantity $K' = 4\pi^2 n^0{}_2 / N_A \lambda_0{}^4$ is the optical constant of the instrument for the vertically polarized light of the laser beam; A_{ii} and A_{jj} are the second virial coefficients (weight scale) characterizing the like pair interactions for biopolymer$_1$ and biopolymer$_2$, respectively; A_{ij} is the cross second virial coefficient (weight scale) characterizing the unlike pair interaction of biopolymers i and j; ν_i and ν_j are the refractive index increments, w_i and w_j are the weight fractions, M_{wi} and M_{wj} are the weight average molar masses, and C_i and C_j are the concentrations of the biopolymers i and j, respectively. In the graphical representation of $K'(C_i+C_j)/[\Delta R_\Theta]_{\Theta=0}$ against (C_i+C_j), the value of the intercept on the y-axis yields a consistency check on the measurements of the molar mass of each biopolymer alone and of their mixture. The cross second virial coefficient (A_{ij}) can be calculated from the initial slope.

It is important to note that the values of the second virial coefficients determined by light scattering are weight-average quantities. The general expression for the second virial coefficient of a polydisperse polymer is as follows (Casassa, 1962):

$$A_2 = \sum_i \sum_j B_{ij} n_i w_i n_j w_j / \langle n \rangle^2 \quad . \tag{5.29}$$

Here w_i and w_j are the weight fractions of the individual polymer species in solution; n_i and n_j are their degrees of polymerization; $\langle n \rangle = \Sigma n_i w_i$ is the weight-average degree of polymerization; and B_{ij} is the second virial coefficient for a single i–j polymer pair in weight units (cm^3 mol g^{-2}). It has been predicted theoretically (Casassa, 1962) that the value of A_2 for a polydisperse polymer in a good solvent is greater than that for a sharp polymer fraction (monodisperse polymer).

The advantage of static light scattering over some other methods of determining the second virial coefficient is the capability to measure both thermodynamic (A_2) and structural parameters (M_w, R_G) in a single experiment. This advantage provides us with the possibility to estimate the contributions from excluded volume effects to the pair interactions between various kinds of biopolymer molecules/particles in aqueous solution. This, in turn, allows an estimation of the contribution from the electrostatic interactions to the value of the second virial coefficient. And this is important because current theories of the second virial coefficient indicate that, in the case of polyelectrolytes (proteins and anionic/cationic polysaccharides), the positive value of the second virial coefficient is mainly determined by the contribution from the thermodynamically excluded volume (A_{ij}^{exc}) of the macroions, as well as by the contribution from the electrostatic forces acting between macroions (A_{ij}^{el}) (Tanford, 1961; Nagasawa and Takahashi, 1972). In this analysis, it is implied that the second virial coefficient may be written as a sum of two terms,

$$A_{ij} = A_{ij}^{\text{exc}} + A_{ij}^{\text{el}} \quad , \tag{5.30}$$

where we set $i = j$ for the case of interactions between pairs of identical macroion species.

To a first approximation, one can assume that the excluded volume term (A_{ij}^{exc}) is determined only by the physical volume occupied by the biopolymer molecules/particles (see equations (5.32), (5.34) and (5.35)). The electrostatic term, $A_{ij}^{\text{el}} = A_{ij}^{0}\Gamma_1\Gamma_2$, is determined by the ideal Donnan contribution, A_{ij}^{0}, corrected by factors to take into account the electrostatic interactions amongst the ions (Γ_1) and the chain-like character of the polyelectrolyte (Γ_2) (Nagasawa and Takahashi, 1972).

The ideal Donnan contribution, A_{ij}^{0}, is a consequence of the Donnan distribution of diffusible ions. This phenomenon arises from a tendency to drive towards fulfilment of the condition of electrical neutrality in the solution (Tanford, 1961; Nagasawa and Takahashi, 1972), leading to the expression

$$A_{ij}^{0} = \frac{1}{2}\frac{Z_i Z_j}{C_{\text{s}}} \quad , \tag{5.31}$$

where the molal units (cm^3/mol) of A_{ij}^{0} are used. The quantities Z_i and Z_j are the total charges on the biopolymer molecules/particles, and C_{s} is the concentration (mol/dm^3) of the supporting electrolyte, *i.e.*, the electrolyte in the 'outside' solution in equilibrium with the polyelectrolyte solution. Whenever the electrical neutrality condition is fulfilled, the ideal Donnan term appears as the leading term in the theoretical equation for the second virial coefficient (Tanford, 1961; Nagasawa and Takahashi, 1972).

Under the conditions of screening of electrostatic interactions between polyions, as occurs at high ionic strength (say, $I \geq 0.1$ mol dm^{-3}), or in solutions containing neutral (non-ionic) polymers, the excluded volume term is the leading term in the theoretical equation for the second virial coefficient. In this latter type of situation, the sizes and conformation/ architecture of the biopolymer molecules/particles become of substantial importance.

It is illuminating to consider some representative examples of effects of biopolymer geometrical structure on the theoretical expression for the excluded volume term of the second virial coefficient on the molal scale (cm^3/mol). The simplest case is that of interacting solid spheres (Tanford, 1961):

$$A_{ij}^{\text{exc}} = 10^{-3}\,\frac{4\pi N_{\text{A}}}{3}\left(R_i + R_j\right)^3 \quad . \tag{5.32}$$

The parameters R_i and R_j in equation (5.32) are the radii of the equivalent hard spheres representing biopolymers i and j, respectively (where $i = j$ for interactions between the same macroions). The equivalent hard sphere corresponds to the space occupied in the aqueous medium by a single biopolymer molecule (or particle) which is completely inaccessible to other biopolymers. In practice, the hard sphere model is a highly satisfactory description for many globular proteins.

For the case of flexible non-rigid particles, a penetration parameter can be used to take into account the softness of the biopolymer interactions. This parameter is defined as the ratio of the equivalent hard sphere radius to the corresponding radius of gyration (Tanford, 1961):

$$\gamma_i' = \frac{R_i}{R_{\text{G}i}} = \frac{\left(3000\,A_{ii} \big/ 32\pi N_{\text{A}}\right)^{1/3}}{R_{\text{G}i}} \quad . \tag{5.33}$$

In equation (5.33), as previously defined, A_{ii} is the second virial coefficient of the i-biopolymer component on the molal scale (cm^3 mol^{-1}); and $R_{\text{G}i}$ is the radius of gyration of the i-biopolymer component. For a flexible polymer chain the value of the penetration parameter γ_i' lies in the range 0.6–0.7, while for a semi-rigid/worm-like polymer we have $\gamma_i' \sim 0.3$ (Tanford, 1961). The shorter the polymer chain, the closer is its effective conformation to that of the equivalent hard sphere.

For interaction of a 'flexible' cylinder and a solid sphere, the excluded volume term of the second virial coefficient is given by (Zvetcov, 1986)

$$A_{ij}^{\mathrm{exc}} = 10^{-3}\pi N_{\mathrm{A}}\left(R_{\mathrm{f}} + R\right)^2 L \quad , \tag{5.34}$$

where R_{f} is the cylindrical radius, L is the total contour chain length of the cylinder, and R is the radius of the equivalent hard sphere. For interaction of a rigid rod and a solid sphere we have (Ogston, 1970)

$$A_{ij}^{\mathrm{exc}} = 10^{-3} N_{\mathrm{A}}\left[2\pi\left(R_i + R_j\right)^2 l_2 + \frac{4\pi}{3}\left(R_i + R_j\right)^3\right] , \tag{5.35}$$

where R_i is the radius of the cylinder and its hemispherical ends, l_2 is the half-length of the cylinder, and R_j is the equivalent hard sphere radius.

Let us now consider some actual numerical data for specific mixed biopolymer systems. Table 5.1 shows a set of examples comparing the values of the cross second virial coefficients obtained experimentally by static laser light scattering with those calculated theoretically on the basis of various simple excluded volume models using equations (5.32) to (5.35). For the purposes of this comparison, the experimental data were obtained under conditions of relatively high ionic strength ($I \geq 0.1$ mol dm^{-3}), *i.e.*, under conditions where the contribution of the electrostatic term (A_{ij}^{el}) is expected to be relatively insignificant.

2.3. *Sedimentation Equilibrium*

In principle, the nature of biopolymer interactions in solution can also be determined from sedimentation equilibrium of a single biopolymer component at constant temperature T and angular velocity ω. The thermodynamic activity on a weight-concentration basis, $z_i = \gamma_i c_i / M_i$, at radial distance r is related to that at a selected reference radial position (r_{f}) by the following formal expression (Deszczynski *et al.*, 2006):

$$M_i z_i(r) = M_i z_i(r_{\mathrm{f}})\Psi_i(r) \quad , \tag{5.36}$$

where

$$\Psi_i(r) = \exp[M_i(1 - v_i'\rho_{\mathrm{s}})\omega^2(r^2 - r_{\mathrm{f}}^2)/(2RT)] \ . \tag{5.37}$$

In equation (5.37), v_i' is the partial specific volume of the biopolymer and ρ_{s} is the solvent density. Making use of the relationship between the thermodynamic activity of the biopolymer in solution and its osmotic second virial coefficient (see chapter 3 for more details), one can get the following relationship (Deszczynski *et al.*, 2006):

Table 5.1 Comparison of the cross second virial coefficients obtained experimentally by static laser light scattering with those calculated from theory on the basis of the excluded volume contribution only.

Biopolymer mixture $(i + j)$	$A_{ij} \times 10^{-5}$ (cm^3/mol)	
	Experiment	Theory
		sphere–sphere
11S globulin + ovalbumin (Wasserman *et al.*, 1997)	0.07 ± 0.007	0.01
11S globulin + dextran (M_w = 48 kDa) (Semenova and Savilova, 1998)	0.09 ± 0.009	0.03
11S globulin + dextran (M_w = 270 kDa) (Semenova and Savilova, 1998)	0.31 ± 0.03	0.29 (γ_j' = 0.7)
11S globulin + dextran (M_w = 2500 kDa) (Semenova and Savilova, 1998)	2.05 ± 0.2	1.90 (γ_j' = 0.65)
11S globulin + pectin (M_w = 250 kDa) (Semenova *et al.*, 1990)	0.27 ± 0.03	sphere–sphere: 0.47 (γ_j' = 0.3); 'flexible' cylinder–sphere: 0.35
fibrinogen + dextran (M_w = 270 kDa) (Semenova and Savilova, 1998)	4.4 ± 0.4	rigid rod–sphere: 4.8

$$c_i(r) = M_i z_i(r_f)\Psi_i(r) - 2(A_{ii}/M_i)[M_i z_i(r_f)\Psi_i(r)]^2 + \dots \quad . \tag{5.38}$$

Nonlinear regression analysis of the dependence of $c_i(r)$ upon $\Psi_i(r)$, a transform of the radial distance r, leads to evaluation of the reference thermodynamic activity, $M_i z_i(r_f)$, and the osmotic second virial coefficient, A_{ii}/M_i, expressed on a weight basis (litre/g) rather than a molar basis (litre/mol). Furthermore, the values of M_i, v_i' and ρ_s can be obtained by curve-fitting the sedimentation equilibrium distribution for low biopolymer concentrations ($M_i z_i(r) \approx c_i$ for all r) to the equations (5.36) and (5.37) in order to deduce the quantity $[M_i(1 - v_i'\rho_s)]$ from the coefficient of the exponent (Winzor *et al.*, 2001; Deszczynski *et al.*, 2006).

2.4. *Self-Interaction Chromatography*

The value of the second virial coefficient A_2 can also be determined from the retention volume that is obtained from self-interaction chromatography (Ahamed *et al.*, 2005; Winzor *et al.*, 2007; Dumetz *et al.*, 2008). The chromatographic retention factor k is calculated from the retention volume V_r using the formula

$$k = \frac{V_r - V_0}{V_0} \quad , \tag{5.39}$$

where V_0 is the retention volume obtained in the absence of interactions. In simple terms, the method is based on measuring the retention of a biopolymer of interest in a column packed with unmodified particles. The value of A_2 is calculated from the measured quantities using the equation (Tessier *et al.*, 2002):

$$A_2 = 1 - \frac{k}{A_{2\mathrm{HS}}\rho_s\phi} \quad . \tag{5.40}$$

Here ρ_s is the biopolymer immobilization density; $A_{2\mathrm{HS}} = 2\pi D^3/3$ is the second virial coefficient based on excluded volume for a biopolymer of equivalent diameter D (a sphere of equal volume) (Neal and Lenhoff, 1995); and $\phi = A_s/V_0$ is the chromatographic phase ratio. The surface area A_s accessible to the biopolymer in the mobile phase is available in the literature, especially for proteins (Tessier *et al.*, 2002; Dumetz *et al.*, 2008).

2.5. *Isothermal Titration/Mixing Calorimetry*

The enthalpy change associated with formation of a thermodynamically ideal solution is equal to zero. Therefore any heat change measured in a mixing calorimetry experiment is a direct indicator of the interactions in the system (Prigogine and Defay, 1954). For a simple biopolymer solution, calorimetric measurements can be conveniently made using titration/flow calorimeter equipment. For example, from isothermal titration calorimetry of solutions of bovine β-casein, Portnaya *et al.* (2006) have determined the association behaviour, the critical micelle concentration (CMC), and the enthalpy of (de)micellization.

With a ternary system of type biopolymer$_i$ + biopolymer$_j$ + solvent, in order to characterize all the different pair interactions, the following heat effects, Q, should be measured in flow mode (Semenova *et al.*, 1991): (i) biopolymer$_i$ solution diluted by pure buffer, $Q_{i\mathrm{b}}$; (ii) biopolymer$_j$ solution diluted by pure buffer, $Q_{j\mathrm{b}}$; and (iii) mixed (biopolymer$_i$ + biopolymer$_j$) solution diluted by pure buffer, $Q_{ij\mathrm{b}}$. The specific enthalpy of interaction between biopolymer$_i$ and biopolymer$_j$ can then be obtained from

$$\Delta H_{ij} = -(Q_{ij\mathrm{b}} - Q_{i\mathrm{b}} - Q_{j\mathrm{b}})/\Delta n \quad , \qquad (5.41)$$

where Δn is the number of moles/grams of the mixed biopolymers.

BIBLIOGRAPHY

Ahamed, T., Ottens, M., van Dedem, G.W.K., van der Wielen, L.A.M. (2005). Design of self-interaction chromatography as an analytical tool for predicting protein phase behaviour. *Journal of Chromatography A,* **1089**, 111–124

Amur, K.S., Harlapur, S.F., Aminabhavi, T.M. (1997). A novel analytical method to estimate molar mass and virial coefficients of polymer from osmometry. *Polymer*, **38**, 6417–6420.

Blandamer, M.J., Cullis, P.M., Engberts, J.B.F.N. (1998). Titration microcalorimetry. *Journal of the Chemical Society, Faraday Transactions*, **94**, 2261–2267.

Burchard, W. (1994) Light scattering. In Ross-Murphy, S.B. (Ed.). *Physical Techniques for the Study of Food Biopolymers*, Glasgow: Blackie, pp. 151–214.

Casassa E.F. (1962). Effect of heterogeneity and molecular weight on the second virial coefficient of polymers in good solvent. *Polymer*, **3**, 625–638.

Clark, A.H. (2000). Direct analysis of experimental tie line data (two polymer–one solvent systems) using Flory–Huggins theory. *Carbohydrate Polymers*, **42**, 337–351.

Deszczynski, M., Harding, S.E., Winzor, D.J. (2006). Negative second virial coefficients as predictors of protein crystal growth: evidence from sedimentation equilibrium studies that refutes the designation of those light scattering parameters as osmotic virial coefficients. *Biophysical Chemistry*, **120**, 106–113.

Dhillon, M.S. (1979). Second virial coefficient and excess thermodynamic functions in benz-ene of a polymer 2-butene-1,4-diol with adipic acid. *Thermochimica Acta*, **31**, 375–379.

Dumetz, A.C., Chockla, A.M., Kaler, E.W., Lenhoff, A.M. (2008). Effects of pH on protein–protein interactions and implications for protein phase behaviour. *Biochimica et Biophys-ica Acta*, **1784**, 600–610.

Edmond, E., Ogston, A. (1968). An approach to the study of phase separation in ternary aqueous systems. *Biochemistry Journal*, **109**, 569–576.

Edsall, J.T., Gutfreund, H. (1983). *Biothermodynamics: The Study of Biochemical Processes at Equilibrium*, Chichester: Wiley.

Edsman, K., Sundelöf, L.-O. (1987). Determination of thermodynamic interaction in mixed polymer solutions by a rapid and precise osmotic method: the system dextran/polyvinyl-pyrrolidone/water. *Polymer*, **28**, 2267–2274.

Evans, J.M. (1972). Manipulation of light scattering data. In Huglin, M.B. (Ed). *Light Scattering from Polymer Solutions*, London: Academic Press, pp. 89–164.

Finkelstein, A.V., Ptitsyn, O.B. (2002). *Protein Physics. A Course of Lectures (Soft Condensed Matter, Complex Fluids and Biomaterials)*, San Diego: Academic Press.

Kaddur, L.O., Strazielle, C. (1986). Experimental investigation of light scattering by a solution of two polymers. *Polymer*, **28**, 459–468.

Kratochvil, P., Sundelöf, L.-O. (1986). Interactions in polymer solutions. *Acta Pharmaceutica Suecica*, **23**, 31–46.

Murphy, R.M. (1997). Static and dynamic light scattering of biological macromolecules: what can we learn? *Current Opinion in Biotechnology*, **8**, 25–30.

Nagasawa, M., Takahashi, A. (1972). Light scattering from polyelectrolyte solutions. In Huglin, M.B. (Ed). *Light Scattering from Polymer Solutions*, London: Academic Press, pp. 671–723.

Neal, B.L., Lenhoff, A.M. (1995). Excluded-volume contribution to the osmotic second virial coefficient for proteins. *AIChE Journal*, **41**, 1010–1014.

Ogston, A.G. (1970). On the interaction of solute molecules with porous networks. *Journal of Physical Chemistry*, **74**, 668–669.

Portnaya, I., Cogan, U., Livney, Y.D., Ramon, O., Shimoni, K., Rosenberg, M., Danino, D. (2006). Micellization of bovine β-casein studied by isothermal titration microcalorimetry and cryogenic transmission electron microscopy. *Journal of Agricultural and Food Chem*istry, **54**, 5555–5561.

Prigogine, I., Defay, R. (1954). *Chemical Thermodynamics*, London: Longmans.

Schaink, H.M., Smit, J.A.M. (2007). Protein–polysaccharide interactions: the determination of the osmotic second virial coefficients in aqueous solutions of β-lactoglobulin and dextran. *Food Hydrocolloids*, **21**, 1389–1396.

Semenova, M.G. (1996). Factors determining the character of biopolymer–biopolymer interactions in multicomponent aqueous solutions modelling food systems. In Parris, N., Kato, A., Creamer, L.K., Pearce, J. (Eds). *Macromolecular Interactions in Food Technology*, ACS Symposium Series No. 650, Washington, DC: American Chemical Society, pp. 37– 49.

Semenova, M.G. (2007). Thermodynamic analysis of the impact of molecular interactions on the functionality of food biopolymers in solution and in colloidal systems. *Food Hydrocolloids*, **21**, 23–45.

Semenova, M., Savilova, L. (1998). The role of biopolymer structure in interactions between unlike biopolymers in aqueous medium. *Food Hydrocolloids*, **12**, 65–75.

Semenova, M.G., Bolotina, V.S., Grinberg, V.Ya., Tolstoguzov, V.B. (1990) Thermodynamic incompatibility of the 11S fraction of soybean globulin and pectinate in aqueous medium. *Food Hydrocolloids*, **3**, 447–456.

Semenova, M.G., Bolotina, V.S., Dmitrochenko, A.P., Leontiev, A.L., Polyakov, V.I., Braudo, E.E., Tolstoguzov, V.B. (1991). The factors affecting the compatibility of serum albumin and pectinate in aqueous medium. *Carbohydrate Polymers*, **15**, 367–385.

Semenova, M.G., Belyakova, L.E., Polikarpov, Y.N., Antipova, A.S., Dickinson, E. (2009). Light scattering study of sodium caseinate + dextran sulfate in aqueous solution: relationship to emulsion stability. *Food Hydrocolloids*, **23**, 629–639.

Su, R., Qi, W., He, Z., Zhang, Y. Jin, F. (2008). Multilevel structural nature and interactions of bovine serum albumin during heat-induced aggregation process. *Food Hydrocolloids*, **22**, 995–1005

Tanford, C. (1961). *Physical Chemistry of Macromolecules*, New York: Wiley.

Tessier, P.M., Lenhoff, A.M., Sandler, S.I. (2002). Rapid measurement of protein osmotic second virial coefficients by self-interaction chromatography. *Biophysical Journal*, **82**, 1620–1631.

Tombs, M.P., Peacocke, A.R. (1974). *The Osmotic Pressure of Biological Macromolecules*, Oxford: Clarendon Press.

Wasserman, L.A., Semenova, M.G., Tsapkina, E.N. (1997). Thermodynamic properties of the 11S globulin of Vicia faba–ovalbumin–aqueous solvent system: phase behaviour and light scattering. Food Hydrocolloids, 11, 327–337.

Wills, P.R., Comper, W.D., Winzor, D.J. (1993). Thermodynamic nonideality in macromolecular solutions: interpretation of virial coefficients. *Archives of Biochemistry and Biophysics*, **300**, 206–212.

Winzor, D.J., Carrington, L.E., Harding, S.E. (2001). Analysis of thermodynamic nonideality in terms of protein solvation. *Biophysical Chemistry*, **93**, 231–240.

Winzor, D.J., Scott, D.J., Wills, P.R. (2007). A simpler analysis for the measurement of second virial coefficients by self-interaction chromatography. *Analytical Biochemistry*, **371**, 21–25.

Zvetcov, V.N. (1986). In Deryagin, B.V. (Ed.). *Rigid Polymer Molecules*, Moscow: Nauka, p. 397 (in Russian).

PART THREE

BIOPOLYMER INTERACTIONS IN THE BULK AQUEOUS MEDIUM OF FOOD COLLOIDS

CHAPTER SIX

SELF-ASSEMBLY OF FOOD BIOPOLYMERS FOR THE DEVELOPMENT OF HEALTH-PROMOTING PROPERTIES OF FOOD COLLOIDS

Proteins and polysaccharides have two main functions in food colloids: nutrition and structure-forming. Both functions are of great importance for the quality of food systems. Our current knowledge suggests that the intermolecular interactions involving biopolymers are central to understanding the biopolymer structure-forming functionality, as well as to controlling the texture/consistency, the physical stability, and the appearance of food colloids. Let us suppose that we can establish quantitative relationships between the structure on the colloidal scale and the interactions of biopolymers with each other and with all the low-molecular-weight components of the system. Then we postulate that we can achieve a fundamental understanding of molecular mechanisms underlying the formation and stabilization of food colloids. This puts us in a position to develop promising ways of creating novel and tasty foods with the optimum compositions for health promotion and sensory pleasure.

Food biopolymers are prone to self-assembly, leading to the formation of supramolecular structures in bulk aqueous medium and at the interfaces of food colloids. Self-assembly is a specific type of aggregation. It is quite close in nature to a polymerization process driven toward a state of minimum free energy, as a result of a combination of different kinds of weak physical bond formation — hydrogen bonding, electrostatic interactions, hydrophobic interactions, van der Waals interactions, and (kinetically labile) metal coordination. While the individual interaction energies are small in magnitude, the effect of the large number of interactions in the final assembly is very significant (Rajagopal and Schneider, 2004). The large entropic cost of self-assembly is only slightly off-set by the favourable enthalpy gained from the combination of all of the weak interactions. Consequently, the ordered biopolymer structures formed are in thermodynamic equilibrium, and the selection of the state of minimum free energy from amongst the many other states of higher free energy is made possible only by the fluctuations of the molecules between those states. The reversible condition means that the self-assembled structure and the average amount of polymer involved are maintained in equilibrium, while the states and positions of individual molecules continue to fluctuate (Oosawa and Asakura, 1975; Kentsis and Borden, 2004; Rajagopal and Schneider, 2004; Graveland-Bikker, 2005; Graveland-Bikker and de Kruif, 2006). In contrast, the process of aggregation is usually regarded as an irreversible self-association reaction leading to formation of amorphous flocs that are generally heterogeneous in structure, and, if relevant, biologically inactive (Kentsis and Borden, 2004). Both covalent and non-covalent interactions are typically involved in aggregation.

Food biopolymer self-assembly is readily controlled and triggered by changes in the environmental conditions — pH, ionic strength, temperature, pressure, and the addition of divalent ions, sugars, surfactants, *etc*. The size of biopolymer aggregates and the character of the interactions between them can be manipulated by changing the conformation of the constituent biopolymers by the action of heat, pH or enzymes. This sensitivity to many external influences makes biopolymer molecules (and their aggregates) very attractive as building blocks for the development of stimuli-sensitive ('switchable') self-assembled structures. This is the basis of the 'bottom-up' nano-biotechnological approach, which attempts to mimic Nature in the creation of new and varied structures with novel applications. The ongoing challenge is to understand what are the key generic factors influencing the formation of different kinds of aggregate morphologies, and ultimately the nanostructure of food colloids under different environmental conditions, whilst at the same time recognizing specific characteristics of the individual biopolymers (Dickinson, 2007).

1. *Self-Assembly due to Specific Kinds of Biopolymer Interactions*

Milk proteins are well studied biopolymers. The self-assembly of caseins (α_{s1}-casein, α_{s2}-casein, β-casein, κ-casein) and the main whey proteins (α-lactalbumin and β-lactoglobulin) has been a central preoccupation of food colloid researchers because of the leading role of milk proteins as structuring agents, emulsifiers and stabilizers. A range of dairy products like yoghurt, quark, cheese and ice-cream contain these mixed proteins and their self-assembled aggregates as their main building blocks.

Caseins are highly disordered proteins having rather limited secondary structure. This is mainly due to the unusually high proline content, which is fairly uniformly distributed along the polypeptide chain. This feature leads to an open extended structure of the casein molecules which differentiates them from the globular whey proteins like α-lactalbumin and β-lactoglobulin. The caseins have been described as 'rheomorphic' proteins (Holt and Sawyer, 1993), indicating that they adopt molecular structures in solution that are dictated by the local environment; that is, their structures are flexible enough to 'go with the flow' (Dickinson, 2006). In solution the caseins typically exist as some form of association colloid. Their flexible and rather open conformation underlies both the high susceptibility of the casein molecules to selective enzymatic hydrolysis and a strong tendency for sticking to hydrophobic surfaces. The different kinds of casein monomer molecules have varying degrees of phosphorylation, and, consequently, a variable sensitivity to calcium ions (Modler, 1985; Dickinson, 2006).

In contrast, the whey proteins are relatively small globular proteins. α-Lactalbumin represents about 20 % of the protein content of bovine whey (3.5 % of total bovine milk protein), and it is the principal protein in human milk (Brew and Grobler, 1992). Nanotube assembly has been discovered in some solutions containing a hydrolysed derivative of this protein. And it appears that the α-lactalbumin nanotube is unique in the sense that it is the only artificial nanotube that has so far been made from a food protein (Graveland-Bikker *et al.*, 2004; Graveland-Bikker and de Kruif, 2006). As for β-lactoglobulin, it has the capacity under certain specific conditions to form nano-fibres in aqueous media (as can various other globular food proteins, such as ovalbumin, soy proteins, and bovine serum albumin) (van der Linden, 2006; Nicolai, 2007).

1.1. *Caseins*

The structure of the naturally created casein particle, the 'casein micelle' (CM), has been the subject of investigation (and controversy) for the best part of a century (Fox and Brodkorb, 2008). Native casein micelles of cow's milk are rather large particles (10^8 Da, mean size ~ 100–300 nm). They are highly polydisperse and have a heterogeneous supramolecular protein structure. The name 'casein micelle' is actually somewhat unfortunate, because in the wider field of colloid science the term 'micelle' routinely refers to a reversible aggregate of small amphiphilic molecules (surfactants). The casein micelle is a complex of calcium phosphate and the four kinds of bovine caseins in the proportions: 38 % α_{s1}-, 10 % α_{s2}-, 36 % β-, and 13 % κ-casein. The individual caseins differ in various respects, including net molecular charge ($\alpha_{s1} > \alpha_{s2} > \beta > \kappa$), sensitivity to precipitation by calcium ions ($\alpha_{s2} > \alpha_{s1} > \beta >> \kappa$), and distribution of hydrophobic and hydrophilic amino acids within their primary structures (Swaisgood, 2003).

The most commonly investigated casein in model colloidal systems is bovine β-casein. It consists of a polypeptide chain of 203–209 amino acids, a net charge of –15*e* at neutral pH, and a molar weight of 24 kDa. The molecule contains five phosphoserine residues distributed towards one end of the molecule. Overall the molecule has a distinctly amphiphilic character. Unlike most proteins, β-casein has no internal covalent cross-links and no tendency to polymerize through intermolecular disulfide bonds (Dickinson *et al.*, 1998; Swaisgood, 2003, Dickinson, 2006).

The casein responsible for colloidal stabilization of the casein micelle is κ-casein. This glycoprotein has a molecular weight of 19 kDa and is composed of 169 amino acids (Swaisgood, 2003). κ-Casein is special in being the only casein component that is insensitive to calcium ions (up to concentrations of 400 mM).

The major casein component of cow's milk is α_{s1}-casein. It consists of a polypeptide chain of 199 amino acids, with a charge of –22*e* at neutral pH and a molecular weight of 24 kDa. The α_{s1}-casein molecule contains eight phosphoserine residues, of which seven are located in the region 43–80, and in addition it bears 12 carboxyl groups. Thus, this section of the protein is extremely polar. As with β-casein, proline residues are distributed along the whole molecule and this hinders the formation of ordered secondary structures. The region 100–199 is strongly non-polar; this is probably mainly responsible for the pronounced tendency of α_{s1}-casein to associate in an aqueous medium. However, the separation of the hydrophilic and hydrophobic residues is statistically more random than for β-casein (Dickinson and Matsumura, 1994). The occurrence of hydrophobic amino acids at both ends of the α_{s1}-casein molecule has certain definite consequences for its self-assembly and adsorption behaviour (Horne, 1998; Dickinson, 1998, 1999a,b, 2006). In the presence of a low concentration of calcium ions, α_{s1}-casein forms insoluble calcium salts (Belitz and Grosch, 1982).

These different casein monomers combine with calcium phosphate to form discrete particles on the nano-size scale. The phosphoserines of the caseins are seemingly clustered for the 'purpose' of linking within the micelle to putative calcium phosphate microcrystallites, also known as 'nanoclusters' (Holt, 1992; Horne, 1998, 2002, 2003, 2006; Holt *et al.*, 2003; Horne *et al.*, 2007). Structural evidence for the existence of such nanoclusters has come from neutron and X-ray scattering (de Kruif and Holt, 2003; Holt *et al.*, 2003; Pignon *et al.*, 2004; Marchin *et al.*, 2007). The presence of nanoclusters allows native casein micelles to be effective natural suppliers of essential calcium salts in the human diet in a readily assimilated functional form. Protein–nanocluster interactions are the central concept of the cross-linking mechanism in Holt's model of casein micellar assembly (Holt *et al.*, 2003; de Kruif and Holt, 2003). Any analogy with conventional soap-like micelles is considered to be inappropriate because of the cementing action of the micellar calcium phosphate (de Kruif *et al.*, 2002). Nevertheless, we note that interaction via the phosphoserine residues is not the only way that the caseins may become self-associated.

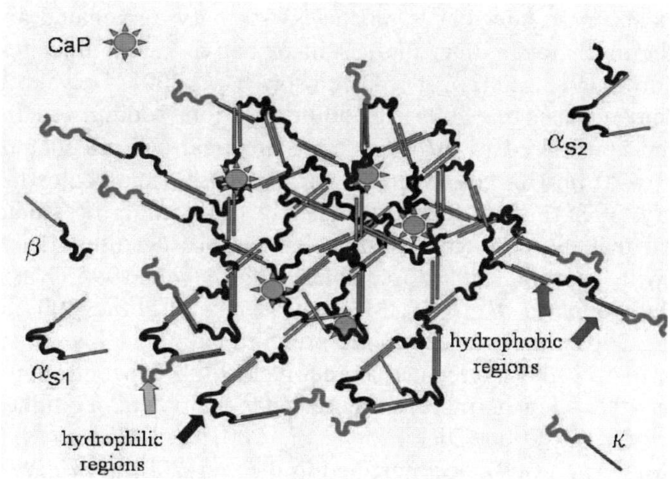

Figure 6.1 The dual binding model of the structure of the casein micelle as built up from the individual caseins (α_{s1}, α_{s2}, β and κ) in combination with calcium phosphate (CaP). Reproduced from Horne (1998) with permission.

Figure 6.1 shows schematically the so-called 'dual-binding' model of casein micelle assembly (Horne, 1998). This model describes an open, irregular CM particle, having the structure of a three-dimensional network extending through space. Individual protein molecules are represented as block copolymers and the micellar assembly is viewed as a polymerization process occurring as a result of hydrophobic interactions or bridging via colloidal calcium phosphate (CCP), with polymer chains terminated by κ-casein, which thus controls the casein micellar size (Horne, 1998; Horne *et al.*, 2007). The model explains how κ-casein forms a hydrophilic coating at the surface of the native casein micelle, hence stabilizing the calcium-sensitive caseins (α_{s1}-, α_{s2}- and β-) against flocculation. This κ-casein surface layer is traditionally termed the 'hairy layer', and it can be considered as a polyelectrolyte brush (de Kruif and Zhulina, 1996). So the instability of casein micelles can be induced by removal of the brush with enzyme (chymosin) or by collapse of the brush in a medium of high ionic strength (Holt and Horne, 1996; de Kruif, 1999). Based on some compelling arguments from the model's proposer, together with recent experimental evidence, it seems that there is now growing support for the Horne dual-binding model over more traditional sub-micellar models (Fox and Brodkorb, 2008).

Lacking the CCP of skim milk, an aqueous dispersion of sodium caseinate exists in a state that is much less strongly aggregated and more molecular in character than the casein of native casein micelles (Holt, 1992; Horne, 1998, 2002, 2006; Dickinson *et al.*, 2001; Belyakova *et al.*, 2003; Semenova *et al.*, 2005). In dilute solution, sodium caseinate has long been considered to exist as small spherical sub-micellar particles (size ~ 10–20 nm) in equilibrium with free casein molecules (Creamer and Berry, 1975). More recently, dynamic light-scattering studies have suggested that the caseinate sub-micellar species ('nanoparticles') may coexist with supramolecular assemblies (Chu *et al.*, 1995; Nash *et al.*, 2002; HadjSadok *et al.*, 2008; Müller-Buschbaum *et al.*, 2007; Semenova *et al.*, 2009); and some osmotic pressure data have been interpreted in terms of rod-like aggregates at relatively high concentration (Farrer and Lips, 1999). Furthermore, combined static and dynamic light scattering (Lucey *et al.*, 2000; Dickinson *et al.*, 2001; Belyakova *et al.*, 2003; Semenova *et al.*, 2009), accompanied in the work of Lucey *et al.* (2000) by size-exclusion chromatography, has indicated a wide range of molecular weight species in commercial sodium caseinate, including some highly non-spherical micelles and polymers of κ-casein.

In a number of studies it has been shown that CCP can be removed without disrupting the micelle if skim milk is acidified at temperatures below ~ 20 °C (Dalgleish and Law, 1988; Lucey *et al.*, 1997; Lucey, 2002). This result is consistent with the Horne dual-bonding model of CM assembly, where the micellar integrity is viewed as being controlled by the balance between attractive and repulsive forces, *i.e.*, a localized excess of hydrophobic attraction over electrostatic repulsion (Horne, 1998, 2002; Lucey, 2002). This type of model also appears to be able to accommodate successfully the responses of native casein micelles and sodium caseinate particles to (i) changes in pH and temperature, (ii) addition of urea, sucrose or calcium ions, and (iii) removal of calcium phosphate by EDTA, thereby rationalizing a broad range of stability and bulk rheological properties of casein dispersions, gels and emulsions (McGann and Fox, 1974; Mora-Gutierrez *et al.*, 1997; Horne, 1998, 2002, 2003; Mora-Gutierrez and Farrell, 2000; Dickinson, *et al.*, 2001; Lucey, 2002; Belyakova *et al.*, 2003; Euston and Horne, 2005; Dauphas *et al.*, 2005; Semenova *et al.*, 2005, 2009). This exceptional degree of responsiveness of sodium caseinate to the solution environmental conditions seems to be determined by casein's amphiphilic nature, as well as by the rather porous structure dictated by the unfolded conformation of the constituent macromolecules (Horne, 2002, 2006; Mora-Gutierrez and Farrell, 2000).

The self-assembly of caseins may be readily manipulated by processing methods that affect the integrity of native casein micelles and the character of the casein interactions in aqueous media. Examples of such procedures are (Dickinson, 2006): (i) acidification toward the isoelectric point (p*I*) (pH ~ 4.6–4.8), leading to a neutralization of the net protein charge; (ii) enzyme action, as exploited in the production of cheeses and fermented milks; (iii) addition of divalent ions, especially, Ca^{2+} ions; (iv) addition of sucrose or ethanol; (v) temperature treatment; and (vi) high-pressure treatment.

In the case of sodium caseinate, combined static and dynamic light scattering studies have indicated (Dickinson *et al.*, 2001) that the extent of casein association is enhanced at neutral pH by the addition of ionic calcium. The increased extent of association is attributed to ionic bridging between sodium caseinate particles, followed by neutralization of the net protein charge, weakening of repulsive electrostatic interactions, and a relative strengthening of the hydrophobic protein–protein interactions. Furthermore, a simultaneous increase in the concentration of calcium ions and in the temperature, along with the acidification of the aqueous medium, can lead to the development of pronounced reversible or even irreversible protein association (Semenova *et al.*, 2005). These changes in the casein self-assembly are considered to be of fundamental importance for the stability and rheological behaviour of emulsions stabilized by sodium caseinate in the presence of low concentrations of calcium ions (as demonstrated in section 4 of this chapter).

In contrast to the above, the extent of sodium caseinate association in solution has been shown to be reduced in the presence of sucrose at a pH slightly above the isoelectric point, *i.e.*, in the pH range from 5.1 to 7.0 (Semenova *et al.*, 2002; Belyakova *et al.*, 2003). This diminished association is reflected in a decrease in molar mass (M_w) and size (R_G, R_h) of the spherical sodium caseinate particles, as shown in Figure 6.2. It is also reflected in an increase in the degree of particle compactness, as indicated by a decrease in the ratio (ρ) of the radius of gyration to the hydrodynamic radius of the sodium caseinate particles. These changes become gradually more pronounced with increasing sucrose concentration in the range from 10% to 78% wt/v.

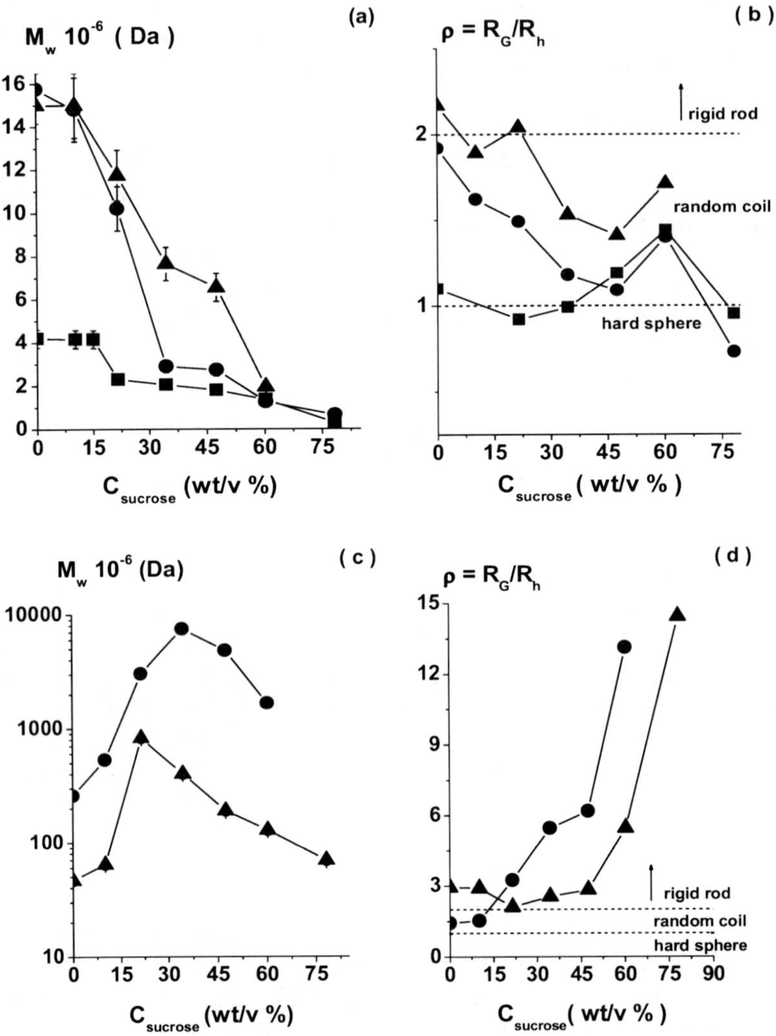

Figure 6.2 Effect of sucrose on the self-assembly / disassembly of sodium caseinate in aqueous medium (ionic strength = 0.01 mol/dm^3, 22 °C) on the basis of combined static and dynamic light scattering. Upper plots refer to pH > p*I*, *i.e.*, (■) pH = 7.0, (●) pH = 6.0, (▲) pH = 5.5: (a) weight-average molar weight, M_w; (b) structure-sensitive parameter ρ. Lower plots refer to pH < p*I*, *i.e.*, (●) pH = 3.9, (▲) pH = 3.5: (c) weight-average molar weight, M_w; (d) structure-sensitive parameter ρ.

The effect of sucrose on casein self-assembly could be attributable to direct hydrogen bonding between the sugar and protein molecules, which weakens bonds in the interior of sodium caseinate micelles and thereby causes protein disassembly. This explanation is supported by other independent experimental investigations: isothermal mixing calorimetry (Antipova and Semenova, 1997; Antipova *et al.*, 1997, 1999); estimation of the free energy of interaction of sugars with proteins in aqueous solution (Jencks, 1969); studies of dynamic and equilibrium moisture take-up, and the positions of carboxylate bands in infra-red spectra (Allison *et al.*, 1999; Tzannis and Prestrelski, 1999); gravimetric estimation of hydration isotherms (López-Diez and Bone, 2000); and hydration studies of the freeze-dried amorphous sugar matrix (Imamura *et al.*, 2001). In addition, it has been suggested (Duan *et al.*, 2001) that the hydrophobic parts of the glucose units in sucrose could also contribute to the interactions with protein and to the disassembly of the protein particles.

Under acidic conditions (below the p*I*), there is a marked increase in casein self-association that manifests itself as a pronounced increase in molar mass M_w, radius of gyration R_G, and the ratio of radius of gyration to hydrodynamic radius of the sodium caseinate particles ($\rho >> 2$) (see Figure 6.2). In addition, an analysis of light-scattering data based on structure-sensitive Kratky plots has shown a transition from Gaussian to worm-like chain/rod behaviour for sodium caseinate on lowering the pH below p*I* (Belyakova *et al.*, 2003). For pH < p*I*, the direct hydrogen bonding between the sugars and the more protonated protein molecules becomes weaker, and so the sugars play the role of a real competitor to protein for the water molecules. This competition leads to protein dehydration followed by protein aggregation (Belyakova *et al.*, 2003).

Owing to its amphipilic character, the sodium caseinate particle can readily interact with amphiphilic/lipophilic food-grade molecules such as surfactants, lipids and vitamins (Istarova *et al.*, 2005; Semenova *et al.*, 2006, 2008; Semenova, 2007; Semo *et al.*, 2007). Therefore the caseinate nanoparticles can perform a dual functionality in food systems. Namely, they can act as promising carriers for hydrophobic nutraceuticals, potentially protecting them against oxidation and degradation (Semo *et al.*, 2007; Semenova *et al.*, 2008), as well as performing a traditional role as structure-forming agents (Semenova *et al.*, 2008). Thus, in the work of Semo *et al.* (2007), it was shown that calcium-containing casein particles that were reassembled from sodium caseinate could provide partial protection against UV-light-induced degradation of vitamin D_2 contained within them. Furthermore, it was found that incorporation of the vitamin D_2 had relatively little effect on the inferred morphology of the reassembled casein micelles.

Mixed systems of sodium caseinate + phosphatidylcholine have been recently investigated (Semenova *et al.*, 2008). It was found that lecithin oxidation was much reduced (or even absent) when the phospholipid was complexed with the protein (see Figure 2.4). In addition, the half-life of foams stabilized by complexes of sodium caseinate (1 wt%) with phosphatidylcholine (10^{-6}–10^{-3} mol/dm^3) showed a four-fold increase, as compared to the protein alone over the range of experimental conditions studied (pH 5.5–7.0; ionic strength 0.001–0.01 M). It was also noted that a solution of pure phosphatidylcholine did not give a fine stable foam for these concentrations and under these experimental conditions.

There have been many studies of the aggregation behaviour of the individual caseins. From measurements of M_w, R_G, R_h, ρ, and the particle scattering factor, $P_z(q)$, as obtained by small-angle neutron scattering and combined static and dynamic light scattering, it has been found that α_{s1}-casein, β-casein and κ-casein each shows a strong tendency towards self-association in aqueous medium (Schmidt, 1982; Thurn *et al.*, 1987a,b; Burchard, 1994; Leclerc and Calmettes, 1997; Dickinson *et al.*, 1998; de Kruif *et al.*, 2002). As with the case of whole casein(ate), this self-association is controlled by a delicate balance of specific electrostatic and hydrophobic interactions (de Kruif *et al.*, 2002; Dickinson, 2006).

Based on a simple view of its primary structure, the single molecule of bovine β-casein can be represented as consisting of a polar head and an apolar tail (see Figure 6.1). Therefore the behaviour of β-casein is expected to be soap-like. In an aqueous solution below 4° C, the protein is monomeric, but above this temperature it exhibits endothermic reversible self-assembly (Belitz and Grosch, 1982; Payens and Vreeman, 1982). This self-assembly appears to be mainly driven by the attractive hydrophobic interactions, and the extent of aggregation increases with increasing temperature and ionic strength (Schmidt and Payens, 1972). A roughly spherical star-like structure of β-casein has been inferred to exist in bulk aqueous media (Schmidt, 1982; Leclerc and Calmettes, 1997; Burchard, 1994), as illustrated in Figure 6.3(a). This is the aggregation pattern expected for a traditional surfactant-like micelle. That is, β-casein micelles are assumed to exhibit monomer–micelle association equilibrium of the 'all or nothing' or closed-association type (Payens and Heremans, 1969; Payens *et al.*, 1969; Kanji *et al.*, 1988; Evans and Wennerstrom, 1999):

$$n \text{ monomers} \leftrightarrows \text{micelle}$$

This description of micelle formation is also called the two-state model. Thus, the solution contains β-casein micelles of about 38–40 monomer units (*i.e.*, the micellization number *n*) when a protein concentration of 4 mg/ml is exceeded at 15 °C and neutral pH. At high concentrations, there is a highly cooperative closed aggregation process. The β-casein micelles start to aggregate at a higher temperatures (> 15 °C), forming structures of complicated architecture (Thurn *et al.*, 1987a). It is interesting to note that removal of just the last 20 residues at the hydrophobic C-terminus has been found to destroy the protein's ability for self-association in aqueous solution (Berry and Creamer, 1975).

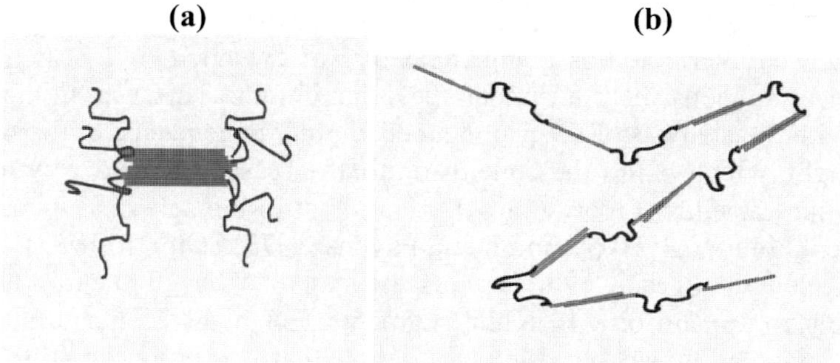

Figure 6.3 Representation of casein self-association structures according to the simple copolymer model: (a) β-casein, (b) α_{s1}-casein. Reproduced from Horne (1998) with permission.

In an alternative interpretation by Mikheeva *et al.* (2003), however, it was shown that β-casein micellization can be described quite adequately using the so-called shell model of Kegeles (1979, 1992). This model describes the association behaviour in terms of the scheme:

$$\beta\text{-cas}_1 + \beta\text{-cas}_1 \leftrightarrows \beta\text{-cas}_2$$
$$\beta\text{-cas}_1 + \beta\text{-cas}_2 \leftrightarrows \beta\text{-cas}_3$$
$$\cdots\cdots\cdots\cdots$$
$$\beta\text{-cas}_1 + \beta\text{-cas}_i \leftrightarrows \beta\text{-cas}_{i+1}$$
$$\beta\text{-cas}_1 + \beta\text{-cas}_{N-1} \leftrightarrows \beta\text{-cas}_N$$

The essence of the shell model is that the association constant of the first step is very much smaller than the association constants of all subsequent steps. It was assumed by de Kruif *et al.* (2002) that, since there is a considerable similarity in various properties of β- and κ-caseins, then the κ-casein micelle can also be described using the shell model. Nevertheless, Vreeman *et al.* (1981) had previously suggested that the structure of the κ-casein aggregates in aqueous medium are more complicated than those of β-casein, being composed of star-like sub-micelles, where each sub-micelle contains nine κ-casein chains and the total degree of association may reach about 140 (Thurn *et al.*, 1987a). Payens and Vreeman (1982) used sedimentation measurements to infer a critical micelle concentration for κ-casein of 0.5 mg/ml.

We turn now to α_{s1}-casein. Based on static and dynamic light-scattering measurements, a constant apparent molar weight of 3.4×10^6 Da was found for this protein over the concentration range from 0.5 to 6.0 mg/ml at 35 °C (Thurn *et al.*, 1987b; Burchard, 1994). The measured value of the ratio of the radius of gyration to the hydrodynamic radius, $\rho = R_G/R_h = 2.78$, was interpreted as being indicative of extended rigid structures formed in aqueous medium under these conditions. This was in agreement with the analysis of the pronounced angular dependence of the scattered light, which led to the conclusion that α_{s1}-casein formed very long worm-like micelles (Figure 6.3(b)), in contrast to the spherical β-casein micelles (Figure 6.3(a)) (Schmidt and Payens, 1972; Schmidt, 1982). The contour length of each cylinder was estimated to be 1600 nm, corresponding to a chain of ~ 12 Kuhn segments. For higher concentrations, lateral aggregation was observed to a degree that was proportional to the concentration. Beyond the overlap concentration, there was a change in shape of the asymptotic scattering curve, which was interpreted as the beginning of reversible gelation (Thurn *et al.*, 1987b; Burchard, 1994).

More recent measurements have investigated the effect of solution conditions on the aggregation of individual caseins. On the basis of static light scattering measurements, it has been established (Dickinson *et al.*, 1998) that the character of the protein–protein interactions and the self-association of both α_{s1}-casein and β-casein vary strongly with alteration of pH over the range from 7.0 to 5.5 and/or ionic strength over the range from 0.01 to 0.05 M.

1.2. *Whey Proteins*

α-Lactalbumin is a globular acidic calcium-binding protein (Brew and Grobler, 1992). The primary structure of α-lactalbumin consists of 123 amino acids (molecular weight 14.2 kDa). Native α-lactalbumin has two domains, a large α-helical domain and a small β-sheet domain, which are connected by a calcium-binding loop (Chrysina *et al.*, 2000). The tertiary structure of α-lactalbumin is stabilized by four disulfide bridges. However, it appears that the overall architecture of the protein fold is essentially determined by the polypeptide sequence itself, as an α-lactalbumin mutant without disulfide bridges has been found to be nearly as compact as wild-type α-lactalbumin at acidic pH (Redfield *et al.*, 1999). The shape of the α-lactalbumin molecule resembles that of a compact prolate ellipsoid, with dimensions of 2.5 × 3.7 × 3.2 nm. The radius of gyration of native holo-α-lactalbumin is 1.57 nm; the acid molten globule has a larger radius of 1.72 nm (Kataoka *et al.*, 1997).

Self-assembly of α-lactalbumin that has been partially hydrolysed by a protease from *Bacillus licheniformis* leads to nanotubular structures (Graveland-Bikker *et al.*, 2006a). These nanotubes are of particular interest as novel ingredients with specific functionality. The enzymatic hydrolysis causes a decrease in mean molar weight, an increase in the number of ionisable groups, and greater exposure of hydrophobic groups (Panyam and Kilara, 1996). Calcium ions play a crucial role as ionic bridges combining the partially hydrolysed protein into the nanotubes (Graveland-Bikker *et al.*, 2004; Graveland-Bikker and de Kruif, 2006) (see Figure 1.1). It appears that the α-lactalbumin nanotubes can only be formed within a rather narrow ion concentration window. Below a molar ratio of calcium to α-lactalbumin of R = 1.5, the ion concentration is considered to be too low to produce enough nuclei, and consequently random aggregates are formed. For a calcium ratio higher than R = 1.5, tubular structures are formed. And the higher is the calcium concentration (up to R = 3), the more the tubular shape is favoured. But a further increase in calcium content has a negative effect, presumably because the increased ionic strength suppresses some important electrostatic interactions, so that only random aggregates are formed at R =10. Calcium ions do not influence the rate of hydrolysis, but they do have a major effect on the kinetics of the self-assembly of partially hydrolysed α-lactalbumin, which is possibly facilitated by nucleation. It was found that the rate of self-assembly increases with the calcium content, and that this could be explained in terms of the nucleation being faster at the higher calcium concentration (Graveland-Bikker *et al.*, 2004; Graveland-Bikker and de Kruif, 2006). These authors have suggested that Ca^{2+} acts as a specific sort of 'glue' between the building blocks involving the many carboxyl groups belonging to aspartic and glutamic acids, as well as those formed as a result of the hydrolysis of peptide bonds. It seems likely that each calcium ion binds between two or more such carboxyl groups, owing to its high affinity for them. The binding of calcium between two specific amino acid residues could then determine the orientation of the building blocks with respect to one another; as a consequence they tend to form a ring structure (see Figure 1.1).

In addition to calcium, we note that divalent manganese and zinc and trivalent aluminium can also induce α-lactalbumin nanotube formation. By using these ions instead of calcium at R = 3, transparent gels (see

Figure 6.18) have been obtained. The structures formed possess tubular structures with diameter of ~ 20 nm and length up to a few micrometres. Hence the role of calcium ions in nanotube formation is not an entirely specific one. Apparently, the 'glue' between two building blocks can be any ion that contains sufficient positive charge to form a bridge between two or more specific carboxyl groups and also fits in with the orientation of these groups (Graveland-Bikker *et al.*, 2004). In the presence of Ba^{2+} or Mg^{2+}, it has been observed that random or fibrillar aggregates are formed instead of nanotubes. Both the precise ion size and its preferred ligand coodination number are therefore considered to play important roles (Graveland-Bikker and de Kruif, 2006).

In the food science context, it is worthwhile to observe the impressive stability of the α-lactalbumin nanotubes in relation to traditional conditions of pasteurization (40 s at 72 °C) and freeze-drying treatment as used in industrial processing. Another important feature in connection with potential applications for controlled release is the possibility of the controlled disassembly of α-lactalbumin nanotubes by several alternative methods: (i) decreasing the Ca^{2+} concentration by dilution in a Ca^{2+}-free buffer; (ii) addition of EDTA (extracting Ca^{2+} from the nanotube); (iii) adding urea (inducing structural change which makes the primary building block unfit for self-assembly); (iv) changing pH to values below 3 or above 9 (*i.e.*, increasing protein–protein repulsion by increasing the net charge on the building blocks) (Graveland-Bikker and de Kruif, 2006).

Atomic force microscopy was used to estimate a value of Young's modulus for α-lactalbumin nanotubes of the order of 0.1 GPa (Graveland-Bikker and de Kruif, 2006; Graveland-Bikker *et al.*, 2006b). This means that the α-lactalbumin nanotubes are substantially stiffer than, for example, myofibrils or casein micelles, which have moduli of the order of 10^{-1} MPa (Nyland and Maughan, 2000; Uricanu *et al.*, 2004). Nevertheless, they are much softer than certain non-proteinaceous structures, such as the well-studied carbon nanotubes, which can have a Young's modulus as high as 1 TPa (Salvetat *et al.*, 1999). Moreover, it has been shown that it is possible to punch a hole in the wall of an α-lactalbumin nanotube, at a precise spot, or to cut the nanotube into pieces, without damaging the complete structure. These findings represent important characteristics for development of nanotube technological applications.

We turn now to the other major whey protein, β-lactoglobulin. This is an acidic globular protein (pI = 5.1) with a molar mass of 18.4 kDa and a radius of about 2 nm (Aymard *et al.*, 1999). The protein can form long semi-flexible fibrils when heated in solution at around or above its denaturation temperature (60–80 °C) at pH = 2 and low ionic strength (Durand *et al.*, 2002; Veerman *et al.*, 2002, 2003a,b). (An example of the fibrils is shown in Figure 1.2). The rate of aggregation is strongly temperature sensitive. The temperature dependence of the rate is considered to be most likely controlled by the temperature dependence of the protein denaturation (Durand *et al.*, 2002). On the basis of data from atomic force microscopy, Ikeda and Morris (2002) estimated that the diameter of the β-lactoglobulin strands is ~ 4 nm.

Other globular food proteins like soy protein (Hermansson, 1985) and egg-white protein (Chamberlain *et al.*, 2000; Eiser *et al.*, 2009) also have the ability to form such fibrils. This suggests that the formation of fibrils is a generic property of proteins. Self-assembly of globular proteins can be induced in different ways — for example, by adding salts, solvents or denaturants, by changing pH, or by heating or applying pressure (Durand *et al.*, 2002; van der Linden, 2006; Nicolai, 2007; Eiser *et al.*, 2009). Where fibril formation occurs, it has been found to be irreversible for β-lactoglobulin and ovalbumin, but reversible for bovine serum albumin (Veerman *et al.*, 2002, 2003b). In general, the structure of these globular protein aggregates can be determined by using the complementary techniques of microscopy and scattering (light, X-ray, neutrons) (Burchard, 1994; Durand *et al.*, 2002; Nicolai, 2007). Types of interactions involved in the assembling include covalent bonding (disulfide bridges), charge interactions, hydrophobic interactions and hydrogen bonding (Bryant and McClements, 1998; Eiser *et al.*, 2009).

It is possible to get a wide range of architecture of protein aggregates depending on the experimental conditions (Clark and Lee-Tuffnell, 1986; Gosal and Ross-Murphy, 2000; Durand *et al.*, 2002; Nicolai, 2007; Eiser *et al.*, 2009). Thus, for example, as a result of partial protein unfolding (typically at acid pH, or in the presence of other chemical denaturants, or as a result of a mutation) globular proteins self-assemble into filamentous aggregates known as 'amyloid fibrils'. These fibrils are characterized by the extensive formation of 'β-sheet' secondary structure, regardless of the initial native structure of the aggregating protein (Chiti and Dobson, 2006). At pH values close to the isoelectric point of the protein or in the presence of salt to screen electrostatic interactions, the partially heat-denatured proteins form large, isotropic, particulate aggregates (Gosal and Ross-Murphy, 2000). In contrast, at pH values away from the isoelectric point and at low salt concentrations, the strong electrostatic repulsion leads to limited hydrophobic contact between partially heat-unfolded proteins, resulting in a proteinaceous 'strings of beads' rather than particulate aggregates (Clark and Lee-Tuffnell, 1986; Eiser *et al.*, 2009). Rigid rods can be formed when the protein is highly charged and ionic strength is low (Aymard *et al.*, 1999; Veerman *et al.*, 2002; Weijers *et al.*, 2002a; Gosal *et al.*, 2004). For example, at pH = 2, linear aggregates of β-lactoglobulin are formed which are very rigid at low ionic strength (Figure 6.4a) (Durand *et al.*, 2002). Under the conditions where repulsive electrostatic forces are weaker, because salt is added or charge density is reduced, flexible linear aggregates are formed (Figure 6.4b) (Aymard *et al.*, 1999; Durand *et al.*, 2002; Weijers *et al.*, 2002b).

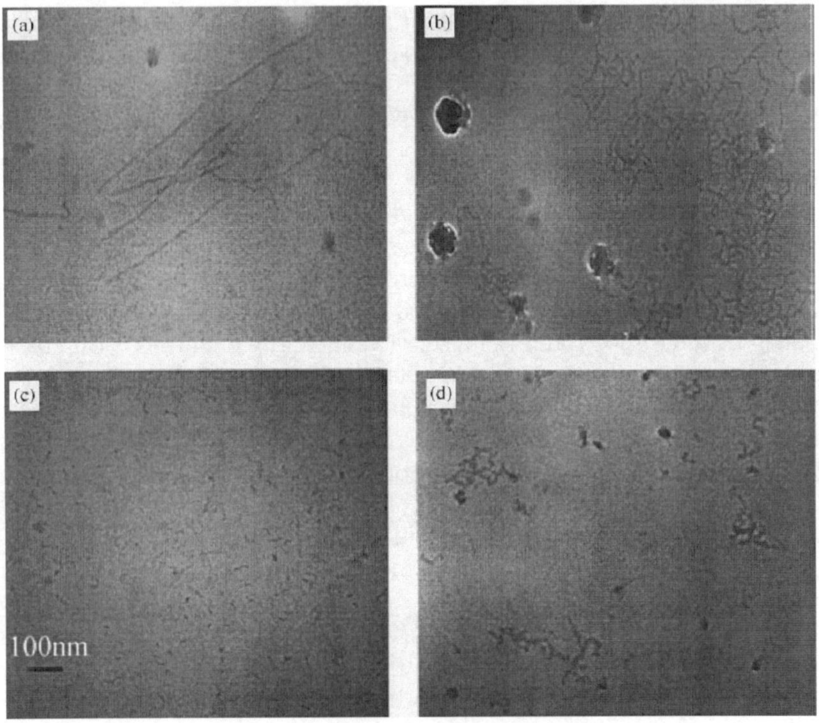

Figure 6.4 Cryo-TEM micrographs of some β-lactoglobulin aggregates formed upon heating: (a) pH = 2, ionic strength = 0.01 M; (b) pH = 2, ionic strength = 0.1 M; (c) pH = 7, ionic strength = 0.001 M; (d) pH = 7, ionic strength = 0.1 M. Reproduced from Durand *et al.* (2002) with permission.

When the electrostatic repulsion becomes negligible, either because the pH is close to the isoelectric point or the ionic strength is high, the aggregates appear more densely branched and their flexibility increases (Koseki *et al.*, 1989; Pouzot *et al.*, 2005). At pH = 7, it is observed that there are small elongated aggregates formed at low ionic strength; but at high ionic strength some larger clusters are observed, which appear to be produced by random association of the small aggregates (Figure 6.4c,d). At the low ionic strength the primary aggregates are apparently inhibited from further association, probably due to electrostatic repulsion. However, at high protein concentrations, Durand *et al.* (2002) have observed further association caused by the presence of protein counter-ions which contribute to the effective ionic strength.

1.3. *Polysaccharides*

It is well known that both neutral and charged polysaccharides have a strong tendency towards self-association in dilute aqueous media, leading to gel network formation at higher concentrations (Harkema, 1998; Kasapis *et al.*, 1993a; Piculell, 1998; Agoub *et al.*, 2009). Because of this behaviour the main stabilizing action of polysaccharides in food colloids is through viscosity modification or gelation in the aqueous continuous phase. Sometimes this behaviour is supplemented by some surface activity. One distinctly identifiable class of surface-active polysaccharides are the hydrophobically modified cellulose derivatives — notably methylcellulose and hydroxypropyl methylcellulose (Karlberg *et al.*, 2005). These polysaccharides can be used as emulsifiers in their own right to prepare stable oil-in-water emulsions. But it is generally found that the droplets produced are considerably coarser than those made using low-molecular-mass emulsifiers or proteins under similar conditions (Darling and Birkett, 1987). This poorer emulsifying behaviour is partly attributable to the relatively high molecular weight of these modified cellulose polymers. Similar arguments can be applied to the emulsifying performance of other surface-active polysaccharides such as hydrophobically modified starch and propylene glycol alginate (Dickinson, 2003).

One particularly common route to polysaccharide self-assembly and gelation involves the coil-to-helix transition. This transition may be induced by cooling a hot solution of the random coil macromolecules. A variety of gel-like states can be produced depending on the type of polysaccharide and on the solvent conditions (Picullel, 1998). For example, imaging techniques such as atomic force microscopy (AFM) and cryo-transmission electron microscopy (cryo-TEM) have revealed rigid thin rods of aggregated helices under gelling conditions for gellan (McIntire and Brant, 1997) and κ-carrageenan (Lundin and Hermansson, 1997).

Figure 6.5 shows experimental data relating to the self-assembly of sodium κ-carrageenan, as induced by cooling in the presence of 0.1 M NaCl, and occurring simultaneously with the coil-to-helix transition for the same polysaccharide (Semenova *et al.*, 1988). In what follows we consider this system in some detail.

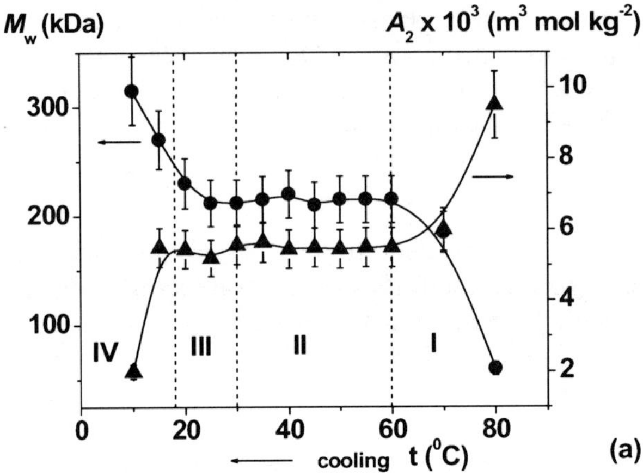

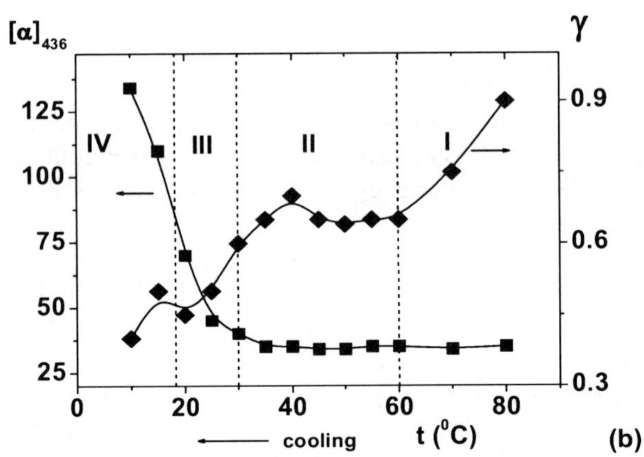

Figure 6.5 Temperature dependence of the characteristics of sodium κ-carrageenan particles dissolved in an aqueous salt solution (0.1 M NaCl). The cooling rate is 1.5 °C min^{-1}. (a) (●) Weight-average molar weight, M_w, and (▲) second virial coefficient, A_2. (b) (■) Specific optical rotation at 436 nm, and (◆) penetration parameter, γ, defined as the ratio of the radius of the equivalent hard sphere to the radius of gyration of the dissolved particles (see equation (5.33) in chapter 5). See the text for explanations of different regions I, II, III and IV.

On the strength of combined experimental data from static light scattering and polarimetry, the self-assembly of sodium κ-carrageenan in 0.1 M NaCl is inferred to occur in four stages (I–IV) (Figure 6.5). In the first region (stage I) a temperature reduction from 80 °C to 60 °C results in a poorer affinity of the polysaccharide for the solvent, as shown by a smaller value of the second virial coefficient A_2. As this takes place, the weight-average molar weight, M_w, increases approximately three-fold (Figure 6.5a). At the same time, as can be seen from the dependence of the specific optical rotation on the temperature, no cooperative conformational coil–helix transition of the κ-carrageenan macromolecules could be observed at this stage (Figure 6.5b). Thus the first stage of the aggregation process occurs under conditions for which macromolecules have a disordered conformation. Actually, at 80 °C, the value of parameter γ (as defined in equation (5.33)) approximates to the theoretical one for isolated molecules of flexible-chain polymers in a good solvent (Berry and Casassa, 1970). The inferred decrease in γ in the high-temperature region (stage I) suggests that there is a strengthening of the attraction between the dissolved particles. With further reduction in temperature from 60 °C to 30 °C (stage II) no aggregation takes place: the second virial coefficient, the optical rotation and the penetration parameter γ all retain their constant values. Beginning at around 25 °C, we can detect helix formation of the macromolecules (stage III), as shown by the polarimetric data (Figure 6.5b). At this stage the constancy of the molecular weight points to the fact that the conformational coil–helix transition occurs inside the aggregates. This coil–helix transition causes a decrease in the value of the penetration parameter γ, reflecting an increase in the radius of gyration of the particles with the retention of the constant values of both M_w and A_2 (Figure 6.5a,b). Further cooling of the solution (stage IV) brings about a decrease in A_2 and an increase in M_w. In other words, it is at this stage (*i.e.*, at a temperature below 20 °C) that the aggregation of the helical macromolecules occurs. The further decrease in γ suggests enhanced permeability of the coils, due to the effect of the helical segments, and stronger attraction between the dissolved particles as demonstrated by the lower value of the second virial coefficient.

This whole analysis is consistent with the commonly accepted model of the double-helical self-assembly of helix-forming polysaccharides (Piculell, 1998; Hjerde *et al.*, 1998a,b), as illustrated schematically in Figure 6.6. Case (a) illustrates that, in the presence of a sufficient amount of the gel-inducing ions, both degraded and intact κ-carrageenan helices associate into thin, rigid 'super-helical rods', which in turn associate into a network structure (Borgström *et al.*, 1996, 1998; Chronakis *et al.*, 1996). Similar rigid rods have also been detected for gellan. Case (b) in

Figure 6.6 shows that helices of intact κ-carrageenan form weak gels at sufficiently high concentrations (> 1%) even in a good solvent such as 0.1 M NaI (Chronakis *et al.*, 1996; Borgström *et al.*, 1998).

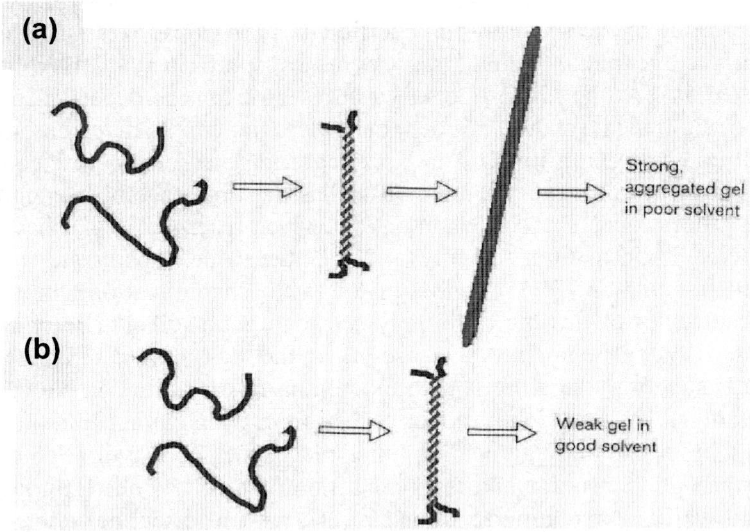

Figure 6.6 Schematic representation of the two types of κ-carrageenan self-assembly: (a) poor solvent conditions; (b) good solvent conditions. Reproduced from Piculell (1998) with permission.

Amongst the anionic polysaccharides, the self-assembly of pectin has attracted very considerable interest because this particular hydrocolloid is commonly used as a gelling and thickening agent in foods (Braudo *et al.*, 1998; Dickinson, 2003; Agoub *et al.*, 2009). A key compositional factor controlling the pectin properties is the proportion of methyl ester groups in the macromolecule, *i.e.*, the degree of esterification (DE). Pectins with DE < 50 % are conventionally known as low-methoxy (LM) pectins, and those with DE > 50 % are called high-methoxy (HM) pectins. Gelation of LM pectin is normally induced by the addition of calcium ions. Intermolecular association has been attributed to formation of 'egg box' junctions (Grant *et al.*, 1973), analogous to those in calcium alginate gels, with extended arrays of site-bound Ca^{2+} ions sandwiched between (1-4)-diaxially linked polyuronate chains in a highly buckled 2-fold conformation (Jarvis and Apperley, 1995; Morris *et al.*, 1978, 1982). Gelation of HM pectin (Rolin, 1993) occurs at acidic pH in the presence of high concentrations (~ 60–65 wt%) of sucrose or other cosolutes (*e.g.*, fructose, polyols). Reduction in pH promotes intermolecular association by converting a proportion of galuronic acid residues from the charged form (COO^-) to the uncharged form (COOH). This suppresses electrostatic repulsion between the HM pectin chains, facilitating hydrophobic attraction of methyl ester groups and thus allowing a network to form.

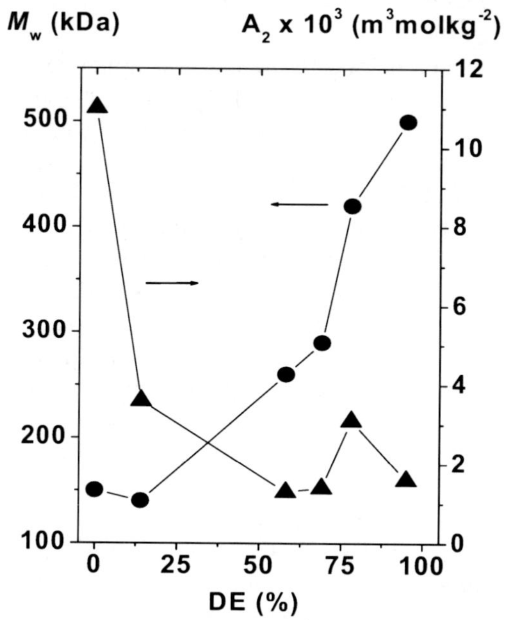

Figure 6.7 Effect of the degree of esterification (DE) of citrus pectin on its extent of self-assembly (M_w) (●) and the character of its intermolecular pair interactions (A_2) (▲) in aqueous solution (0.09 M NaCl and 0.01 M NaF) (Braudo *et al.*, 1998).

It was established by Braudo and coworkers (1998) that an increase in the degree of pectin esterification leads to a greater extent of pectin association in aqueous solution (an increase in M_w). This is illustrated in Figure 6.7. The strength of the attractive interactions between the polysaccharide associates increases with the degree of pectin esterification, as reflected in the lowering of the value of the second virial coefficient.

The formation of long hydrogen-bonded junctions between the polar groups of the pectins, which can favour pectin association on cooling, has also been suggested (Tsoga *et al.*, 2004a; Agoub *et al.*, 2009). One obvious effect of incorporating large amounts of cosolute is to decrease the amount of water available to maintain the polymer in the solvated (solution) state. Nevertheless, it appears that different cosolutes have widely varying ability to promote the formation of HM pectin gels (May, 1990; May and Stainsby, 1986; Tsoga *et al.*, 2004a,b).

2. *Surfactant-Based Self-Assembly of Proteins*

Most food surfactants belong to the class of lipids consisting of ionic or non-ionic derivatives of long-chain fatty acids. These are the phospholipids (lecithin) and the esters of fatty acids and glycerol (or other polyvalent alcohols), which are often combined with organic hydroxy-acids (acetic, lactic, diacetyl, tartaric, or citric) (Krog, 1997). Owing to their high surface activity and their ability to take part in the stabilization of food emulsions and foams, surfactants are widely used as the principal functional ingredients in food colloids along with biopolymers. Therefore, in order to be able to control the quality of food colloids or design new structures, it is essential to take into account the role of biopolymer–surfactant interactions in determining the functional properties of both these kinds of ingredients, including their self-assembly.

Let us consider the various possible types of biopolymer–surfactant interactions. We first note that, because of the amphiphilic nature of both biopolymers and surfactants, it can be envisaged that the mechanistic interpretation could be based on attractive or repulsive interactions acting between the original biopolymer and surfactant molecules/particles or between biopolymer particles modified by the surfactants. For example, attractive interactions could arise from:

- (i) hydrophobic interactions between hydrocarbon chains of surfactant molecules and the non-polar patches and/or functional groups of biopolymers;
- (ii) hydrophobic interactions between hydrocarbon chains of surfactant molecules attached to one biopolymer particle and the non-polar patches and/or functional groups of another;
- (iii) hydrophobic interactions between hydrocarbon chains of surfactant molecules attached to different biopolymer particles;
- (iv) electrostatic attractions between opposite charges on surfactant and biopolymer functional groups;
- (v) electrostatic attractions between opposite charges on biopolymer particles modified by surfactant attachment;

(vi) multiple hydrogen bonding between oxygen atom (acceptor) and hydrogen atom (donor) on relevant functional groups (hydroxyl, carboxyl, ester, or ether) of surfactants and biopolymers; or
(vii) multiple hydrogen bonding between oxygen atom (acceptor) and hydrogen atom (donor) on relevant functional groups (hydroxyl, carboxyl, ester, or ether) of biopolymer particles modified by surfactant attachment.

And repulsive interactions could arise from:

(i) electrostatic repulsions between the like charges of surfactant and biopolymer functional groups;
(ii) electrostatic repulsions between the like charges on biopolymer particles modified by surfactant attachment;
(iii) effects of excluded volume (steric repulsions) between surfactant and biopolymer molecules/particles; or
(iv) effects of excluded volume (steric repulsions) between biopolymer particles modified by surfactant attachment.

Analysis of recent experimental data has shown that there are various issues that have to be considered when interpreting the complex character of the interactions in mixed protein + surfactant systems (Kelley and McClements, 2003; Il'in *et al.*, 2004, 2005; Malhotra and Coupland, 2004; Istarova *et al.*, 2005). Below we identify these general issues as factors A, B, C and D.

(A) The protein conformation does or does not provide access to the protein interior for the surfactant molecules, with implications for the number of available binding sites on each protein.

We illustrate this factor with experimental data for the enthalpy of the protein–surfactant interactions for the globular protein legumin, which has a compact native conformation (M_w = 330 kDa, radius 3–5 nm (Derbyshire *et al.*, 1976; Lawrence *et al.*, 1994)). It seems that the interior of the native legumin globule is inaccessible to the rather long hydrocarbon chains of anionic surfactant molecules such as CITREM and SSL. These food emulsifiers are generally based on equimolar mixtures of the esters of stearic and palmitic acids in combination with citric acid or lactic acid, respectively (Krog, 1997). Following thermal denaturation, however, the aggregates of legumin (M_w = 6300 kDa, R_h = 58 nm (Il'in *et al.*, 2004)) and sodium caseinate nanoparticles (M_w = 4000 kDa, R_h = 117 nm (Il'in *et al.*, 2005)) were found to exhibit enormously greater (two orders of magnitude larger) enthalpy of interactions with the same anionic surfactants at the same concentrations (Semenova, 2007). This experimental observation can be primarily attributable to a greater accessibility for these surfactant molecules of the rather porous structure of either the sodium caseinate nanoparticles or the heat-induced legumin aggregates. This enhanced accessibility correlates with the rather open conformation of the constituent polypeptide chains of these proteins, and incidentally also with the much larger molar masses of these protein nanoparticles, compared with the native compact globular legumin molecules.

(B) The size and charge of the surfactant head-group is favourable (when both are small) or unfavourable (when they are large) for penetration of whole surfactant molecules into the interior of protein nanoparticles.

This factor concerns the probability of realizing the different kinds of protein–surfactant interactions in the protein interior between polar and non-polar parts of the interacting protein and surfactant molecules. In the case of a high penetration ability of the surfactant, and thus a high probability of realization of the different kinds of interactions, the overall energy of the interactions between polar functional groups on interacting protein and surfactant (electrostatic forces between opposite charges and hydrogen bonding) could generally exceed to those between their non-polar parts (hydrophobic forces) (Jenks, 1969; Finkelstein and Ptitsyn, 2002). This hypothesis is supported by the greater exothermic heat effect for protein–surfactant interactions measured with SSL as compared to CITREM: the comparatively lower charge and smaller size of the SSL head-group provides easier penetration of SSL molecules into the protein interior (Semenova *et al.*, 2006).

(C) When charges on surfactant and protein are of similar sign, the electrostatic repulsion between them hinders the penetration of the surfactant head-groups into the protein interior and therefore reduces the probability of polar surfactant–protein interactions.

Where this factor plays a role, the hydrophobic interaction between the hydrocarbon chains of the surfactant and the non-polar parts of protein functional groups are predominant. An example of this effect is the marked endothermic character of the interactions between the anionic CITREM and sodium caseinate at pH = 7.2 (Semenova *et al.*, 2006), and also between sodium dodecyl sulfate (SDS) and soy protein at pH values of 7.0 and 8.2 (Nakai *et al.*, 1980). It is important here to note that, when the character of the protein–surfactant interactions is endothermic (*i.e.*, involving a positive contribution from the enthalpy to the change in the overall free energy of the system), the main thermodynamic driving force is considered to be an increase in the entropy of the system due to release into bulk solution of a great number of water molecules. This entropy change arises from the dehydration of the surfactant and protein molecules consequent upon their intimate hydrophobic interactions.

(D) The molecular state of the surfactant in the aqueous medium is of micellar form (concentration above critical micelle concentration (cmc)) or non-micellar form (concentration below cmc) (Belyakova *et al.*, 1999; Chen *et al.*, 1998; Fang and Dalgleish, 1997; Il'in *et al.*, 2005; Kelley and McClements, 2003; Semenova *et al.*, 2003, 2006).

The situation is complicated by the fact that the cmc value determined in the pure surfactant solution generally differs from that found in the presence of biopolymer. This is mainly because the surfactant–biopolymer interactions can shift the equilibrium between free surfactant molecules and their micelles, leading to a change in the effective cmc of surfactant molecules in the biopolymer system (Kelley and McClements, 2003; McClements, 2000; Thongngam and McClements, 2005).

As a result of interaction of protein with individual surfactant molecules (below the cmc), a marked degree of protein aggregation is generally observed. This is reflected in changes in the weight-average molar mass, M_w, and the radius of gyration of the protein, R_G, as determined by laser light scattering (Il'in *et al.*, 2004, 2005; Istarova *et al.*, 2005) and turbidity measurements (Kelley and McClements, 2003). It is as if the individual surfactant molecules were fulfilling the role of powerful cross-linking agents.

To illustrate some of the structural features of the aggregates formed, we can make use of an empirical relationship between M_w and R_G which takes the form of a power-law equation (Burchard, 1994):

$$M_w = K R_G^{d^*} \quad . \tag{6.1}$$

Here the exponent d^* has the meaning of a fractal dimensionality which assists in describing the aggregate structure formed according to some well-defined geometrical model (Burchard, 1994); and K is a pre-factor which becomes larger if the elementary unit of the self-similar structure is denser (Nicolai, 2007). On the basis of equation (6.1), it seems safe to assume that under comparison of the extent of protein aggregation, k_1 (ratio of weight-average molar mass of protein + surfactant, M_w^c, to that of pure protein, M_w^{pr}), with the extent of increase in size of the protein nano-particles, k_2 (ratio of radius of protein + surfactant, R_G^c, to that of pure protein, R_G^{pr}), the following equation will apply:

$$k_1 = k_2^{d^*} \quad . \tag{6.2}$$

Table 6.1 shows some results of calculations based on light scattering data, in which the fractal-type exponent d^* is fitted to different types of polymer aggregation (Semenova *et al.*, 2006). Values of the quantity $k_2^{d^*}$ are listed for four different kinds of model aggregate structure: a rigid rod ($d^* = 1$), a random aggregate ($d^* = 2.5$), a shell-like aggregate ($d^* = 3.0$), and a structure with contracted units ($d^* = 3.5$). The experimental values of k_1 are presented for mixed aqueous solutions (pH = 7.2, ionic strength = 0.05 M) of sodium caseinate or heat-denatured legumin with two ionic surfactants (CITREM and SSL) and one non-ionic surfactant (PGE) at concentrations well below their cmc values.

In the case of the rather porous and flexible structure of sodium caseinate nanoparticles, the data show that the interaction with surfactants causes a tendency towards the shrinkage of the aggregates, most likely due to the enhanced cross-linking in their interior as a result of the protein–surfactant interaction. This appears most pronounced for the case of the anionic surfactants (CITREM and SSL) interacting with the sodium caseinate nanoparticles. Consistent with this same line of interpretation, a surfactant-induced contraction of gelatin molecules of almost 30% has been demonstrated as a result of interaction with the anionic surfactant α-olefin sulfonate (Abed and Bohidar, 2004).

It is pertinent to note here the observed increase in the value of the structure-sensitive parameter ρ from 1 to 2. This implies that the architecture of the sodium caseinate aggregates, as modified by interaction with the surfactant, becomes generally more open, despite the inferred collapse of their constituent protein nanoparticles. In contrast, a shell-like aggregation structure can be inferred for the self-assembly of sodium caseinate as a result of its interaction with the non-ionic surfactant PGE (this surfactant is based on a mixture of the esters of stearic and palmitic acids in chemical combination with polyglycerol (Krog, 1997)).

Figure 6.8 shows a schematic illustration of the proposed molecular mechanism of self-assembly of sodium caseinate induced by interaction with surfactant, including inferred changes in the distribution of bound water molecules arising from the formation of protein–surfactant interactions. This mechanistic interpretation is based on combined data from mixing calorimetry and light scattering, using CITREM as an example at a concentration of 5 mg/L (0.3 cmc) (Il'in *et al.*, 2004). The original sodium caseinate nanoparticles were characterized by the following set of molecular and thermodynamic parameters: $M_w = 4 \times 10^6$ Da, $R_G = 117$ nm, $A_2 = 0.93$ m^3 mol^{-1} and $\rho \approx 1$ (Semenova *et al.* 2006). This inferred value of $\rho \approx 1$ is characteristic of a rather dense sphere.

Table 6.1 Theoretical estimation of the types of structures of protein self-assembly induced by interactions of food proteins (0.5 % w/v) with food-grade surfactants in aqueous solution (pH = 7.2, ionic strength = 0.05 M, 293 K) below the cmc (CITREM cmc = 15 mg/L; SSL cmc = 3.5 mg/L; PGE cmc = 1.0 mg/L (Il'in *et al.*, 2005)). See the text for definitions of k_1 and k_2.

		$k_2^{\ d*}$			
		Aggregate structure			
System	k_1	rigid rod $d^* = 1$	random $d^* = 2.5$	shell -like $d^* = 3.0$	with contracted units $d^* = 3.5$
Sodium caseinate					
Protein + CITREM (0.3 cmc)	7.3	1.8	4.3	5.8	7.8
Protein + SSL (0.3 cmc)	6.3	1.8	4.3	5.8	7.8
Protein + PGE (0.5 cmc)	3.9	1.6	4.3	4.1	5.2
Heat-denatured legumin					
Protein + CITREM (0.4 cmc)	3.0	2.7	11.9	19.7	32.3
Protein + SSL (0.3 cmc)	8.7	6.8	121	314	820
Protein + PGE (0.5 cmc)	1.4	3.0	15.6	27	46.8

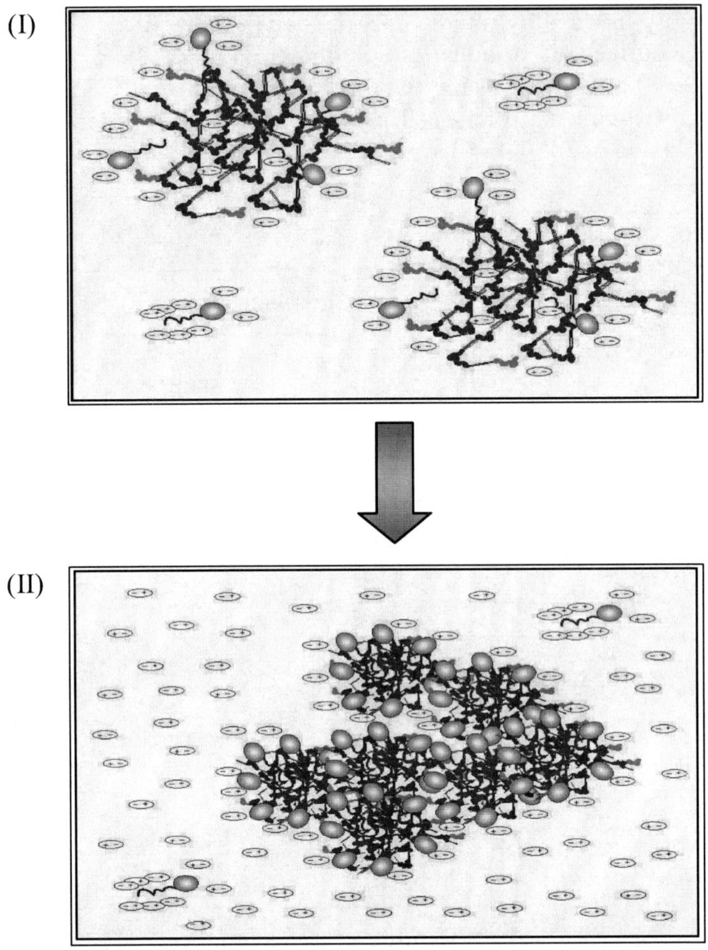

Figure 6.8 Sketch of proposed molecular mechanism of protein–surfactant interaction for CITREM + sodium caseinate (0.5 % w/v in aqueous medium (pH = 7.2, ionic strength = 0.05 M) at 293 K. Picture (I) shows the water molecules bound with polar groups of the protein and surfactant, as well as water molecules structured as a result of hydrophobic hydration around the hydrocarbon chain of the surfactant. (For clarity, the free water molecules are not shown.) Picture (II) demonstrates the release of bound and structured water molecules resulting from the predominantly hydrophobic interactions between protein and surfactant. Reproduced from Semenova *et al.* (2006) with permission.

Due to interaction with CITREM, modified protein aggregates were formed with $M_w = 29 \times 10^6$ Da, $R_G = 211$ nm and $A_2 = 188$ m^3/mol. This practically two orders of magnitude increase in A_2 is attributable to the presence of charged head-groups of CITREM around the protein aggregates. Such an interpretation agrees well with the predominantly hydrophobic character of the protein–CITREM interactions as manifest in the endothermic enthalpy measured by mixing calorimetry (Il'in *et al.*, 2005; Semenova *et al.*, 2006). We infer that the contracted protein nanoparticles modified by interaction with surfactant are combined into complex particles of more open architecture (ρ goes up from 1 to 2). The main driving force for the spontaneous hydrophobic interactions, characterized by heat taken up by the system ($\Delta H > 0$), is an entropy increase ($\Delta S > 0$) due to release of water molecules, as induced by dehydration of surfactant and protein under the influence of their interaction (see Figure 6.8).

For the case of heat-denatured legumin, the data presented in Table 6.1 suggest that there is a one-dimensional self-assembly of the original protein particles into rigid rods. This inferred structural change agrees well with the marked increase in the value of the parameter ρ from 0.7 up to 3.1 for the case of interaction with SSL (Semenova *et al.*, 2006). It was recently shown by a number of methods — isothermal mixing calorimetry, static and dynamic light scattering, and viscometry of dilute solutions — that the manipulation of protein self-assembly can be easily performed by changing the concentration of the charged anionic surfactant molecules in the solution (Semenova *et al.*, 2007). Figure 6.9 shows, by way of example, how sodium caseinate self-assembly, in the presence of an anionic surfactant (CITREM) at pH = 7.2, may be modified to give structures with different molecular and thermodynamic parameters. In discussing this behaviour, we distinguish three concentration ranges.

In region I, at rather low surfactant concentration ($C_{CITREM} < 1.75$ mg/L), we have the largest (M_w, R_G) and most linear ($\rho >> 2$) protein–surfactant nanoparticles. The balance between attractive and repulsive interactions of the protein nanoparticles, as modified by the surfactant complexation, favours linear structuring in aqueous solution. Under these conditions, the observed marked increase in thermodynamic affinity of the protein associates for the aqueous medium, as mirrored by the increase in the molal A_2 (Figure 6.9c), is most likely determined by the pronounced increase in their size (Figures 6.9a and 6.9b) and hence in their excluded volume, owing to their linear form (Figure 6.9b) (Tanford, 1961; Nagasawa and Takahashi, 1972). This explanation is instead of the increase in their hydrophilicity due to attachment of charged carboxylic groups of CITREM to the protein.

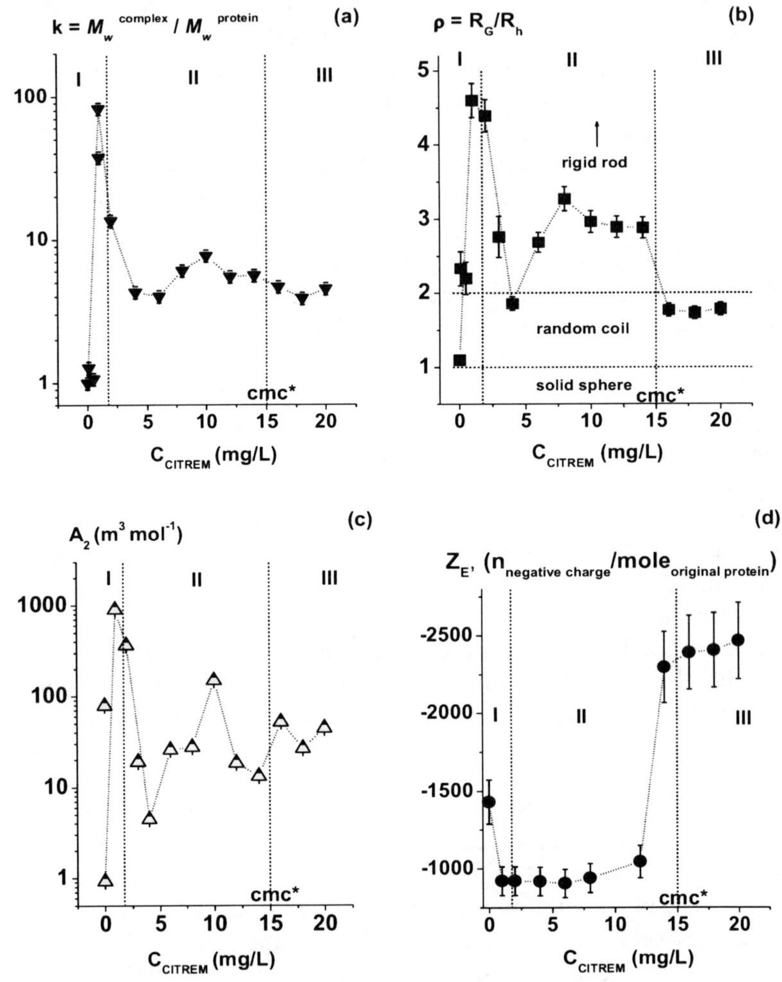

Figure 6.9 Effect of CITREM concentration on the molecular and thermodynamic parameters of complex protein–surfactant nanoparticles in aqueous medium (phosphate buffer, pH = 7.2, ionic strength = 0.05 M; 20 °C): (a) extent of protein association, $k = M_w^{complex}/M_w^{protein}$; (b) structure-sensitive parameter, ρ; (c) second virial coefficient, A_2 (molal scale); (d) effective charge, Z_E (net number *n* of moles of negative charges per mole of original sodium caseinate nanoparticles existing at pH = 7.2 ($M_w = 4\times10^6$ Da)). The indicated cmc* value refers to the pure CITREM solution. Reproduced from Semenova *et al.* (2007) with permission.

Measurements of the effective protein charge, Z_E, indicate a reduction in its value in this concentration region (Figure 6.9d). This observation points to the important contribution of electrostatic attraction between the negatively charged head-groups of the CITREM and the positively charged groups of sodium caseinate to their interactions in this concentration range. Such charge neutralization could be expected to facilitate and enhance the marked protein association found in this concentration region (Figure 6.9a), owing to the additional hydrophobic attractions involving the alkyl chains of CITREM molecules already bound electrostatically on different protein nanoparticles.

In region II, the higher surfactant concentration (1.75 mg/L < C_{CITREM} < 15 mg/L (cmc of pure CITREM)) leads to a reduction in the extent of protein self-association. The linear protein aggregates possess a smaller average molar mass (Figure 6.9a) and lower thermodynamic affinity for the aqueous medium (Figure 6.9c). This can be attributed to a strengthening of the electrostatic repulsive forces acting between protein particles, as modified by the increasing number of charged head-groups of the surfactant molecules added to the surface of the protein particles. This explanation agrees well with the predominantly hydrophobic character of the protein–CITREM interactions, as indicated by the endothermic character of the interactions measured by the isothermal mixing calorimetry under similar experimental conditions (Il'in *et al.*, 2005; Semenova *et al.*, 2006). As a consequence there is disassembly of the complex particles. The independence of the effective protein charge (Figure 6.9d) with respect to surfactant concentration in region II could be attributable to compensative contributions to the value of Z_E from the neutralization and growth of the protein charge due to interactions between oppositely charged functional groups of CITREM and the sodium caseinate, and the addition of negative charge from attached CITREM molecules. Additionally, we infer that the appearance of local minimum and maximum values of measured molecular and thermodynamic parameters in region II (Figure 6.9a–c) most likely indicates a shift in the fine balance of interparticle forces towards predominant repulsion or attraction, respectively, between the complex nanoparticles.

In the third concentration regime (region III), defined as being in the close vicinity of the pure surfactant cmc ($C_{CITREM} \geq 15$ mg/L), complex nanoparticles of slightly lower molar mass and more compact structure are formed. An additional sharp contraction of the complex nanoparticles occurs at surfactant concentrations just beyond the cmc (Figure 6.9b) due to surfactant cluster formation within the interior of the complex nanoparticles. This interpretation is based on information from a combination of experimental methods: light scattering, isothermal mixing calorimetry, and viscometry of dilute solutions (Semenova *et al.*, 2006). The explanation agrees well with previous literature data (Chen *et al.*, 1995; Vasilescu *et al.*, 1999; Abed and Bohidar, 2004). It is likely that the cluster formation reduces the electrostatic repulsion between surfactant head-groups, due to their non-uniform and restricted location in the form of clusters in the interior of the protein nanoparticles. This could strengthen the attractive forces between internal surfactant-free non-polar patches of the protein nanoparticles, inducing their contraction. Previously, based on data from dynamic light scattering, Abed and Bohidar (2004) had observed that the presence of the apparently similar anionic surfactant, α-olefin sulfonate (AOS), causes the contraction of gelatin molecules at 30 °C by almost 30% up to the cmc, and induces some molecular expansion beyond that concentration. But the great difference in the structure and flexibility of a gelatin chain as compared with a sodium caseinate nanoparticle obviously prevents us from inferring the same sort of necklace-bead-type model of protein–surfactant interactions as was suggested for the gelatin–AOS complex by Abed and Bohidar (2004). Hence, it seems that the validation of the putative mechanism of the contraction of sodium caseinate nanoparticles, as induced by surfactant cluster formation in their interior, requires some further confirmatory evidence.

One additional peculiarity of the properties of the complex nanoparticles in region III is their larger effective charge as compared with the values of Z_E of the complex nanoparticles existing in regions I and II (Figure 6.9d). This finding can be attributed to the moderately large quantity of the negatively charged head-groups of surfactant molecules attached to each protein nanoparticle, which seems to be enough for both the partial neutralization of the positively charged functional groups of the protein and the generation of some excess negative charge. In addition, in line with the marked increase in the values of Z_E for the complex nanoparticles in region III (Figure 6.9d), we have found an increase in their thermodynamic affinity for the aqueous medium in passing from region II to region III, as reflected in the higher values of A_2 (Figure 6.9c). The fact that this increase is found despite the evident decline in the expected contribution to A_2 from the excluded volume of collapsed complex nanoparticles implies that the increase in A_2 is caused by the greater hydrophilicity of the nanoparticles, as a result of CITREM charge addition to the net protein charge.

In line with the general increase in thermodynamic affinity of surfactant–protein particles for the aqueous medium, a marked increase in the solubility of soy protein has been observed in response to interactions with SDS (Malhotra and Coupland, 2004). In contrast, however, the self-assembly of the globular protein legumin, as modified by the anionic surfactants CITREM and SSL, was found to be associated with a clearly defined increase in the relative hydrophobicity ($A_2 < 0$) of the protein surface (Il'in *et al.*, 2004). This result accords with reported insoluble complex formation between bovine serum albumin and ionic surfactants of the same charge sign (Kelley and McClements, 2003).

It would appear that the changes found in the hydrophilic/lipophilic properties of the protein surface in response to interactions with surfactants are most likely dictated by the ultimate mutual spatial arrangement of the hydrophobic and hydrophilic parts of the protein and surfactant molecules, as determined by the predominant nature of their local interactions. We may assume that interaction with surfactant causes gradual denaturation of the native protein. Hence, the peculiar features of the changes in molecular properties of native globular legumin — in particular the increase in the protein surface hydrophobicity ($A_2 < 0$) — occur in parallel with a decrease in conformational stability (a partial unfolding) of the protein globule, as a consequence of the interactions with either anionic surfactant (CITREM or SSL) or non-ionic surfactant (PGE) (Il'in *et al.*, 2004).

When a protein is interacting with surfactant micelles, the following intrinsic features of the micelles are suggested as being of general importance in relation to their influence on the character of the protein self-assembly (Il'in *et al.*, 2005):

(i) the thermodynamic stability of surfactant micelles, as expressed in terms of ΔG_{mic} (the Gibbs free energy of micellization), which may be favourable or otherwise towards the participation of the hydrophobic 'core' of the surfactant micelles in intimate interaction with the protein;

(ii) the net charge and the associated relative hydrophilic/lipophilic balance of the micellar surface properties; and

(iii) the size of the surfactant micelles, which affects the impact of the excluded volume effects on the various types of repulsive intermolecular interactions (protein–micelle, micelle–micelle, and complex–complex).

Based on a recent isothermal mixing calorimetric study (Il'in *et al.*, 2005) it was postulated that, when both surfactant micelles and protein carry a rather high like net charge, then the micelles cannot take part in interactions with protein on account of the strong electrostatic repulsion between them. Hence, in this case, the interactions occur only between the protein and the individual surfactant molecules which are in equilibrium with their micelles. This behaviour is reflected in the lack of any change in the thermodynamic character (heat effects in the case of isothermal mixing calorimetry) of protein–surfactant interactions in passing through the cmc.

In contrast, let us consider experimental conditions more favourable to weak electrostatic repulsion: a pH closer to the protein p*I* and a change from charged to uncharged micelles. Now the surfactant micelles themselves can interact with the protein (Il'in *et al.*, 2005, Semenova *et al.*, 2006). In this case, if the surfactant micelles are of low thermodynamic stability, they are disrupted by encounters with protein particles. This disruption is accompanied by a simultaneous release of a great number of surfactant molecules, which can therefore increase significantly the number of surfactant–protein contacts. This, in turn, is revealed by the isothermal mixing calorimetry. In such measurements, a sharp increase in the value of the enthalpy of formation of interactions between surfactant molecules and protein has been observed when passing through the cmc. On the other hand, if the surfactant micelles are of high thermodynamic stability they can interact with the protein as single entities. This behaviour manifests itself in a dramatic change in the enthalpic character of the protein–surfactant interactions when passing through the cmc (Il'in *et al.*, 2005; Semenova *et al.*, 2006).

The greatest degree of affinity between the protein and the surfactant micelles is found under conditions of availability of the greatest quantity of opposite charges on their surfaces (Istarova *et al.*, 2005). In addition, the overall amount of surfactant micelles that can be accommodated by protein is dictated by the protein molar mass, *i.e.*, by the number of binding sites (Istarova *et al.*, 2005). It has been observed that each polymer coil can accommodate only a limited number of micelles. Typically, in practice, 1 g of protein would be capable of binding up to around 1.5 g of surfactant micelles (Evans and Wennerstrom, 1999).

Combined static and dynamic light scattering data indicate that the values of the molecular and thermodynamic properties of proteins modified by surfactant micelles depend on the specific features of their interactions (Il'in *et al.*, 2005; Istarova *et al.*, 2005; Semenova *et al.*, 2006). For instance, intensive intermolecular and intramolecular protein crosslinking may occur due to the participation of numerous surfactant molecules from the hydrophobic 'core' of micelles following their disruption as a result of interactions with protein. As a consequence of this, large protein aggregates of much higher hydrophilicity and a more compact spherical architecture can be formed, as compared with the pure proteins. This type of situation is illustrated in Table 6.2 for the case of sodium caseinate + CITREM ($\Delta G_{\mathrm{mic}} = -4\ kT$).

Table 6.2 Schematic representation of nanoscale structure and experimental data relating to self-assembly of sodium caseinate induced by interactions of the protein (1.0 % w/v) with micelles of food-grade surfactants (CITREM and SSL) in an aqueous medium (pH = 5.5, ionic strength = 0.05 M, 293 K) above the surfactant cmc.

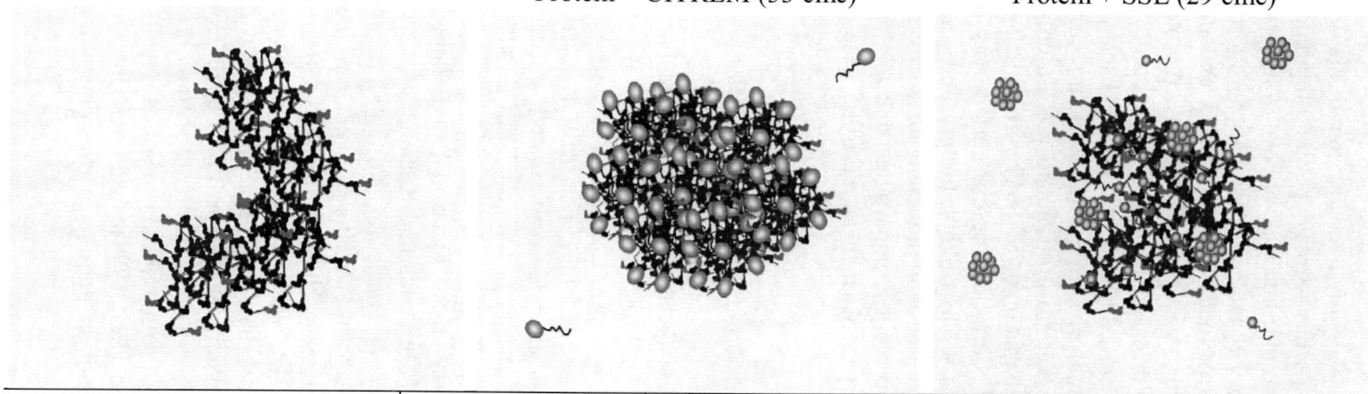

System	$C_{surfactant}$ (mg L^{-1})	$M_w \times 10^{-6}$ (Da)	A_2 (m^3 mol^{-1})	R_G (nm)	ρ
Protein	0	15	9	200	2.2
Protein + CITREM (33 cmc)	500	123	517	186	0.6
Protein + SSL (29 cmc)	100	18	38	187	1.4

When whole micelles are involved in interactions with protein their influence on protein properties seems depend on the balance between the attractive forces (electrostatic between opposite charges) and the repulsive forces (electrostatic between similar charges, and excluded volume effects). These forces may act between protein and micelles, between micelles closely attached to protein, and/or between protein nanoparticles modified by micelles. For example, in the presence of added SSL (with $\Delta G_{mic} = -6\ kT$ (Il'in *et al.*, 2005)), it could be observed that molecular parameters of the original protein particles change insignificantly (see Table 6.2). That is, there is no pronounced protein cross-linking, while there is an apparent enhancement of the hydrophilicity of the micellar surface compared to the original protein surface (as indicated by the more positive value of A_2). Most likely, the presence of some additional charges on the protein arising from attached surfactant micelles prevents further aggregation of protein nanoparticles due to the greater electrostatic repulsion between them (Il'in *et al.*, 2005; Istarova *et al.*, 2005; Semenova *et al.*, 2006).

3. *Surfactant-Based Self-Assembly of Polysaccharides*

Interactions of food-grade surfactants with polysaccharides has attracted considerable recent research interest (Semenova *et al.*, 2001a; Lundqvist *et al.*, 2002a,b; Wangsakan *et al.*, 2001, 2003; Anokhina *et al.*, 2005; Grant *et al.*, 2006; Mun *et al.*, 2006). The main motivation for this activity is the fact that polysaccharides are frequently incorporated into food systems containing surfactants for the control of gravity-induced separation/creaming and rheological behaviour (Dickinson, 2003).

Maltodextrins are convenient functional ingredients for use in food emulsion formulations (Chronakis and Kasapis, 1995; Chronakis, 1997; Dokic-Baucal *et al.*, 2004; Hardas *et al.*, 2000; Hogan *et al.*, 2001; Klinkesorn *et al.*, 2004). Maltodextrins can vary enormously in terms of their average molecular size. They are classified on the basis of their dextrose equivalence (DE), which is a measure of the reducing power of the starch-derived polysaccharide/oligosaccharide as compared with that for D-glucose as calculated on a dry-weight basis (Wang and Wang, 2000). Conventionally, maltodextrins are defined as having DE < 20, in order to distinguish them from corn syrup solids with DE > 20 (Reineccius, 1991; Shahidi and Han, 1993). Maltodextrins with the same DE value may also have rather different properties depending on such factors as the hydrolysis procedure, the original source of the starch (maize, rice, potato, *etc.*), and the ratio of amylose to amylopectin (Dokic-Baucal *et al.*, 2004).

Let us consider the interaction of CITREM with potato maltodextrin of low dextrose equivalence (DE = 2), *i.e.*, with a relatively high extent of polymerization. It has been established (Anokhina *et al.*, 2007) that a pronounced self-assembly of the maltodextrin in solution is induced by increasing the molar ratio R of the anionic surfactant to glucose (the monomer unit of the maltodextrin macromolecules). This self-assembly manifests itself in increases in weight-average molar mass (Figure 6.10a) and radius of gyration (Figure 6.10b) of the surfactant-modified maltodextrin, as measured by static laser light scattering. We have found that only 3 moles of CITREM per 10^4 moles of glucose units were required to maximize the weight-average molar mass and sizes of the surfactant-modified maltodextrin associates below the cmc of the surfactant. It is inferred that this result could only be reached if the inclusion complex formation between the CITREM and the linear maltodextrin molecules induces helix formation, which is then followed by a process of helix–helix aggregation.

In fact, it is well established that the central event in intermolecular association and gelation of starch-based polymers is association of 1,4-linked α-D-glucan chains into co-axial double helixes, which then associate further by lateral aggregation. Average helix length in amylose gels appears to be around 50–70 glucose residues (Kasapis *et al.*, 1993a), and this is a chain length that the maltodextrin SA-2 (DE = 2) can apparently just provide. It would appear that such a mechanism can facilitate the involvement also of branched chains of maltodextrin in the aggregate formation (Kasapis *et al.*, 1993b). Furthermore, it has been demonstrated that small-molecule surfactants can bind to maltodextrins by inserting their non-polar tails into the interior of the helical coil formed by the polysaccharide chain (Wangsakan *et al.*, 2001, 2003). The contributions from hydrophobic interactions between the alkyl chains of the surfactant molecules attached to the maltodextrin molecules can further enhance the extent of maltodextrin self-association in the aqueous medium (Lindman *et al.*, 1993).

An increasing concentration of anionic surfactant can strengthen the electrostatic repulsive forces between the maltodextrin associates, as modified by the addition of the negatively charged head-groups of the surfactant to the neutral molecules of polysaccharide. The consequence of this effect is a reduction in the extent of maltodextrin association. In combination with the other effects, this leads to parameter dependency with a local maximum and minimum below the cmc of the surfactant (see Figures 6.10a and 6.10b).

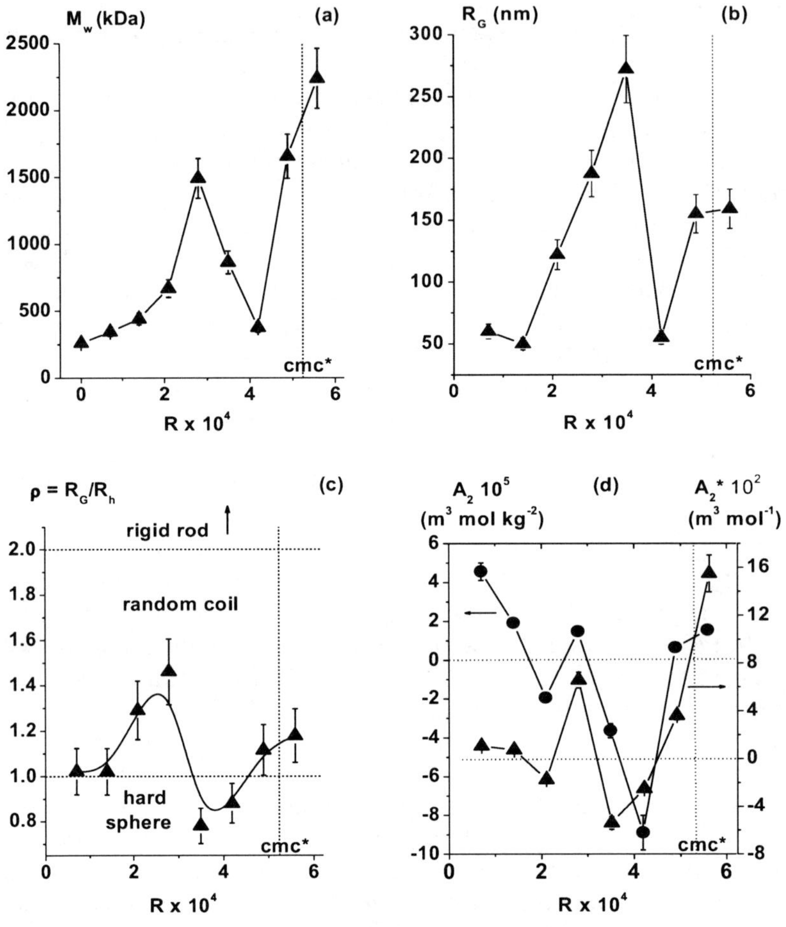

Figure 6.10 Effect of CITREM on the molecular and thermodynamic parameters of maltodextrin SA-2 (DE = 2) in aqueous medium (phosphate buffer, pH = 7.2, ionic strength = 0.05 M; 20 °C): (a) weight average molar mass, M_w; (b) radius of gyration, R_G; (c) structure sensitive parameter, ρ, characterizing the architecture of maltodextrin associates; (d) second virial coefficient, A_2 or A_2*, on the basis of the weight (●) and molal (▲) scales, respectively. The parameter R is defined as the molar ratio of surfactant to glucose monomer units in the polysaccharide. The indicated cmc* value refers to the cmc of the pure CITREM solution. Reproduced from Anokhina *et al.* (2007) with permission.

We can observe a pronounced increase in the values of M_w and R_G in the cmc region (Figures 6.10a and 6.10b). It is as if the most effective cross-linking occurs between maltodextrin molecules and highly charged CITREM micelles as a result of effective hydrogen bonding between them. The structure sensitive parameter (ρ) shows a similar dependence on the molar ratio *R* (Figure 6.10c). The behaviour is indicative of the formation of maltodextrin associates in the shape of spheres (Burchard, 1994), which are less compact in the local maximum (large aggregates) and *vice versa* in the local minimum (small aggregates).

In turn, the concentration dependence of the second virial coefficient, A_2, shows a progressive reduction in the thermodynamic affinity for the aqueous medium of the maltodextrin particles with increasing concentration of CITREM below the cmc. This result seems indicative of a steady increase in the strength of attraction between surfactant-modified maltodextrin associates, which could be attributable to the addition of hydrophobic hydrocarbon tails of the surfactant to the relatively hydrophilic surface of the original maltodextrin associates. Because of this effect, the expected repulsive contribution from the charged head-groups of the CITREM molecules attached to the polysaccharide aggregates seems not so dominant. Nevertheless, this contribution manifests itself most likely in the marked increase in the positive values of the weight and molal second virial coefficients in the close vicinity of the cmc, where a strong contribution of highly charged CITREM micelles to the intermolecular interactions is expected (Figure 6.10d).

Let us consider now the case of a specific ionic polysaccharide. The unique properties of complexes of the cationic chitosan with non-ionic sorbitan esters provides an interesting example. Grant and co-workers (2006) have established that mixtures of chitosan and surfactant form emulsion-like solutions and/or creams, where the surfactant component is present as droplets or micelle-like particles and the chitosan solution acts as the system's continuous phase. It was established that the length and the degree of saturation of the surfactant hydrocarbon chain have a significant impact on the development of the chitosan–surfactant complexes. Moreover, an optimal distance between the chitosan's protonated amine groups is required for effective interactions to occur between the polysaccharide and the sorbitan esters.

According to Groot (2000), the mechanism of interaction between a polymer and surfactant may be deduced by considering parameters such as polymer size, mode of surfactant adsorption (continuous or discrete micelles), and possible sites of interaction (head group or tail). For the case of the mechanism of the interaction between chitosan and sorbitan esters, the polymer concentration (dilute, semi-dilute, concentrated) of the formulation must also be considered, since different structural conformations occur in various concentration regimes (Grant *et al.*, 2006). In a dilute formulation, small localized chitosan–surfactant droplets that do not interact with each other are distributed along the polymer chains. Grant and co-workers (2006) postulated that formation of the primary polymer–surfactant complex could occur in two stages. First, the sorbitan esters interact with chitosan by hydrogen bonding. This interaction may occur between the amine, ammonium, and hydroxyl groups of chitosan and the hydroxyl and carbonyl groups of the sorbitan ester head-group. Secondly, the hydrophobic interactions between the sorbitan ester tails reinforce the complex structures. In the semi-dilute formulations (above the overlap concentration), the droplets can expand in size with increasing concentration of the chitosan + sorbitan ester mixtures. In this way, the localized droplets become closer together or even superimposed, leading to the formation of globalized polymer–surfactant complexes. The hydrophobic interaction between the droplets enhances the stability of the complex structures. In the concentrated state, the chitosan chains are entangled, which restricts the expansion of the droplets. Therefore, depending on the diffusivity of sorbitan ester within the chitosan matrix and the degree of chitosan–surfactant interaction, some kind of enhanced system-spanning complexes may be established. In connection with this work, Grant *et al.* (2006) have suggested that the chitosan–sorbitan monooleate cream could be used to develop more stable emulsions for applications in the food and drug-delivery industries.

There are many other examples of the potential use of mixed systems of chitosan + surfactants to be found in the recent research literature. For instance, it has been established that chitosan–surfactant interactions can be successfully used for preparing emulsions stabilized by multilayer adsorbed films (Mun *et al.*, 2006; Chuah *et al.*, 2009).

4. *Manipulation of Food Colloid Properties through Biopolymer Self-Assembly*

There is now a solid body of available knowledge to indicate that the general features of biopolymer self-assembly in bulk aqueous solutions can account for various detailed aspects of the stability, rheology and microstructure of oil-in-water emulsions (and foams) stabilized by the same kinds of biopolymers (Dickinson, 1997, 1998; Casanova and Dickinson, 1998; Dickinson *et al.*, 1997, 1998; Semenova *et al.*, 1999, 2006; van der Linden, 2006; Semenova, 2007; Ruis *et al.*, 2007). In particular, the richness of the self-assembly and surface-active properties of the milk proteins (especially caseins) makes these common food ingredients especially suitable for formulation of food colloids, as shown in the following set of examples.

The two main molecular components of bovine casein, α_{s1}-casein and β-casein, can each be used to make and stabilize fine oil-in-water emulsions (Dickinson *et al.*, 1988). The character of their self-assembly dictates, on the one hand, the thickness/extension of the protein adsorbed layers at the surface of the emulsion droplets (through M_w and R_G), and, on the other hand, the strength of the protein–protein attraction between the colloidal particles covered by the protein layers (through A_2) (Dickinson *et al.*, 1998). Based on the experimental data, it would seem that these two types of parameters act in opposition with respect to determining the stability of protein-stabilized emulsions. Large protein aggregates can provide a rather thick adsorbed protein layer around the colloidal particles, thereby conferring upon them a high level of steric stabilization. But strong protein–protein attraction, as reflected in a large negative value of A_2, can induce attraction between protein adsorbed layers on different emulsion droplets, implying enhanced flocculation and coalescence. This interpretation is consistent with experimental data (Dickinson *et al.*, 1998) showing the effect of ionic strength on molecular (M_w) and thermodynamic (A_2) parameters of α_{s1}-casein or β-casein entities in a bulk aqueous medium of pH = 5.5, and the corresponding droplet size (average diameter $d_{43} = \Sigma d^4/\Sigma d^3$) and creaming stability (% serum layer formation) of oil-in-water emulsions stabilized by these same proteins (Figures 6.11 and 6.12). In fact, within the experimental uncertainty, no discernible effect of salt on either the droplet size or the creaming kinetics was found for the β-casein emulsions (Figure 6.11b). This behaviour is attributable, on the one hand, to the formation of a thick and, consequently, sterically stable adsorbed layer on the emulsion droplets arising from the adsorption of rather large protein particles (large M_w), and, on the other hand, to the rather weak and salt-insensitive protein–protein interactions (small A_2), as shown in Figure 6.11a.

In contrast with the salt-stability of the model β-casein emulsions, the corresponding α_{s1}-casein emulsions show a pronounced increase in d_{43} accompanied by a greatly enhanced degree of serum layer separation at ionic strength above ~ 0.05 M, as illustrated in Figure 6.12b. This behaviour can be attributed to droplet flocculation in the α_{s1}-casein emulsions at ionic strength > 0.05 M, as induced by a strengthening of the protein–protein attractive interactions (an increasingly negative value of A_2) between thinning protein adsorbed layers (decreasing M_w), as is illustrated in Figure 6.12a.

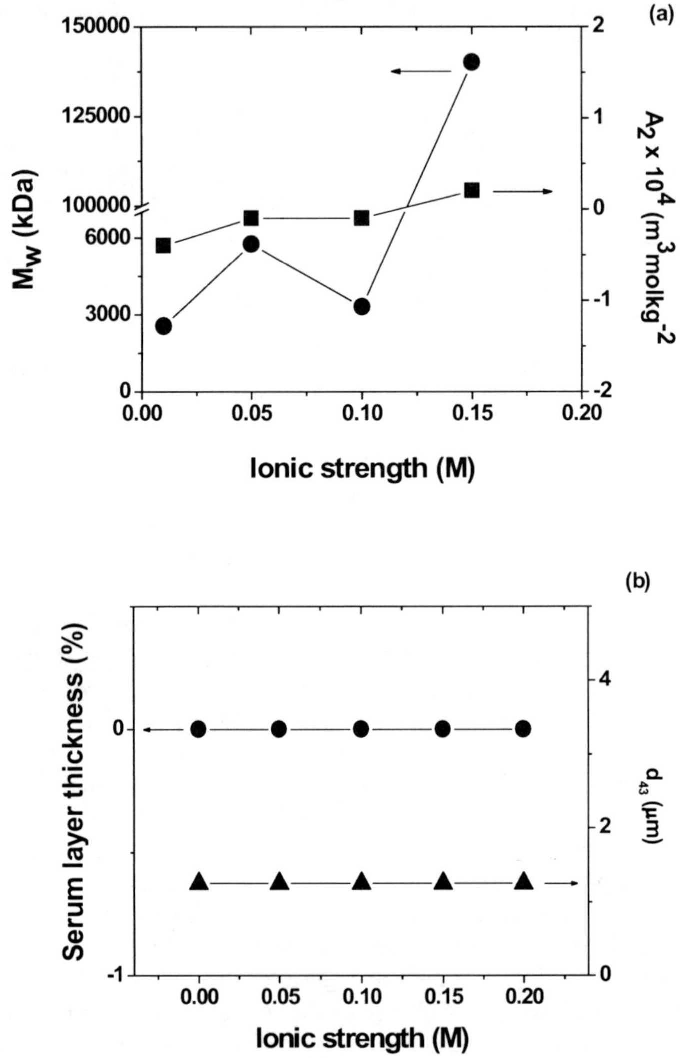

Figure 6.11 Effect of ionic strength on (a) weight-average molar weight, M_w (●), and second virial coefficient, A_2 (■), of β-casein in solution at pH = 5.5 and 22 °C, as determined by static light scattering; and (b) average droplet diameter, d_{43} (▲), and extent of gravity creaming (●) of β-casein-stabilized emulsion (11 vol% oil, 0.6 wt% protein, pH = 5.5, 22 °C).

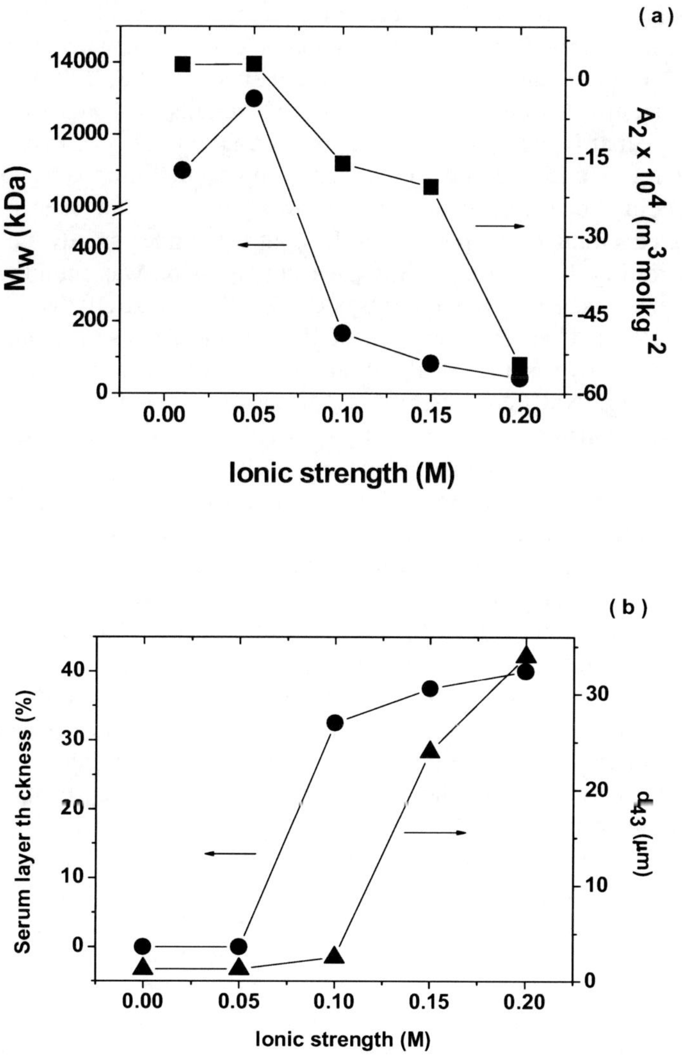

Figure 6.12 Effect of ionic strength on (a) weight-average molar weight, M_w (●), and second virial coefficient, A_2 (■), of α_{s1}-casein in solution at pH = 5.5 and 22 °C, as determined by static light scattering; and (b) average droplet diameter, d_{43} (▲), and extent of gravity creaming (●) of α_{s1}-casein-stabilized emulsion (11 vol% oil, 0.6 wt% protein, pH = 5.5, 22 °C).

It is pertinent to recall here that flocculation, by whatever mechanism, typically leads to a reduced creaming stability, except in concentrated emulsions where formation of a coherent gel-like network can actually have a positive stabilizing influence (Dickinson, 1998, 2006). Generally speaking, the flocculation of a non-dilute emulsion leads to a type of rheology that is more characteristic of a weak gel network than a viscous dispersion. Hence a large increase in complex modulus G^* (1 Hz) with change in electrolyte concentration was observed for the α_{s1}-casein-stabilized emulsion at pH = 5.5. In contrast, under steady-state viscometry testing, the corresponding β-casein emulsion was found to behave like a low-viscosity Newtonian system at all ionic strengths (Dickinson *et al.*, 1998). Consistent with these data, a mixed casein emulsion that was rich in α_{s1}-casein was also found to be highly susceptible to flocculation at high NaCl concentrations (Casanova and Dickinson, 1998). (We note that emulsions based on sodium caseinate, containing α_{s1}-casein and β-casein in roughly equal amounts, are stable to flocculation by salt.)

Another way to interpret the above observations would be in terms of the general principle that effective steric stabilization of polymer-coated droplets requires that the continuous phase be a good quality solvent for the polymeric stabilizer. Under poor quality solvent conditions (α_{s1}-casein at high ionic strength), the required entropic stabilizing repulsion of the adsorbed protein layer is converted into a destabilizing polymer-mediated attraction (Dickinson and Stainsby, 1982; Dickinson, 2006).

The different aggregation behaviour of β-casein and α_{s1}-casein with respect to increasing ionic strength of the solution (Figures 6.11a, and 6.12a) may indicate that the mechanism of α_{s1}-casein aggregation is dictated by attraction between opposite charges, while β-casein aggregation is more influenced by repulsion between like charges, as suggested by Dickinson *et al.* (1998). Anyway, these results have been found to correlate well with the irreversible sticking of individual α_{s1}-casein-stabilized emulsion droplets, as detected by means of the shear-driven formation of particle doublets at a wall using a colloidal particle scattering apparatus (Figure 3.5). In contrast, a complete lack of sticking was found for β-casein-stabilized emulsion droplets (Whittle *et al.*, 2000; Semenova *et al.*, 2001b). Obviously, if aggregation involves the entanglement and adhesion (friction) between protein layers on colliding droplets, then this can be expected to increase when there is a greater degree of attractive interaction between the adsorbed protein molecules.

It seems worthwhile to note here also, that these experimental results on α_{s1}/β-casein systems all agree semi-quantitatively with self-consistent field calculations (Leermakers *et al.*, 1996; Dickinson *et al.*, 1997; Dickinson, 1998, 1999b) of interaction potentials for a pair of plane surfaces coated with either α_{s1}-casein or β-casein under the same solution conditions. The calculations predict a stronger repulsive interaction for adsorbed β-casein layers than for α_{s1}-casein layers, due to the greater effective thickness of the β-casein 'tail' configuration as compared with the α_{s1}-casein 'loop' configuration (see Figure 6.1). Thus, at pH = 5.5, when the net charge on the β-casein molecules is low but still significant, it seems likely that it is the extended thickness of the adsorbed protein layer that mainly contributes to the (steric) repulsive forces.

Biopolymer self-assembly can promote flocculation of emulsions, not only through the nature of the interactions between different biopolymer adsorbed layers, but also via the presence of unadsorbed biopolymers in the aqueous continuous phase. A well-studied case is the flocculation by sodium caseinate nanoparticles (Dickinson and Golding, 1997; Dickinson *et al.*, 2001; Radford and Dickinson, 2004). Light scattering analysis has shown that, through the regulation of the self-assembly of sodium caseinate particles using calcium ions, a manipulation of the negative depth of the Gibbs free energy of depletion flocculation can be achieved (Dickinson *et al.*, 2001). (See Figure 3.6 and associated text in chapter 3 for more details.) Using standard colloid science theory, it could be estimated (Radford and Dickinson, 2004) what is the optimum size of non-adsorbed protein nanoparticles for the generation of depletion flocculation in caseinate-stabilized emulsions. It turns out that this optimum size[1] is the result of a compromise between (i) the nanoparticles being small enough to generate a high enough particle number density, and hence a sufficiently large osmotic pressure, and (ii) the nanoparticles being large enough to produce a substantial depletion layer thickness (eq. (3.41)).

The study of the thermodynamics of sodium caseinate self-assembly (Semenova *et al.*, 2005) allows us also to get insight into the molecular basis of the heat-induced gelation of sodium caseinate-based emulsions occurring in the presence of specific concentrations of Ca^{2+} at particular pH values (Eliot and Dickinson, 2003; Eliot *et al.*, 2003). Figure 6.13 compares light-scattering data measured for protein solutions with rheological data measured for protein-stabilized emulsions. We can observe that the most pronounced increase in the extent of caseinate aggregation with increasing temperature, in the range from 35 to 45 °C, was observed for the experimental conditions (pH = 5.1, 15 mM Ca^{2+}) corresponding

---

[1] These calculations also demonstrate the general theoretical principle, which has been confirmed in practice for various dairy-type emulsions, that the depletion interaction is of insufficient magnitude to induce flocculation when the non-adsorbed protein species are too small (*e.g.*, individual protein molecules) or too large (*e.g.*, native casein micelles).

to solid-like emulsions (category C) forming gels below 35 °C. In contrast, the least temperature-sensitive aggregation behaviour was found for conditions (pH = 6.5, 15 mM Ca^{2+}) corresponding to liquid-like emulsions remaining liquid on heating (category A). An intermediate extent of temperature sensitivity of caseinate aggregation was found at the pH values of 6.0 and 5.5 (15 mM Ca^{2+}), corresponding to liquid-like emulsions becoming converted into emulsion gels on heating to 35–43 °C (category B). Furthermore, there was found to be a systematic lowering of the aggregation temperature with decreasing pH at fixed Ca^{2+} content and with increasing Ca^{2+} content at fixed pH. This behaviour correlates well with the systematic lowering of the gelation temperature through the set of emulsion categories A → B → C (Dickinson and Casanova, 1999; Eliot and Dickinson, 2003; Eliot *et al.*, 2003; Semenova *et al.*, 2005). Moreover, different conditions of protein aggregation, in terms of the temperature reversibility (category A and B systems) and irreversibility (category C behaviour), could be established from the light scattering measurements (Figure 6.13). These same trends are seen to correlate well with corresponding degrees of thermo-reversibility / irreversibility found for the heat-induced emulsion gels.

The light-scattering data on the caseinate solutions also reinforce the tentative molecular explanation (Dickinson and Casanova, 1999; Eliot and Dickinson, 2003; Eliot *et al.*, 2003) for this heat-induced gelation phenomenon in terms of temperature-dependent changes in the interactions of adsorbed and unadsorbed caseins. That is, there is an increase in relative strength of the hydrophobic protein–protein attraction in the presence of calcium ions, both between casein-coated emulsion droplets and within casein aggregates in the bulk aqueous phase. Hence, the data in Figure 6.13 demonstrate an increase in the thermodynamic affinity between caseinate nanoparticles in the aqueous medium over the whole temperature cycle of heating and subsequent back-cooling, which is mirrored in progressively decreasing values of A_2 from positive to negative. This behaviour is most likely to be attributable to an increase in the total extent of nanoparticle surface hydrophobicity.

The high level of intrinsic hydrophobicity of caseins, together with their specific charge distributions, explains the characteristically strong tendency of the caseins towards hydrophobically-driven self-association in aqueous solution, as is well known from the extensive dairy science literature (Walstra and Jenness, 1984; Leman and Kinsella, 1989; Rollema, 1992). An additional contribution from interparticle Ca^{2+} bridging to the overall caseinate aggregation mechanism is also to be expected (Dickinson *et al.*, 1992, 2001; Horne and Leaver, 1995; Ye and Singh, 2000a; Antipova *et al.*, 2002).

(a)

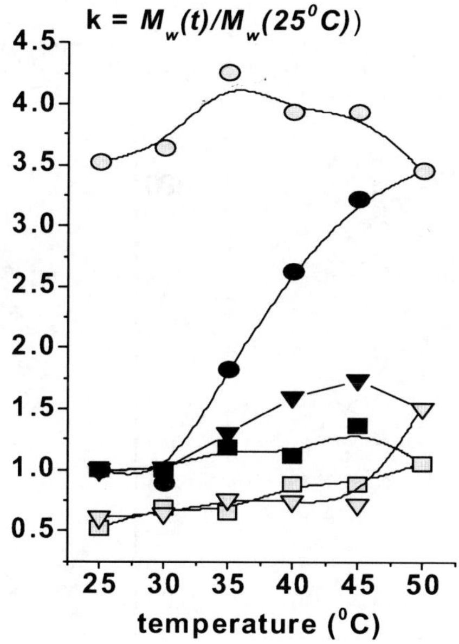
k = M_w(t)/M_w(25°C))
4.5
4.0
3.5
3.0
2.5
2.0
1.5
1.0
0.5
25 30 35 40 45 50
temperature (°C)

(b)

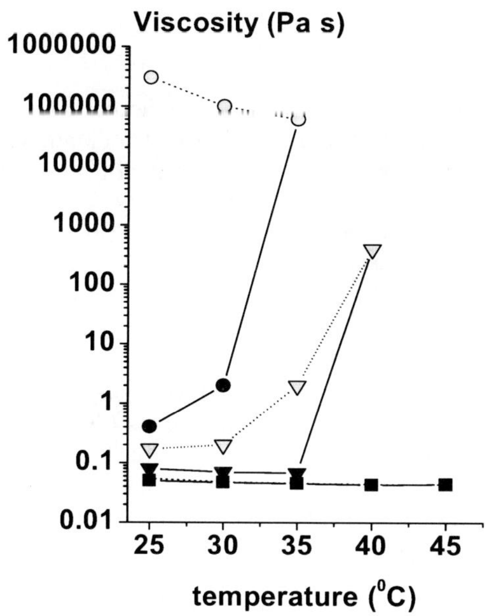
Viscosity (Pa s)
1000000
100000
10000
1000
100
10
1
0.1
0.01
25 30 35 40 45
temperature (°C)

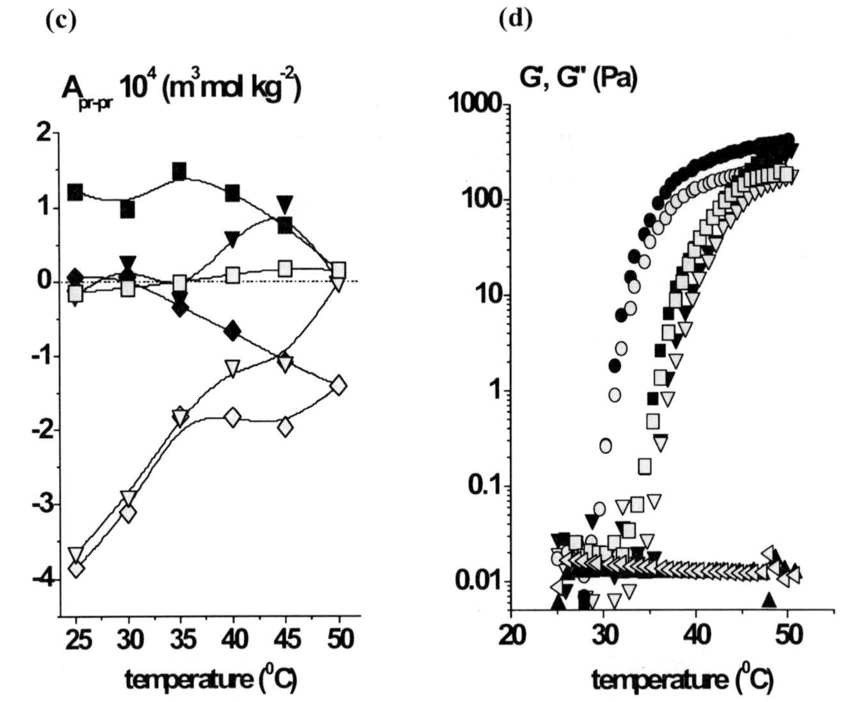

Figure 6.13 The effect of heating (filled symbols) and subsequent back-cooling (open symbols) on behaviour of systems containing sodium caseinate. (a) State of sodium caseinate aggregation in aqueous solution: (●) pH = 5.1, 15 mM Ca^{2+}; (▼) pH = 5.5, 15 mM Ca^{2+}; (■) pH = 6.5, 15 mM Ca^{2+}. (b) Steady state viscosity of caseinate-stabilized emulsions (30 vol% vegetable oil, 4 wt% protein): (●) pH = 5.1, 15 mM Ca^{2+}; (▼) pH = 5.5, 15 mM Ca^{2+}; (■) pH = 6.5, 15 mM Ca^{2+}. (c) Character of protein–protein pair interaction in aqueous solution, as described by second virial coefficient, $A_{pr\text{-}pr}$: (■) pH = 6.0, 15 mM Ca^{2+}; (▼) pH = 5.5, 15 mM Ca^{2+}; (◆) pH = 5.5, 10 mM Ca^{2+}. (d) Storage modulus G' (filled symbols) and loss modulus G'' (open symbols) of caseinate-stabilized emulsions (30 vol% vegetable oil, 4 wt% protein): (●) pH = 5.1, 15 mM Ca^{2+}; (▼) pH = 5.5, 15 mM Ca^{2+} ; (■) pH = 6.0, 15 mM Ca^{2+}; (▲) pH = 5.5, 5 mM Ca^{2+} ; (◁) pH = 6.5, 15 mM Ca^{2+}. Plots (a), (c) and (d) are reproduced from Semenova *et al.* (2005), and plot (b) from Eliot and Dickinson (2003), with permission.

The pronounced strengthening of the protein–protein attractive interactions under back-cooling is shown clearly in Figure 6.13c by the change in sign of the second virial coefficient. This feature can explain the failure of the heat-induced flocculated network of protein-coated emulsion droplets to disrupt spontaneously on cooling, *i.e.*, under the influence of Brownian motion alone. We can postulate the triggering of some changes in the interactions of the surface of caseinate nanoparticles as a result of the rearrangement of their constituent elements, involving both intramolecular and intermolecular Ca^{2+} cross-linking induced by the heating and subsequent back-cooling (Semenova *et al.*, 2005). Based on the light scattering measurements, there was inferred to be a marked transformation in size (R_G) and architecture (ρ) of the caseinate aggregates under heating and subsequent back-cooling. These molecular changes in solution would be expected to be reflected in the properties of the protein adsorbed layers (thickness/density), and consequently in the delicate balance between the strengths of repulsive interactions (steric, electrostatic) and attractive interactions (hydrophobic, polymer bridging) involving the protein-stabilized emulsion droplets. Consistent with this explanation is evidence elsewhere in the literature (Vetier *et al.*, 1997) of a reduction in average hydrodynamic diameter of aggregating primary particles of caseinate with increasing temperature. The higher particle voluminosity observed at the lower temperatures could be due to weaker hydrophobic interactions in the interior of the caseinate nanoparticles (Chen and Dickinson, 2000).

Hence, from the previously described light-scattering study of caseinate self-assembly in solution, we can postulate that heating/cooling not only alters the nature and strength of the physical (hydrophobic) interactions between emulsion droplets covered by caseinate. It most likely also transforms the nanoscale structural characteristics of the protein network in the bulk and at the interface, thereby affecting the viscoelastic and microstructural properties of the emulsions.

A further illustrative example of the role of protein self-assembly on the structure-forming ability of sodium caseinate concerns the effect of a cosolute (sucrose) on the microstructure and rheology of acid-induced sodium caseinate gels (Dickinson and Matia-Merino, 2002). As shown in Figures 6.14 and 6.15, there was found to be a marked enhancement in the shear viscoelasticity and degree of homogeneity of the acid-induced sodium caseinate gels in the presence of sucrose (Dickinson and Matia-Merino, 2002). On the basis of light scattering data, this enhancement was attributed primarily to the disaggregation of the original sodium caseinate nanoparticles at pH $\geq$ p*I* (Belyakova *et al.*, 2003) (Figure 6.2), as a result of direct hydrogen bonding between sodium caseinate and sucrose

(Semenova *et al.*, 2002). This disaggregation leads to a marked rise in the effective number density of smaller protein particles. Following further acidification of the protein solution below the p*I*, the light scattering data are consistent with a substantial degree of aggregation of sodium caseinate nanoparticles (Belyakova *et al.*, 2003) (Figure 6.2), as induced by protein dehydration caused in part by the competition between the sodium caseinate and sucrose for available water molecules (Semenova *et al.*, 2002).

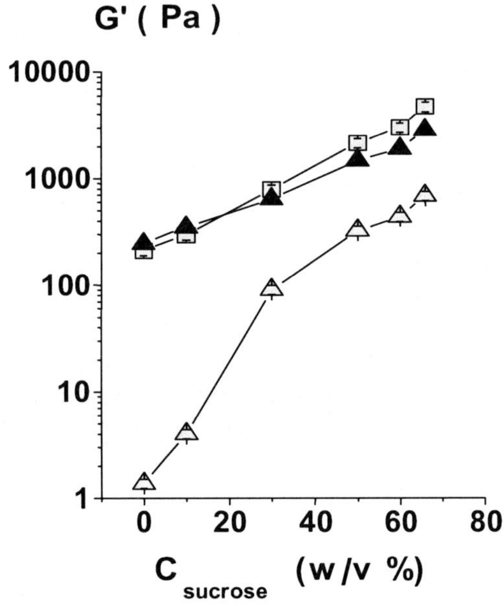

Figure 6.14 Effect of sucrose on the storage modulus *G'* at 1 Hz and 25 °C of acid-induced sodium caseinate gels (3 wt%): (Δ) pH = 5.5; (▲) pH = 5.1; (□) pH = 3.9. Reproduced from Belyakova *et al.* (2003) with permission.

The combined processes of protein self-assembly and disassembly in the presence of sucrose form the basis of the explanation of the observed enhancements in the viscoelasticity (Figure 6.14) and structural homogeneity (Figure 6.15) of acid-induced caseinate gels containing sucrose (Dickinson and Matia-Merino, 2002). That is, there is the attainment of a greater degree of interconnectivity of the gel microstructure as a result of the larger number of structurally important bonds between the protein nanoparticles. This is caused by the marked rise in the effective number density of smaller interacting protein particles, and also in the strength of the protein–protein attractive interactions below p*I* (Belyakova *et al.*, 2003). Image analysis of the two micrographs shown in Figure 6.15 has indicated (Pugnaloni *et al.*, 2005) that the mean pore size of the casein gel microstructure is 5–10 times smaller in the system containing 60 wt% sucrose than in the sugar-free system.

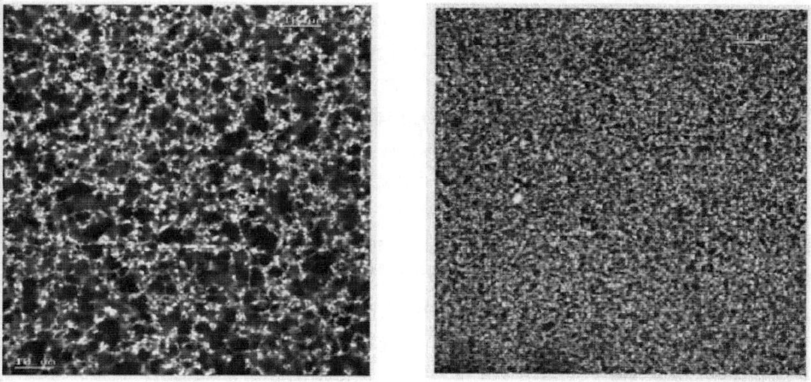

(a) **(b)**

Figure 6.15 Confocal micrographs (100 μm × 100 μm) showing the microstructure of acid-induced sodium caseinate gels (3 w/v%): (a) without any sucrose at pH = 5.2; (b) in the presence of 60 w/v% sucrose at pH = 4.9. Reproduced from Belyakova *et al.* (2003) with permission.

Let us now consider how the nature of protein–surfactant interactions can affect the surface behaviour of self-assembled protein particles. Generally speaking, we can report that good qualitative agreement has been observed between light scattering data and tensiometry data: that is, the greater is the hydrophobicity of protein particles modified by surfactant, the higher is the surface activity of their mixtures at the air–water interface and *vice versa* (Il'in *et al.*, 2004). Examples of both negative and positive effects of protein self-assembly, as induced by surfactants, are shown in Figure 6.16 for foams stabilized by complexes of sodium caseinate with micellar forms of the anionic surfactants CITREM and SSL (Semenova *et al.*, 2006). We can observe that the enhanced protein self-assembly in the case of CITREM leads to a significant decrease in the number density of nanoparticles that have the potential to adsorb on the newly formed air bubbles. As a consequence, there are not enough surface-active species present to form a stable foam. By way of contrast, in the case of SSL, for which the number density of modified nanoparticles practically does not change, while their hydrophilicity increases, and consequently so does the hydration of the protein adsorbed layer, we can observe that the foam stability is enhanced also.

In support of the possibility to manipulate foam stability by changing the nature of protein assembly in the presence of surfactant, Table 6.3 shows a correlation between molecular parameters of protein–phospholipid complexes and the visual appearance of foams stabilized by them in solutions of different pH. The data indicate that the foams stabilized by complexes of phospholipid liposomes with sodium caseinate exhibit a dramatic increase in stability as compared to the corresponding pure protein foams. (The phospholipid sample by itself did not make fine stable foams at any of the concentrations investigated).

Based on the light-scattering data, it is evident that the nature of the synergistic effect found for mixtures of phospholipids + sodium caseinate is dependent on pH. Therefore, at pH = 7.2, the synergistic effect is mainly attributable to the enhancement of the steric characteristics of the protein adsorbed layers and their hydrophilicity (Prins and van Kalsbeek, 1998; Rouimi *et al.*, 2005; Sanchez and Rodriguez Patino, 2005), as reflected in the degree of protein association (increase in M_w) and the greater thermodynamic affinity of complexes for the aqueous medium (increase in A_2), respectively (see Table 6.3). The increase in effective protein hydrophilicity can be attributed to the attachment of the polar head-groups of the phospholipids to the protein. In contrast, however, at lower pH values (6.0 and 5.1), under conditions where the original sodium caseinate is itself more associated, the attachment of the phospholipid liposomes to the protein leads to protein dissociation. This is probably caused by the repulsion between charged phospholipid head-groups attached close together on the protein. It is most dramatic at the pH value (5.1) in the closest vicinity to the protein's isoelectric point. Under these conditions, the synergistic effect in the mixture of phospholipid + sodium caseinate is caused by the increase in the number density of the modified caseinate nanoparticles (due to protein dissociation) and also by the increase in the hydrophilicity of the protein nanoparticles.

In an O/W emulsion system containing a mixture of surfactant + polysaccharide, the stability behaviour will generally depend on two sets of factors: (i) the nature of the surfactant–polysaccharide interactions at the surface of the emulsion droplets, and (ii) the surfactant–polysaccharide interactions in the aqueous medium between the droplets (Dickinson *et al.*, 1993; Dickinson, 2003; Aoki *et al.*, 2005; Klinkesorn *et al.*, 2004; Chuah *et al.*, 2009).

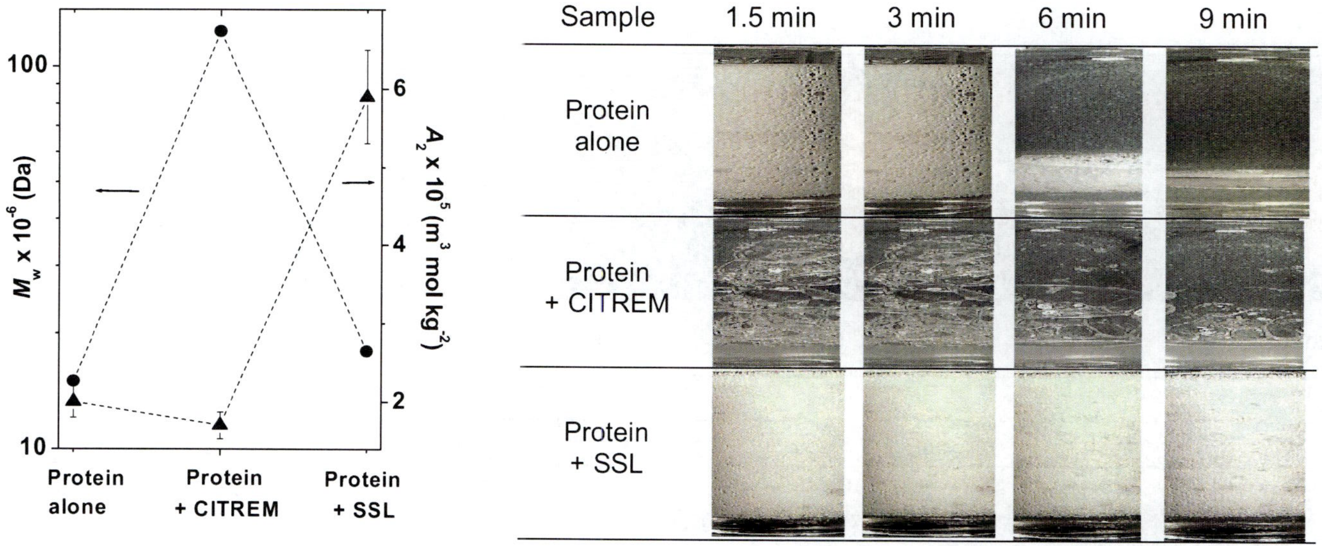

Figure 6.16 Effect on foam stability of the self-assembly of sodium caseinate (1% wt/v) induced by interactions with micelles of food-grade anionic surfactants CITREM (500 mg/L) and SSL (100 mg/L) in aqueous medium (pH = 5.5, ionic strength = 0.05 M, 20 °C). On the left is a plot of the weight-average molar mass, M_w (●), and the second virial coefficient, A_2 (▲), of the protein with and without added surfactant. On the right are images of the foams after various times. The foam samples (25 ml) had been generated through a glass membrane (1 μm) using the bubbling method. The velocity of supplied air was constant. The volume of the foaming solutions and the protein concentration were maintained constant at 5 ml and 1 wt/v%, respectively. In each of the images the volume of the glass vessel containing the foam is 10 ml.

Table 6.3 Effect of protein self-assembly, induced by interaction with lecithin, on the stability of foams stabilized by complexes of sodium caseinate (1 % w/v) with soy phospholipids Lipoid S-21 (10^{-5} M) (Istarova *et al.*, 2005; Semenova, 2007). Values of M_w and A_2 are presented for the protein with and without surfactant at three pH values. Also shown are photographs of foams recorded 9 minutes following foam preparation. In each of the images the volume of the glass vessel containing the foam is 10 ml.

	pH = 7.2		pH = 6.0		pH = 5.1	
Sample	$M_w \times 10^{-6}$ (Da)	$A_2 \times 10^5$ (m^3 mol kg^{-2})	$M_w \times 10^{-6}$ (Da)	$A_2 \times 10^5$ (m^3 mol kg^{-2})	$M_w \times 10^{-6}$ (Da)	$A_2 \times 10^5$ (m^3 mol kg^{-2})
protein	4	2.9	16	4.7	1023	8.9
complex	14	596	6	58	11	119
	protein	complex	protein	complex	protein	complex

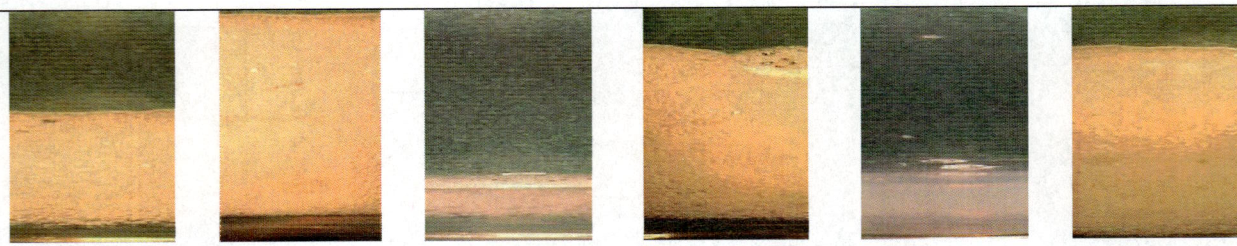

A recently reported example of the enhancement of the stability of a surfactant-based emulsion through addition of polysaccharide is provided by the case of a soybean oil-in-water emulsion stabilized by lecithin in the presence of chitosan (Chuah *et al.*, 2009). It was established that, in the presence of chitosan, there was good long-term stability of the emulsions against aggregation at acidic pH values (pH < 6) and at moderately low ionic strengths (< 0.5 M NaCl). Moreover, addition of chitosan conferred extra benefits on the emulsions in terms of thermal stability (see Figure 6.17), freeze–thaw stability and long-term shelf-life. It appears that the primary stabilizing mechanism here is the establishment of additional lecithin–chitosan interactions at the surface of the emulsion droplets, as well as the formation of a three-dimensional polysaccharide network surrounding the droplets.

We have seen earlier in this chapter how the self-assembly of casein systems is sensitively affected by temperature. Another thermodynamic variable that can affect protein–protein interactions in aqueous media is the hydrostatic pressure. Static high-pressure treatment causes the disintegration of casein micelles due to the disruption of internal hydrophobic interactions and the dissociation of colloidal calcium phosphate. This phenomenon has been used to modify the gelation ability of casein without acidification as a consequence of exposure of hydrophobic parts of the casein molecules into the aqueous medium from the interior of the native casein micelles (Dickinson, 2006). High-pressure treatment leads to a reduction in the casein concentration required for gelation under neutral conditions, especially in the presence of cosolutes such as sucrose (Abbasi and Dickinson, 2001, 2002, 2004; Keenan *et al.*, 2001).

In the case of the self-assembly of partially hydrolysed α-lactalbumin into nanotubes in the presence of various concentrations of calcium ions, the resulting protein gel may exhibit different kinds of microstructure, appearance and rheological behaviour (Graveland-Bikker *et al.*, 2004), as shown in Figures 6.18 and 6.19. Below $R = 1.5$ (calcium/α-lactalbumin molar ratio), white turbid gels are produced with structure consisting of a random arrangement of amorphous aggregates. For $1.5 < R < 6$, translucent gels are formed consisting of a fine-stranded network of tubules with large spaces between them. The fine-stranded gel structure scatters less light than the more amorphous aggregate structure. A high calcium content ($R = 10$) has a negative effect on tubule formation, leading again to amorphous turbid structures. The optimum calcium ion concentration for nanotube formation is therefore $R \approx 3$. By increasing the calcium concentration, the rate of self-assembly is enhanced, and so gelation proceeds faster (Graveland-Bikker *et al.*, 2004; Graveland-Bikker and de Kruif, 2006).

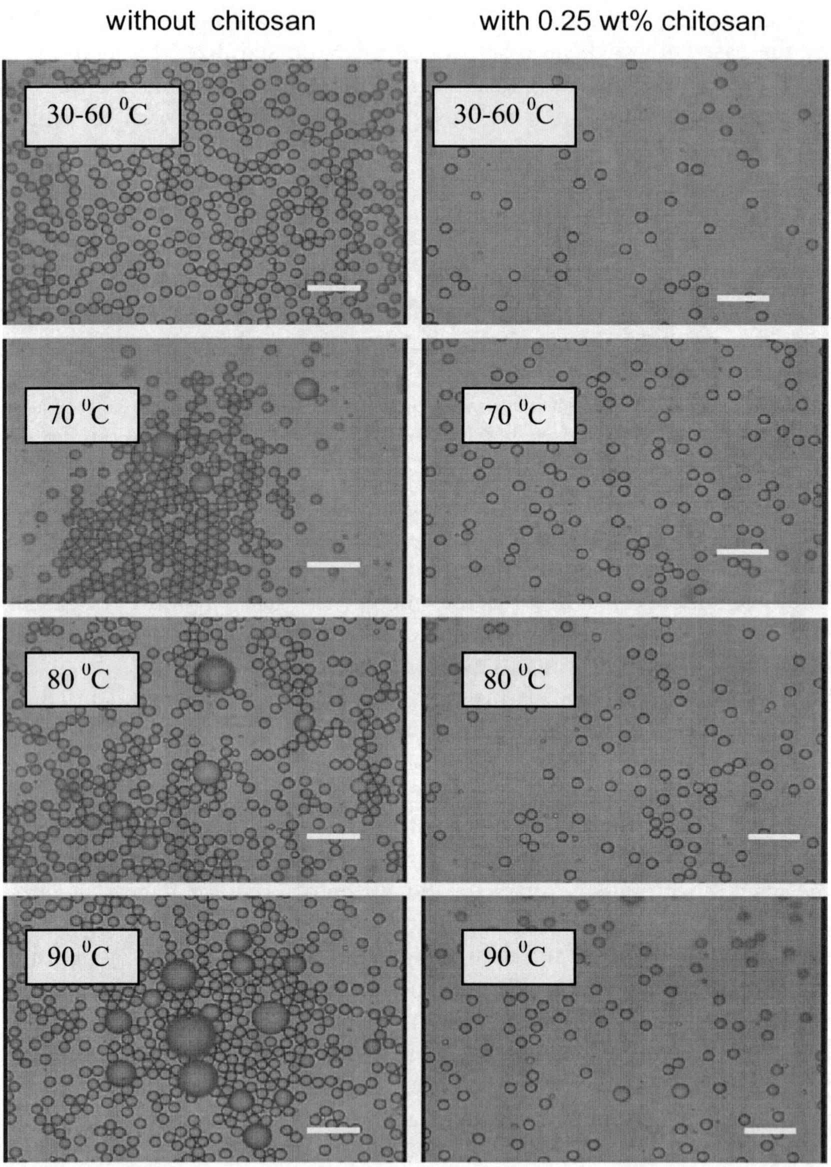

Figure 6.17 Microscopy images of 1 wt% soybean oil-in-water emulsion (0.25 wt% modified lecithin, 0 or 0.25 wt% chitosan, pH = 3.0) following heat treatment (30–90 °C, 30 min). The scale bars correspond to 100 μm. Reproduced from Chuah *et al.* (2009) with permission.

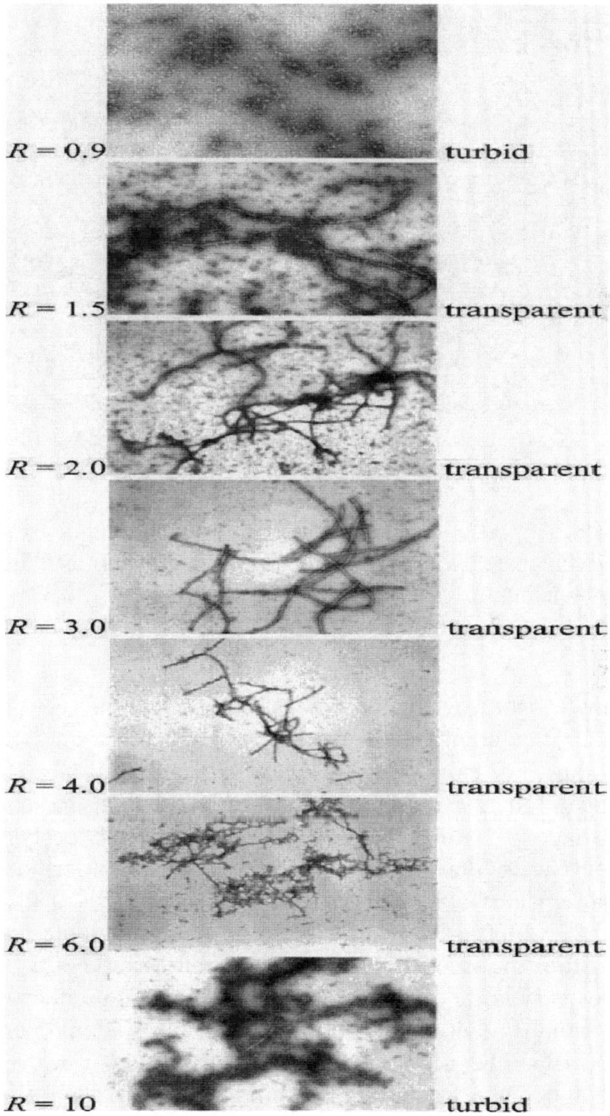

Figure 6.18 Transmission electron micrographs of protein structures from solutions of 3 wt% α-lactalbumin incubated with 4 wt% *Bacillus licheniformis* at 50 °C (75 mM Tris-HCl, pH = 7.5) at different values of the calcium/α-lactalbumin molar ratio *R*. The image dimensions are 2100 nm × 2600 nm. Reproduced from Graveland-Bikker *et al.* (2004) with permission.

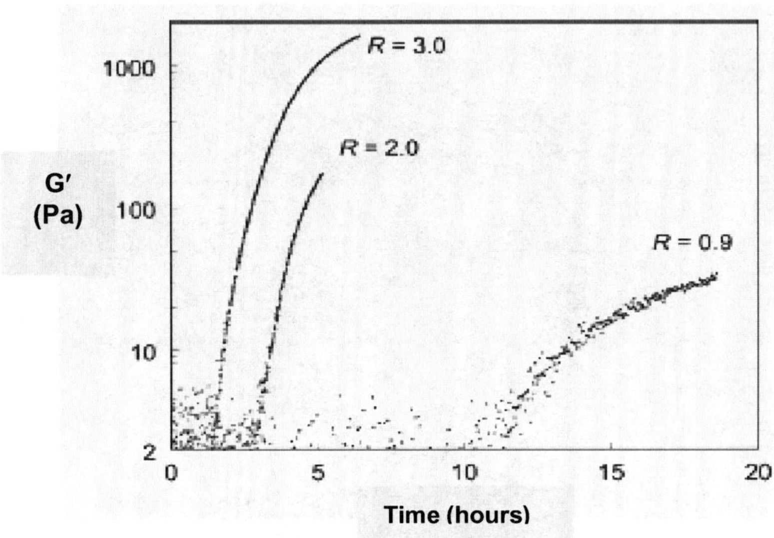

Figure 6.19 Time-dependent storage modulus *G′* of solutions of 3 wt% α-lactalbumin incubated with 4 wt% *Bacillus licheniformis* as a function of the calcium/α-lactalbumin molar ratio *R* (75 mM Tris-HCl, pH = 7.5, 50 °C). Reproduced from Graveland-Bikker *et al.* (2004) with permission.

Gels made from α-lactalbumin nanotubes have a very high rigidity (Figure 6.19), as compared with most other protein gels at equivalent concentrations. For example, at 10 wt% α-lactalbumin, the storage modulus has been found to be up to 100 times larger than that obtained for a 10 wt% β-lactoglobulin gel induced by enzymatic hydrolysis (Ipsen *et al.*, 2001). The modulus value is also about two orders of magnitude higher than that of acid gels of heated skim milk (10% total solids) (Bikker *et al.*, 2000). Despite the strength of the nanotube gel at deformations within the linear regime, it has been observed to be relatively weak under very large deformations, such as can be produced by shaking firmly by hand. In fact, it is reported that simply shaking the cuvette immediately makes the gel become completely liquid; however, the gelation could apparently be reversed within hours (Graveland-Bikker and de Kruif, 2006), which appears to be a potentially interesting property for certain applications. In fact, there are several desirable characteristics of gels formed from α-lactalbumin nanotubes which are noteworthy: (i) substantial rigidity; (ii) reversibility of the gelation; (iii) gel transparency; and (iv) structures that can be broken down in a controllable way, *e.g.*, by changing pH to acidic values. The combination of these properties is indicative of a biopolymer gelling agent with especially novel functional properties. Additionally, because of the existence of a cavity (diameter ~ 8.7 nm (Graveland-Bikker *et al.*, 2006a)) inside the α-lact-albumin nanotube, and its ability for controlled disassembly, the system has real potential as delivery vehicles for drugs, vitamins and enzymes.

Globular food proteins, such as β-lactoglobulin, ovalbumin, soy proteins and bovine serum albumin, when assembled into fibrils, can also form gels at bulk concentrations as low as 0.5 % (Veerman *et al.*, 2002, 2003a,b; van der Linden, 2006; Eiser *et al.*, 2009). Making use of the irreversible assembly into fibrils, Veerman *et al.* (2003a) have produced a β-lactoglobulin gel at a concentration as low as 0.07 % protein. The fibrils were formed at pH = 2, where the protein monomer is highly positively charged (net charge 20*e*), and then after preparation the pH was adjusted to 7–8, where the net protein charge is negative. Transmission electron micrographs showed that, upon changing the pH from 2 to 7 or 8, the fibrils remained stable, retaining their original contour and persistence length. Subsequent addition of calcium ions (from $CaCl_2$) induced gelation by screening electrostatic interactions and forming ion bridges.

Van der Linden and Sagis (2001) have suggested that, once a critical concentration for gel formation is known, one can predict, for example, the dependence of the gel elasticity on the concentration of protein. The minimum gelation concentration may be expressed in terms of the basic fibre characteristics like the stiffness and also as a function of salt concentration (Veerman, 2004; Sagis *et al.*, 2004). It is pointed out by van der Linden (2006) that there are two important factors which allow the manipulation of this minimum gelation concentration to an extremely low value. The first factor is that the nano-fibre should be robust against dilution. And the second is that the fibre should robust also with respect to other treatments, especially the adjustment of the pH. It would appear that each of these conditions is satisfied.

Another potential advantage of protein nano-fibres for novel applications is the ability to orient themselves along the direction of the applied flow (van der Linden, 2006). This preferred orientation implies that more fibres per unit of volume are required in a flow field in order to interact to the same extent as occurs at the gel point at rest. Thus, the minimum gelation concentration increases with flow rate. This implies that one might prepare, under flowing conditions, a fibre solution that does not show gelation (under flow), but which does exhibit gelation as soon as it stops flowing, *e.g.*, once the solution is poured into a container.

More that twenty years ago, Doi and Kitabatake (1989) established that the transition between transparent gels and opaque gels occurs over a relatively narrow range of salt concentrations or pH for different types of globular proteins. So, for example, a β-lactoglobulin gel formed at pH = 7 is quite transparent if < 50 mM NaCl is present, but is highly opaque in 200 mM NaCl. In the complete absence of salt, the gel is transparent at pH = 6.2 and opaque at pH = 5.8. The strong growth in turbidity with slight reduction in pH is caused by the loss of order on the nanoscale due to a weakening of the repulsive interactions between the aggregates.

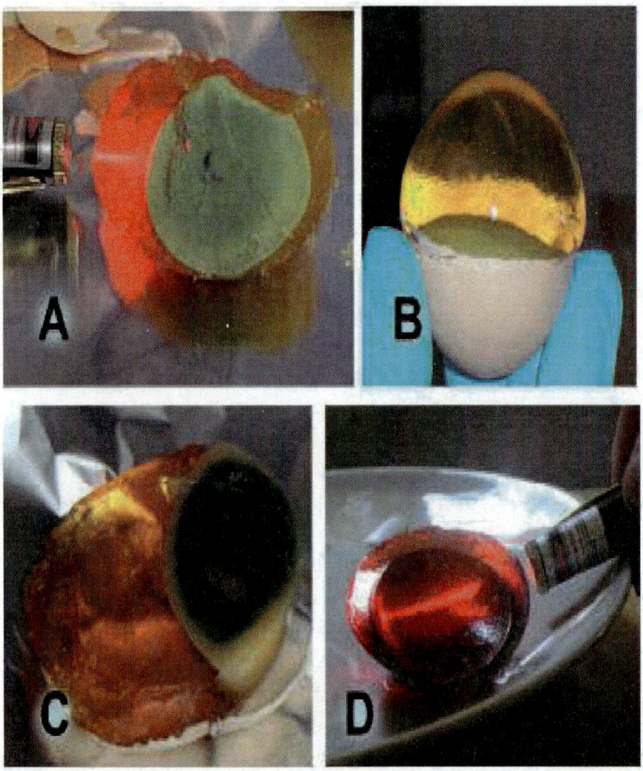

Figure 6.20 Photographs of variously prepared egg white gels. (A) A hard-boiled fresh egg pickled for 23 days in 0.9 M NaOH + 0.5 M NaCl solution. The laser light (λ = 660 nm) shows that the initial white egg gel becomes transparent. (B) A fresh chicken egg incubated in a similar NaOH–NaCl solution for 15 days. (C) Another fresh chicken egg incubated for 15 days in a 0.9 M KOH + 0.5 M KCl solution. (D) A slice of a fresh chicken egg gel prepared in the same way as before which remains transparent after boiling for 10 min in H_2O. Reproduced from Eiser *et al.* (2009) with permission.

Eiser *et al.* (2009) have recently reported on the pH-induced transformation of a hard-boiled egg from a white, brittle particulate gel to a transparent, elastic polymer gel (see Figure 6.20). Eggs were incubated in their hard protective shells for up to 26 days in a strong alkaline solution (0.9 M NaOH + 0.5 M NaCl, pH ~ 12) at room temperature. These harsh experimental conditions are apparently rather similar to those used in a traditional Chinese method developed over two thousand years ago as a way of preserving eggs so that they would remain safely edible for many months.

In attempting to explain the physico-chemical mechanism behind this egg-white gelation process, Eiser *et al.* (2009) have suggested that the gel transformation is driven by a balance between a short-range intermolecular attraction, arising from hydrophobic patches exposed on the partially denatured egg-white protein, and a long-range electrostatic repulsion caused by the high net charge on the protein under conditions of alkaline pH. Such interactions result in the formation of dense strands of particles which, at high enough particle densities, lead to a stable, transparent and elastic network.

Particulate globular protein gels tend to be rather heterogeneous in their microstructure. Durand *et al.* (2002) have shown that, when aggregates of β-lactoglobulin become very large (*i.e.*, under conditions of high ionic strength, pH close to the isoelectric point, or with added polysaccharide) they tend to phase separate, leading to the formation of protein-rich micro-domains. In such a situation the overall morphology of the protein gel is determined by competition between phase separation, on the one hand, and aggregation and gelation, on the other.

Globular protein aggregation can influence the microstructure and the stability of emulsions. The data of Ye and Singh (2000b) on the effect of addition of calcium ions on the properties of emulsions stabilized by whey protein concentrate (WPC) provides an illustrative example (see Figures 6.21 and 6.22) of how whey protein self-association can affect emulsion stability properties. The authors established that the addition of $CaCl_2$ to a WPC solution prior to emulsion formation increases dramatically the average particle size (d_{43}) and reduces the creaming stability of the stored emulsions. Also there was found to be an increase in protein surface concentration from 2 to 7 mg m^{-2} with increasing $CaCl_2$ content from 0 to 20 mM in the 3.0 wt% WPC solution.

It was suggested by Ye and Singh (2000b) that binding of Ca^{2+} induces aggregation of whey proteins in solution, resulting in a relatively lower amount of protein being available to cover the oil droplet surface. This causes a decrease in emulsifying capacity and ultimately results in the coalescence of small droplets. It was observed subsequently, however, that any further coalescence seems to be prevented by the steric repulsion associated with the larger protein particles that have already become adsorbed. In addition, it was found that a relatively high concentration of added $CaCl_2$ induces protein-based bridging flocculation between the emulsion droplets. The change in the microstructure of these emulsions prepared with different concentrations of calcium ions is illustrated in the set of images in Figure 6.21.

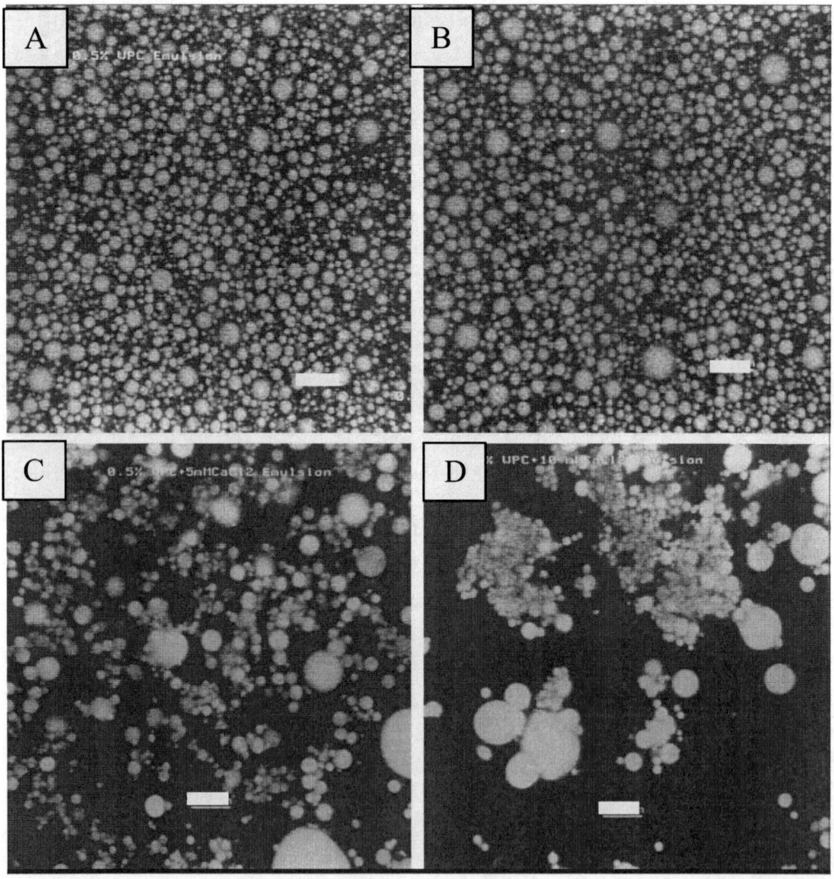

Figure 6.21 Confocal micrographs of samples of 0.5 wt% WPC emulsions containing $CaCl_2$ added prior to emulsification: A, no added $CaCl_2$; B, 3 mM; C, 5 mM; D, 10 mM. Scale bars correspond to 10 μm. Reproduced from Ye and Singh (2000b) with permission.

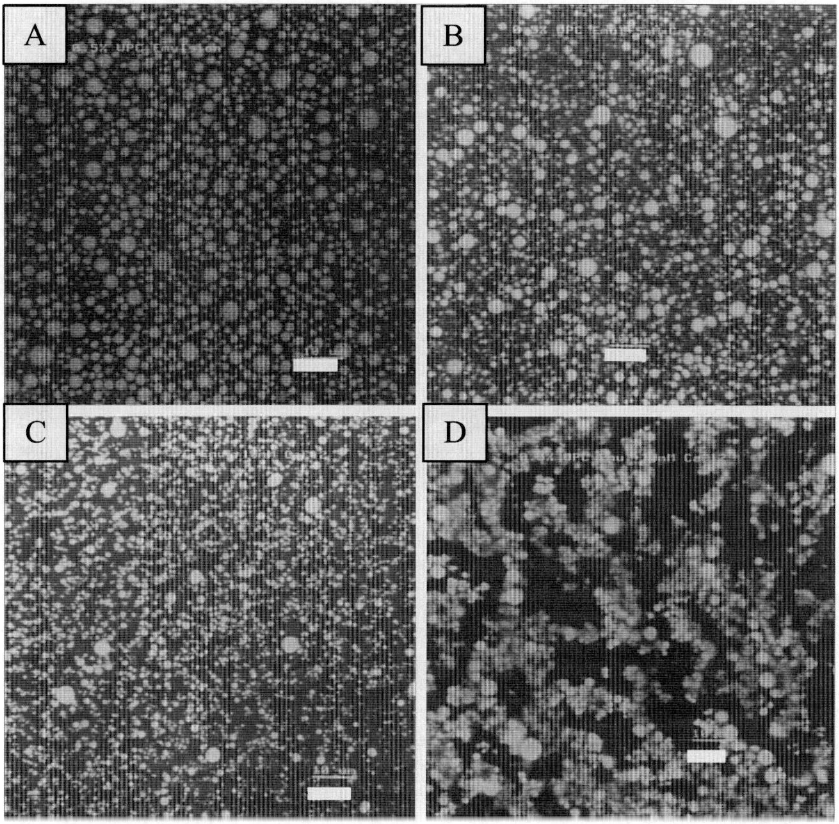

Figure 6.22 Confocal micrographs of samples of 0.5 wt% WPC emulsions with $CaCl_2$ added after emulsification: A, no added $CaCl_2$; B, 5 mM; C, 10 mM; D, 30 mM. Scale bars correspond to 10 μm. Reproduced from Ye and Singh (2000b) with permission.

In contrast to the behaviour when ionic calcium was present before emulsification, the addition of $CaCl_2$ to WPC emulsions *after* preparation was found to have no significant effect on protein surface concentration, although it did lead to aggregation of emulsion droplets and formation of a droplet network structure by the bridging mechanism (see Figure 6.22). The formation of a network structure was deemed to be responsible for inhibiting further creaming of these whey protein emulsions. A similar type of behaviour occurs with caseinate-stabilized emulsions containing added calcium ions (Dickinson and Golding, 1998, Dickinson, 2006).

Concluding Remark

On the strength of all the examples presented in this chapter, the reader should be convinced that variations in self-assembly of food biopolymers in aqueous media can have an enormous influence on food colloid stability, rheology and microstructure. It therefore seems reasonable to infer that further study and understanding of the molecular mechanisms of self-assembly and interactions of biopolymers in aqueous solution should provide increased opportunities for the creation of new classes of structured soft materials with potential application for incorporation in a wide range of new food and pharmaceutical products.

BIBLIOGRAPHY

Abbasi, S., Dickinson, E. (2001). Influence of sugars on high-pressure induced gelation of skim milk dispersions. *Food Hydrocolloids*, **15**, 315–319.

Abbasi, S., Dickinson, E. (2002). High-pressure-induced rheological changes of low-methoxyl pectin plus micellar casein mixtures. *Journal of Agricultural and Food Chemistry*, **50**, 3559–3565.

Abbasi, S., Dickinson, E. (2004). Gelation of ι-carrageenan and micellar casein mixtures under high hydrostatic pressure. *Journal of Agricultural and Food Chemistry*, **52**, 1705–1714.

Abed, M.Ah., Bohidar, H.B. (2004). Gelatin-alpha olefin sulfonate interactions studied by dynamic light scattering. *International Journal of Biological Macromolecules*, **34**, 49–54.

Agoub, A.A., Giannouli, P., Morris, E.R. (2009). Gelation of high methoxy pectin by acidification with D-glucono-δ-lactone (GDL) at room temperature. *Carbohydrate Polymers*, **75**, 269–281.

Allison, S.D., Chang, B., Randolph, T.W., Carpenter, J.F. (1999). Hydrogen bonding between sugar and protein is responsible for inhibition of dehydration-induced protein unfolding. *Archives of Biochemistry and Biophysics*, **365**, 289–298.

Anokhina, M.S., Il'in, M.M., Semenova, M.G., Belyakova, L.E., Polikarpov, Yu.N. (2005). Calorimetric investigation of the thermodynamic basics of the effect of maltodextrins on the surface activity of legumin in the presence of small-molecule surfactants. *Food Hydrocolloids*, **19**, 455–466.

Anokhina, M.S., Semenova, M.G., Belyakova, L.E., Polikarpov, Yu.N. (2007). The modification of the molecular and thermodynamic parameters of the low-DE potato maltodextrin in an aqueous medium through the interactions with anionic small-molecule surfactants. *Food Hydrocolloids*, **21**, 693–703.

Antipova, A.S., Semenova, M.G. (1997). Effect of neutral carbohydrate structure in the set glucose/sucrose/maltodextrin/dextran on protein surface activity at the air/water interface. *Food Hydrocolloids*, **11**, 71–77.

Antipova, A., Semenova, M., Gauthier-Jacques, A. (1997). Effect of neutral carbohydrate structure on protein surface activity at air–water and oil–water interfaces. In Dickinson, E., Bergenståhl, B. (Eds). *Food Colloids: Proteins, Lipids and Polysaccharides*, Cambridge, UK: Royal Society of Chemistry, pp. 245–258.

Antipova, A.S., Semenova, M.G., Belyakova, L.E. (1999). Effect of sucrose on the thermodynamic properties of ovalbumin and sodium caseinate in bulk solution and at air–water interface. *Colloids and Surfaces B: Biointerfaces*, **12**, 261–270.

Antipova, A.S., Dickinson, E., Murray, B.S., Semenova, M.G. (2002). On the effect of calcium ions on the sticking behaviour of casein-coated particles in shear flow. *Colloids and Surfaces B: Biointerfaces*, **27**, 123–131.

Aoki, T., Decker, E.A., McClements, D.J. (2005). Influence of environmental stresses on stability of O/W emulsions containing droplets stabilized by multilayered membranes produced by a layer-by-layer electrostatic deposition technique. *Food Hydrocolloids*, **19**, 209–220.

Aymard, P., Nicolai, T., Durand, D. (1999). Static and dynamic scattering of β-lactoglobulin aggregates formed after heat-induced denaturation at pH 2. *Macromolecules*, **35**, 2542–2552.

Belitz, H.D., Grosch, W. (Eds) (1982). *Lehrbuch der Lebensmittelchemie*, Heidelberg: Springer-Verlag.

Belyakova, L.E., Semenova, M.G., Antipova, A.S. (1999). Effect of small molecule surfactants on molecular parameters and thermodynamic properties of legumin in a bulk and at the air–water interface depending on a protein structure in an aqueous medium. *Colloids and Surfaces B: Biointerfaces*, **12**, 271–285.

Belyakova, L.E., Antipova, A.S., Semenova, M.G., Dickinson, E., Matia-Merino, L., Tsapkina, E.N. (2003). Effect of sucrose on molecular and interaction parameters of sodium caseinate in aqueous solution: relationship to protein gelation. *Colloids and Surfaces B: Biointerfaces*, **31**, 31–46.

Berry, G.C., Casassa, E.F. (1970). Thermodynamic and hydrodynamic behaviour of dilute polymer solutions. *Macromolecular Reviews*, **4**, 1–66.

Berry, G.P., Creamer, L.K. (1975). The association of bovine β-casein. *Biochemistry*, **14**, 3542–3545.

Bikker, J.F., Anema, S.G., Li, Y., Hill, J.P. (2000). Rheological properties of acid gels prepared from heated milk fortified with whey protein mixtures containing the A, B and C variants of β-lactoglobulin. *International Dairy Journal*, **10**, 723–732.

Borgström, J., Piculell, L., Viebke, C., Talmon, Y. (1996). On the structure of aggregated kappa-carrageenan helices: a study by cryo-TEM, optical rotation and viscometry. *International Journal of Biological Macromolecules*, **18**, 223–229.

Borgström, J., Egermayer, M., Sparrman, T., Quist, P.O., Piculell, L. (1998). Liquid crystallinity versus gelation of κ-carrageenan in mixed salts: effects of molecular weight, salt composition and ionic strength. *Langmuir*, **14**, 4935–4938.

Braudo, E.E., Plashchina, I.G., Semenova, M.G., Tolstoguzov, V.B. (1998). Structure formation in liquid solutions and gels of polysaccharides. *Food Hydrocolloids*, **12**, 253–261.

Brew, K., Grobler, J.A. (1992). α-Lactalbumin. In Fox, P.F. (Ed.). *Advanced Dairy Chemistry — Vol. 1: Proteins*, Barking: Elsevier Science.

Bryant, M.C., McClements, D.J. (1998). Molecular basis of protein functionality with special consideration of cold-set gels derived from heat-denatured whey. *Trends in Food Science and Technology*, **9**, 143–151.

Burchard, W. (1994). Light scattering. In Ross-Murphy, S.B. (Ed.). *Physical Techniques for the Study of Food Biopolymers*, Glasgow: Blackie, pp. 151–214.

Casanova, H., Dickinson, E. (1998). Influence of protein interfacial composition on salt stability of mixed casein emulsions. *Journal of Agricultural and Food Chemistry*, **46**, 72–76.

Chamberlain, A.K., MacPhee, C.E., Zurdo, J., Morozova-Roche, L.A., Allen, H., Hill, O., Dobson, C.M., Davis, J.J. (2000). Ultrastructural organization of amyloid fibrils by atomic force microscopy. *Biophysical Journal*, **79**, 3282–3293.

Chen, J., Dickinson, E. (1998) Viscoelastic properties of protein-stabilized emulsions: effect of protein–surfactant interactions. *Journal of Agricultural and Food Chemistry*, **46**, 91–97.

Chen, J., Dickinson, E. (2000). On the temperature reversibility of the viscoelasticity of acid-induced sodium caseinate emulsion gels. *International Dairy Journal*, **10**, 541–549.

Chen, A., Wu, D., Johnson, C.S. (1995). Determination of the binding isotherm and size of the bovine serum albumin–sodium dodecyl sulphate complex by diffusion-ordered 2D NMR. *Journal of Physical Chemistry*, **99**, 828–834.

Chiti, F., Dobson, C.M. (2006). Protein misfolding, functional amyloid, and human disease. *Annual Review of Biochemistry*, **75**, 333–366.

Chronakis, I.S. (1997). Structural–functional and water-holding studies of biopolymers in low fat content spreads. *Lebensmittel-Wissenschaft und-Technologie*, **30**, 36–44.

Chronakis, I.S., Kasapis, S. (1995). Preparation and analysis of water-continuous very low fat spreads. *Lebensmittel-Wissenschaft und-Technologie*, **28**, 488–494.

Chronakis, I.S., Piculell, L., Borgström, J. (1996). Rheology of kappa-carrageenan in mixtures of sodium and cesium iodide: two types of gels. *Carbohydrate Polymers*, **31**, 215–225.

Chrysina, E.D., Brew, K., Acharya, K.R. (2000). Crystal structure of apo- and holo-bovine α-lactalbumin at 2.2-Å resolution reveals an effect of calcium on inter-lobe interactions. *Journal of Biological Chemistry*, **275**, 37021–37029.

Chu, B., Zhou, Z., Wu, G., Farrell, Jr., H.M. (1995). Laser light scattering of model casein solutions: effects of high temperature. *Journal of Colloid and Interface Science*, **170**, 102–112.

Chuah, A.M., Kuroiwa, T., Kobayashi, I., Nakajima, M. (2009). Effect of chitosan on the stability and properties of modified lecithin stabilized oil-in-water monodisperse emulsion prepared by microchannel emulsification. *Food Hydrocolloids*, **23**, 600–610.

Clark, A.H., Lee-Tuffnell, C.D. (1986). In Mitchell, J.R., Ledward, D.A. (Eds). *Functional Properties of Food Macromolecules*, London: Elsevier Applied Science, pp. 203–272.

Creamer, L.K., Berry, G.P. (1975). A study of the properties of dissociated bovine casein micelles. *Journal of Dairy Research*, **42,** 169–183.

Dalgleish, D.G., Law, A.J.R. (1988). pH-Induced dissociation of casein micelles. 1. Analysis of liberated caseins. *Journal of Dairy Research*, **55**, 529–538.

Darling, D.F., Birkett, R.J. (1987). Food colloids in practice. In Dickinson, E. (Ed.). *Food Emulsions and Foams*, London: Royal Society of Chemistry, pp.1–29.

Dauphas, S., Mouhous-Riou, N., Metro, B., Mackie, A.R., Wilde, P.J., Anton, M., Riaublanc, A. (2005). The supramolecular organization of β-casein: effect on interfacial properties. *Food Hydrocolloids*, **19**, 387–393.

de Kruif, C.G. (1999). Casein micelle interactions. *International Dairy Journal*, **9**, 183–188.

de Kruif, C.G., Holt, C. (2003). Casein micelle structure, functions and interactions. In Fox, P.F., McSweeney, P.L.H. (Eds). *Advanced Dairy Chemistry, Vol. 1 — Proteins*, 3rd edn, New York: Kluwer Academic/Plenum, part A, pp. 233–276.

de Kruif, C.G., Zhulina, E.B. (1996). κ-Casein as a polyelectrolyte brush on the surface of casein micelles. *Colloids and Surfaces A: Physicochemical and Engineering Aspects*, **117**, 151–159.

de Kruif, C.G., Tuinier, R., Holt, C., Timmins, P.A., Rollema, H.S. (2002). Physicochemical study of κ- and β-casein dispersions and the effect of cross-linking by transglutaminase. *Langmuir*, **18**, 4885–4891.

Derbyshire, E., Wright, D.J., Boulter, D. (1976). Legumin and vicilin, storage proteins of legume seeds. *Phytochemistry*, **15**, 3–24.

Dickinson, E. (1997). Flocculation and competitive adsorption in a mixed polymer system: relevance to casein-stabilized emulsions. *Journal of the Chemical Society, Faraday Transactions*, **93**, 2297–2301.

Dickinson, E. (1998). Proteins at interfaces and in emulsions: stability, rheology and interactions. *Journal of the Chemical Society, Faraday Transactions*, **94**, 1657–1669.

Dickinson, E. (1999a). Caseins in emulsions: interfacial properties and interactions. *International Dairy Journal*, **9,** 305–312.

Dickinson, E. (1999b). Adsorbed protein layers at fluid interfaces: interactions, structure and surface rheology. *Colloids and Surfaces B: Biointerfaces*, **15**, 161–176.

Dickinson, E. (2003). Hydrocolloids at interfaces and the influence on the properties of dispersed systems. *Food Hydrocolloids*, **17**, 25–39.

Dickinson, E. (2006). Structure formation in casein-based gels, foams, and emulsions. *Colloids and Surfaces A: Physicochemical and Engineering Aspects*, **288**, 3–11.

Dickinson, E. (2007). Food colloids... how do interactions of ingredients control structure, stability and rheology? *Current Opinion in Colloid and Interface Science*, **12**, 155–157.

Dickinson, E., Casanova, H. (1999). A thermoreversible emulsion gel based on sodium caseinate. *Food Hydrocolloids*, **13**, 285–289.

Dickinson, E., Golding, M. (1997). Depletion flocculation of emulsions containing unadsorbed sodium caseinate. *Food Hydrocolloids*, **11**, 13–18.

Dickinson, E., Golding, M. (1998). Influence of calcium ions on creaming and rheology of emulsions containing sodium caseinate. *Colloids and Surfaces A: Physicochemical and Engineering Aspects*, **144**, 167–177.

Dickinson, E., Matia-Merino, L. (2002). Effect of sugars on the rheological properties of acid caseinate-stabilized emulsion gels. *Food Hydrocolloids*, **16**, 321–331.

Dickinson, E., Matsumura, Y. (1994). Proteins at liquid interfaces: role of the molten globule state. *Colloids and Surfaces B*: *Biointerfaces*, **3**, 1–17.

Dickinson, E., Stainsby, G. (1982). *Colloids in Food*, London: Applied Science.

Dickinson, E., Goller, M.I., Wedlock, D.J. (1993). Creaming and rheology of emulsions containing polysaccharide and non-ionic or anionic surfactants. *Colloids and Surfaces A: Physicochemical and Engineering Aspects*, **75**, 195–201.

Dickinson, E., Hunt, J.A., Horne, D.S. (1992). Calcium induced flocculation of emulsions containing adsorbed β-casein or phosvitin. *Food Hydrocolloids*, **6**, 359–370.

Dickinson, E., Pinfield, V.J., Horne, D.S., Leermakers, F.A.M. (1997). Self-consistent-field modelling of adsorbed casein: interaction between two protein-coated surfaces. *Journal of the Chemical Society, Faraday Transactions*, **93**, 1785–1790.

Dickinson, E., Semenova, M.G., Antipova, A.S. (1998). Salt stability of casein emulsions. *Food Hydrocolloids*, **12**, 227–235.

Dickinson, E., Semenova, M.G., Belyakova, L.E., Antipova, A.S., Il'in, M.M., Tsapkina, E.N., Ritzoulis, C. (2001). Analysis of light scattering data on the calcium ion sensitivity of caseinate solution thermodynamics: relationship to emulsion flocculation. *Journal of Colloid and Interface Science*, **239**, 87–97.

Dickinson, E., Whyman, R.H., Dalgleish, D.G. (1988). Colloidal properties of model oil-in-water food emulsions stabilized separately by α_{s1}-casein, β-casein and κ-casein. In Dickinson, E. (Ed.). *Food Emulsions and Foams*, London: Royal Society of Chemistry, pp. 40–51.

Doi, E., Kitabatake, N. (1989). Structure of glycinin and ovalbumin gels. *Food Hydrocolloids*, **3**, 327–337.

Dokic-Baucal, L., Dokic, P., Jakovljevic, J. (2004). Influence of different maltodextrins on properties of O/W emulsions. *Food Hydrocolloids*, **18**, 233–239.

Duan, X.Q., Hall, J.A., Nikaido, H., Quiocho, F.A. (2001). Crystal structures of the maltodextrin/maltose-binding protein complexed with reduced oligosaccharides: flexibility of tertiary structure and ligand binding. *Journal of Molecular Biology*, **306**, 1115–1126.

Durand, D., Gimel, J.Ch., Nicolai, T. (2002). Aggregation, gelation and phase separation of heat denatured globular proteins. *Physica A*, **304**, 253–265.

Eiser, E., Miles, C.S., Geerts, N., Verschuren, P., MacPhee, C.E. (2009). Molecular cooking: physical transformations in Chinese 'century' eggs. *Soft Matter*, **5**, 2725–2730.

Eliot, C., Dickinson, E. (2003). Thermoreversible gelation of caseinate stabilized emulsions at around body temperature. *International Dairy Journal*, **13**, 679–684.

Eliot, C., Radford, S.J., Dickinson E. (2003). Effect of ionic calcium on the flocculation and gelation of sodium caseinate oil-in-water emulsions. In Dickinson, E., van Vliet, T.

(Eds). *Food Colloids: Biopolymers and Materials*, Cambridge, UK: Royal Society of Chemistry, pp. 234–242.

Euston, S.R., Horne, D.S. (2005). Simulating the self-association of caseins. *Food Hydrocolloids*, **19,** 379–386.

Evans, D.F., Wennerstrom, H. (1999). *The Colloidal Domain*, New York: Wiley.

Fang, Y., Dalgleish, D. (1997). Interactions between sodium caseinate and dioleoylphosphatidylcholine on oil–water interfaces and in solution. In Dickinson, E., Bergenståhl, B. (Eds). *Food Colloids: Proteins, Lipids and Polysaccharides*, Cambridge, UK: Royal Society of Chemistry, pp. 67–76.

Farrer, D., Lips, A. (1999). On the self-assembly of sodium caseinate. *International Dairy Journal*, **9**, 281–286.

Finkelstein A.V., Ptitsyn, O.B. (2002). *Protein Physics. A Course of Lectures (Soft Condensed Matter, Complex Fluids and Biomaterials)*, California, USA: Academic Press.

Fox, P.F., Brodkorb, A. (2008). The casein micelle: historical aspects, current concepts and significance. *International Dairy Journal*, **18**, 677–684.

Gosal, W.S., Ross-Murphy, S.B. (2000). Globular protein gelation. *Current Opinion in Colloid and Interface Science*, **5**, 188–194.

Gosal, W.S., Clark, A.H., Ross-Murphy, S.B. (2004). Fibrillar β-lactoglobulin gels. Part 1. Fibril formation and structure. *Biomacromolecules*, **5**, 2408–2419.

Grant, G.T., Morris, E.R., Rees, D.A., Smith, P.J.C., Thom, D. (1973). Biological interactions between polysaccharides and divalent cations: the egg-box model. *FEBS Letters*, **32**, 195–198.

Grant, J., Cho, J., Allen, C. (2006). Self-assembly and physicochemical and rheological properties of a polysaccharide–surfactant system formed from the cationic biopolymer chitosan and nonionic sorbitan esters. *Langmuir*, **22**, 4327– 4335.

Graveland-Bikker, J. F. (2005). Self-assembly of hydrolysed α-lactalbumin into nanotubes. PhD Thesis, Utrecht University, the Netherlands.

Graveland-Bikker, J.F., de Kruif, C.G. (2006). Unique milk protein-based nanotubes: food and nanotechnology meet. *Trends in Food Science and Technology*, **17**, 196–203.

Graveland-Bikker, J.F., Ipsen, R., Otte, J., de Kruif, C.G. (2004). Influence of calcium on the self-assembly of partially hydrolyzed α-lactalbumin. *Langmuir*, **20**, 6841–6846.

Graveland-Bikker, J.F., Fritz, G., Glatter, O. (2006a). Growth and structure of α-lactalbumin nanotubes. *Journal of Applied Crystallography*, **39**, 180–184.

Graveland-Bikker, J.F., Schaap, I.A.T., Schmidt, C.F., de Kruif, C.G. (2006b). Structural and mechanical study of a self-assembling protein nanotube. *Nano Letters*, **6**, 616–621.

Groot, R.D. (2000). Mesoscopic simulation of polymer–surfactant aggregation. *Langmuir*, **16**, 7493–7502.

HadjSadok, A., Pitkowski, A., Nicolai, T., Benyahia, L., Moulai-Mostefa, N. (2008). Characterization of sodium caseinate as a function of ionic strength, pH and temperature using static and dynamic light scattering. *Food Hydrocolloids*, **22**, 1460–1466.

Hardas, N., Danviriyakul, S., Foley, J.L., Nawar, W.W., Chinachoti, P. (2000). Accelerated stability studies of microencapsulated anhydrous milk fat. *Lebensmittel-Wissenschaft und-Technologie*, **33**, 506–513.

Harkema, Ir.J. (1998). Paselli SA-2 and paselli excel. In Dalzell, J.M. (Ed.). *Ingredients Handbook: Fat Substitutes*, London: Leatherhead Food RA, pp. 103–133.

Hermansson, A.-M. (1985). Structure of soya glycinin and conglycinin gels. *Journal of the Science of Food and Agricultutre*, **36**, 822–832.

Hjerde, T., Smidsrød, O., Christensen, B.E. (1998a). Analysis of the conformational properties of κ- and ι-carrageenan by size-exclusion chromatography combined with low-angle laser light scattering. *Biopolymers*, **49**, 71–80.

Hjerde, T., Smidsrød, O., Christensen, B.E. (1998b). Acid hydrolysis of κ- and ι-carrageenan in the disordered and ordered conformations: characterization of partially hydrolyzed samples and single-stranded oligomers released from the ordered structures. *Macromolecules*, **31**, 1842–1851.

Hogan, S.A., McNamee, B. F., O'Riordan, E. D., O'Sullivan, M. (2001). Emulsification and microencapsulation properties of sodium caseinate/carbohydrate blends. *International Dairy Journal*, **11**, 137–144.

Holt, C. (1992). Structure and stability of the bovine casein micelle. In Anfinsen, C.B., Edsall, J.D., Richards, F.R., Eisenberg D.S. (Eds). *Advances in Protein Chemistry*, vol. 43, San Diego: Academic Press, pp. 63–151.

Holt, C., Horne, D.S. (1996). The hairy casein micelle: evolution of the concepts and its implication for dairy technology. *Netherlands Milk and Dairy Journal*, **50**, 85–111.

Holt, C., Sawyer, L. (1993). Caseins as rheomorphic proteins: interpretation of the primary and secondary structures of the α_{s1}-, β- and κ-caseins. *Journal of the Chemical Society, Faraday Transactions*, **89**, 2683–2692.

Holt, C., de Kruif, C.G., Tuinier, R., Timmins, P.A. (2003). Substructure of bovine casein micelles by small-angle X-ray and neutron scattering. *Colloids and Surfaces A: Physicochemical and Engineering Aspects*, **213**, 275–284.

Horne, D.S. (1998). Casein interactions: casting light on the black boxes, the structure in dairy products. *International Dairy Journal*, **8**, 171–177.

Horne, D.S. (2002). Casein structure, self-assembly and gelation. *Current Opinion in Colloid and Interface Science*, **7**, 456–461.

Horne, D.S. (2003). Casein micelles as hard spheres: limitations of the model in acidified gel formation. *Colloids and Surfaces A: Physicochemical and Engineering Aspects*, **213**, 255–263.

Horne, D.S. (2006). Casein micelle structure: models and muddles. *Current Opinion in Colloid and Interface Science*, **11**, 148–153.

Horne, D.S., Leaver, J. (1995). Milk proteins on surfaces. *Food Hydrocolloids*, **9**, 91–95.

Horne, D.S., Anema, S., Zhu, X., Nicholas, K.R., Singh, H. (2007). A lactational study of the composition and integrity of casein micelles from the milk of the tammar wallaby (*Macropus eugenii*). *Archives of Biochemistry and Biophysics*, **467**, 107–118.

Ikeda, S., Morris, V.J. (2002). Fine-stranded and particulate aggregates of heat-denatured whey proteins visualized by atomic force microscopy. *Biomacromolecules*, **3**, 382–389.

Il'in, M.M., Semenova, M.G., Belyakova, L.E., Antipova, A.S., Polikarpov, Yu.N. (2004). Thermodynamic and functional properties of legumin in the presence of small-molecule surfactants: effect of temperature and pH. *Journal of Colloid and Interface Science*, **278**, 71–80.

Il'in, M.M., Anokhina, M.S., Semenova, M.G., Belyakova, L.E., Polikarpov, Yu.N. (2005). Calorimetric study of the interactions between small-molecule surfactants and sodium caseinate with reference to the surface activity of their binary mixtures. *Food Hydrocolloids*, **19**, 441– 453.

Imamura, K., Iwai, M., Ogawa, T., Sakiyama, T., Nakanishi, K. (2001). Evaluation of hydration states of protein in freeze-dried amorphous sugar matrix. *Journal of Pharmaceutical Sciences*, **90**, 1955–1963.

Ipsen, R., Otte, J., Qvist, K.B. (2001). Molecular self-assembly of partially hydrolysed alpha-lactalbumin resulting in strong gels with a novel microstructure. *Journal of Dairy Research*, **68**, 277–286.

Istarova, T.A., Semenova, M.G., Sorokoumova, G.M., Selishcheva, A.A., Belyakova, L.E., Polikarpov, Yu.N. (2005). Effect of pH on caseinate interactions with soy phospholipids in relation to surface activity of their mixtures. *Food Hydrocolloids*, **19**, 429–440.

Jarvis, M.C., Apperley, D.C. (1995). Chain conformation in concentrated pectic gels: evidence from ^{13}C NMR. *Carbohydrate Research*, **275**, 131–145.

Jenks, W.P. (1969). *Catalysis and Chemistry in Enzymology*, New York: McGraw-Hill, part II, chap. 8.

Kanji, K., Niki, R., Urakawa, H., Hiragi, Y., Donkai, N., Nagura, M. (1988). Micellar structure of β-casein observed by small-angle X-ray scattering. *Biochimica et Biophysica Acta*, **955**, 128–134.

Karlberg, M., Thuresson, K., Lindman, B. (2005). Hydrophobically modified ethyl (hydroxyethyl)cellulose as stabilizer and emulsifying agent in macroemulsions. *Colloids and Surfaces A: Physicochemical Engineering Aspects*, **262**, 158–167.

Kasapis, S., Morris, E.R., Norton, I.T., Clark, A.H. (1993a). Phase equilibria and gelation in gelatine/maltodextrin systems. Part I. Gelation of individual components. *Carbohydrate Polymers*, **21**, 243–248.

Kasapis, S., Morris, E.R., Norton, I.T., Gidley, M.J. (1993b). Phase equilibria and gelation in gelatine/maltodextrin systems. Part II. Polymer incompatibility in solution. *Carbohydrate Polymers*, **21**, 249–259.

Kataoka, N., Kuwajima, K., Tokunaga, F., Goto, Y. (1997). Structural characterization of the molten globule of α-lactalbumin by solution X-ray scattering. *Protein Science*, **6**, 422–430.

Keenan, R.D., Young, D.J., Tier, C.M., Jones, A.D., Underdown, J. (2001). Mechanism of pressure-induced gelation of milk. *Journal of Agricultural and Food Chemistry*, **49**, 3394–3402.

Kegeles, G. (1979). A shell model for the size distribution in micelles. *Journal of Physical Chemistry*, **83**, 1728–1732.

Kegeles, G. (1992). The critical micelle condition revisited. *Indian Journal of Biochemistry and Biophysics*, **29**, 97–102.

Kelley, D., McClements, D.J. (2003). Interactions of bovine serum albumin with ionic surfactants in aqueous solutions. *Food Hydrocolloids*, **17**, 73–85.

Kentsis, A., Borden, K.L.B. (2004). Physical mechanisms and biological significance of supramolecular protein self-assembly. *Current Protein and Peptide Science*, **5**, 125–134.

Klinkesorn, U., Sophanodora, P., Chinachoti, P., McClements, D.J. (2004). Stability and rheology of corn oil-in-water emulsions containing maltodextrin. *Food Research International*, **37**, 851–859.

Koseki, T., Kitabatake, N., Doi, E. (1989). Irreversible thermal denaturation and formation of linear aggregates of ovalbumin. *Food Hydrocolloids*, **3**, 123–134.

Krog, N. (1997). Food emulsifiers and their chemical and physical properties. In Friberg, S.E., Larsson, K. (Eds). *Food Emulsions*, 3rd edn, New York: Marcel Dekker, pp. 141–187.

Lawrence, M.C., Izard, T., Beuchat, M., Blagrove, R.J., Colman, P.M. (1994). Structure of phaseolin at 2.2 Å resolution. Implication for a common vicilin/legumin structure and the genetic engineering of seed storage proteins. *Journal of Molecular Biology*, **238**, 748–776.

Leclerc, E., Calmettes, P. (1997). Interactions in micellar solutions of β-casein. *Physical Review Letters*, **78**, 150–153.

Leermakers, F.A.M., Atkinson, P.J., Dickinson, E., Horne, D.S. (1996). Self-consistent-field modeling of adsorbed β-casein: effects of pH and ionic strength on surface coverage and density profile. *Journal of Colloid and Interface Science*, **178**, 681–693.

Leman, J., Kinsella, J.E. (1989). Surface activity, film formation, and emulsifying properties of milk proteins. *Critical Reviews in Food Science and Nutrition*, **28**, 115–138.

Lindman, B., Carlsson, A., Gerdes, S., Karlstroem, G., Piculell, L., Thalberg, K., *et al.* (1993). Polysaccharide–surfactant systems: interactions, phase diagrams and novel gels. In Dickinson, E., Walstra, P. (Eds). *Food Colloids and Polymers: Structure and Dynamics*, Cambridge, UK: Royal Society of Chemistry, pp. 113–125.

López-Driez, E.C., Bone, S. (2000). An investigation of the water-binding properties of protein + sugar systems. *Physics in Medicine and Biology*, **45**, 3577–3588.

Lucey, J.A. (2002). Formation and physical properties of milk protein gels. *Journal of Dairy Science*, **85**, 281–294.

Lucey, J.A., Dick, C., Singh, H., Munro, P.A. (1997). Dissociation of colloidal calcium-phospate depleted casein particles as influenced by pH and concentration of calcium and phosphate. *Milchwissenschaft*, **52**, 603–606.

Lucey, J.A., Srinivasan, M., Singh, H., Munro, P.A. (2000). Characterization of commercial and experimental sodium caseinate by multiangle laser light scattering and size-exclusion chromatography. *Journal of Agricultural and Food Chemistry*, **48**, 1610–1616.

Lundin, L., Hermansson, A.-M. (1997). Rheology and microstructure of Ca- and Na-κ-carrageenan and locust bean gum gels. *Carbohydrate Polymers*, **34**, 365–375.

Lundqvist, H., Eliasson, A.-Ch., Olofsson, G. (2002a) Binding of hexadecyl trimethyl-ammonium bromide to starch polysaccharides. Part I. Surface tension measurements. *Carbohydrate Polymers*, **49**, 43–55.

Lundqvist, H., Eliasson, A.-Ch., Olofsson, G. (2002b) Binding of hexadecyl trimethyl-ammonium bromide to starch polysaccharides. Part II. Calorimetric study. *Carbohydrate Polymers*, **49**, 109–120.

Malhotra, A., Coupland, J.N. (2004). The effect of surfactants on the solubility, zeta potential, and viscosity of soy protein isolates. *Food Hydrocolloids*, **18**, 101–108.

Marchin, S., Putaux, J.L., Pignon, F., Leonil, J. (2007). Effects of the environmental factors on the casein micelle structure studied by cryo-transmission electron microscopy and small-angle X-ray scattering/ultra-small-angle X-ray scattering. *Journal of Chemical Physics*, **126**, 45–101.

May, C.D. (1990). Industrial pectins: sources, production and applications. *Carbohydrate Polymers*, **12**, 79–99.

May, C.D., Stainsby, G. (1986). Factors affecting pectin gelation. In Phillips, G.O., Wedlock, D.J., Williams, P.A. (Eds). *Gums and Stabilisers for the Food Industry 3*, London: Elsevier Applied Science, pp. 515–523.

McClements, D.J. (2000). Isothermal titration calorimetry study of pectin–ionic surfactant interactions. *Journal of Agricultural and Food Chemistry*, **48**, 5604–5611.

McGann, T.C.A., Fox, P.F. (1974). Physico-chemical properties of casein micelles reformed from urea-treated milk. *Journal of Dairy Research*, **41**, 45–53.

McIntire, T.M., Brant, D.A. (1997). Imaging of individual biopolymers and supramolecular assemblies using noncontact atomic force microscopy. *Biopolymers*, **42**, 133–146.

Mikheeva, L.M., Grinberg, N.V., Grinberg, V.Ya., Khokhlov, A.R., de Kruif, C.G. (2003). Thermodynamics of micellization of bovine β-casein studied by high-sensitivity differential scanning calorimetry. *Langmuir*, **19**, 2913–2921.

Modler, H.W. (1985). Functional properties of nonfat dairy ingredients — a review. Modification of products containing casein. *Journal of Dairy Science*, **68**, 2195–2205.

Mora-Gutierrez, A., Farrell, Jr., H.M. (2000). Sugar–casein interaction in deuterated solutions of bovine and caprine casein as determined by oxygen-17 and carbon-13 nuclear magnetic resonance: a case of preferential interactions. *Journal of Agricultural and Food Chemistry*, **48**, 3245–3255.

Mora-Gutierrez, A., Kumosinski, T.F., Farrell, Jr., H.M. (1997). Oxygen-17 nuclear magnetic resonance studies of bovine and caprine casein hydration and activity in deuterated sugar solutions. *Journal of Agricultural and Food Chemistry*, **45**, 4545–4553.

Morris, E.R., Powell, D.A., Gidley, M.J., Rees, D.A. (1982). Conformations and interactions of pectins. I. Polymorphism between gel and solid states of calcium polygalacturonate. *Journal of Molecular Biology*, **155**, 507–516.

Morris, E.R., Rees, D.A., Thom, D., Boyd, J. (1978). Chiroptical and stoichiometric evidence of a specific, primary dimerization process in alginate gelation. *Carbohydrate Research*, **66**, 145–154.

Müller-Buschbaum, P., Gebhardt, R., Roth, S.V., Metwalli, E., Doster, W. (2007). Effect of calcium concentration on the structure of casein micelles in thin films. *Biophysical Journal*, **93**, 960–968.

Mun, S., Decker, E.A., McClements, D.J. (2006). Effect of molecular weight and degree of deacetylation of chitosan on the formation of oil-in-water emulsions stabilized by surfactant–chitosan membranes. *Journal of Colloid and Interface Science*, **296**, 581–590.

Nagasawa, M., Takahashi, A. (1972). Light scattering from polyelectrolyte solutions. In Huglin, M.B. (Ed.). *Light Scattering from Polymer Solutions*, London: Academic Press, pp. 671–723.

Nakai, S., Ho, L., Tung, M.A. (1980). Solubilization of rapeseed, soy and sunflower protein isolates by surfactants and proteinase treatments. *Canadian Institute of Food Science and Technology Journal*, **13**, 14–22.

Nash, W., Pinder, D.N., Hemar, Y., Singh, H. (2002). Dynamic light scattering investigation of sodium caseinate and xanthan mixtures. *International Journal of Biological Macromolecules*, **30**, 269–271.

Nicolai, T. (2007). Structure of self-assembled globular proteins. In Dickinson, E., Leser, M.E. (Eds). *Food Colloids: Self-assembly and Material Science*, Cambridge, UK: Royal Society of Chemistry, pp. 35–56.

Nyland, L.R., Maughan, D.W. (2000). Morphology and transverse stiffness of Drosophila myofibrils measured by atomic force microscopy. *Biophysical Journal*, **78**, 1490–1497.

Oosawa, F., Asakura, S. (1975). *Thermodynamics of the Polymerization of Proteins*, London: Academic Press.

Panyam, D., Kilara, A. (1996). Enhancing the functionality of food proteins by enzymatic modification. *Trends in Food Science and Technology*, **7**, 120–125.

Payens, T.A.J., Heremans, K. (1969). Effect of pressure on the temperature-dependent association of β-casein. *Biopolymers*, **8**, 335–345.

Payens, T.A.J., Vreeman, H. (1982). Casein micelles and micelles of β- and κ-casein. In Mittal, K.L., Fendler, E.J. (Eds). *Solution Behaviour of Surfactants*, New York: Plenum, vol. 1, pp. 543–571.

Payens, T.A.J., Brinkhuis, J.A., van Markwijk, B.W. (1969). Self-association in nonideal systems. Combined light scattering and sedimentation measurements in β-casein solutions. *Biochimica et Biophysica Acta*, **175**, 434–437.

Piculell, L. (1998). Gelling polysaccharides. *Current Opinion in Colloid and Interface Science*, **3**, 643–650.

Pignon, F., Belina, G., Narayanan, T., Paubel, X., Magnin, A., Gezan-Guiziou, G. (2004). Structure and rheological behaviour of casein micelle suspensions during ultrafiltration process. *Journal of Chemical Physics*, **121**, 8138–8146.

Pouzot, M., Nicolai, T., Visschers, R.W., Weijers, M. (2005). X-ray and light scattering study of the structure of large protein aggregates at neutral pH. *Food Hydrocolloids*, **19**, 231–238.

Prins, A., van Kalsbeek, H.K.A.I. (1998). Foaming behaviour and dynamic surface properties of liquids. *Current Opinion in Colloid and Interface Science*, **3**, 639–642.

Pugnaloni, L.A., Matia-Merino, L., Dickinson, E. (2005). Microstructure of acid-induced caseinate gels containing sucrose: quantification from confocal microscopy and image analysis. *Colloids and Surfaces B: Biointerfaces*, **42**, 211–217.

Radford, S.J., Dickinson, E. (2004). Depletion flocculation of caseinate-stabilized emulsions: what is the optimum size of the non-adsorbed protein nano-particles? *Colloids and Surfaces A: Physicochemical and Engineering Aspects*, **238**, 71–81 .

Rajagopal, K., Schneider, J.P. (2004). Self-assembling peptides and proteins for nanotechnological applications. *Current Opinion in Structural Biology*, **14**, 480–486.

Redfield, C., Schulman, B.A., Milhollen, M.A., Kim, P.S., Dobson, C.M. (1999). Alpha-lactalbumin forms compact molten globule in the absence of disulfide bonds. *Nature Structural and Molecular Biology*, **10**, 948–952.

Reineccius, G.A. (1991). Carbohydrates for flavor encapsulation. *Food Technology*, **45**(3), 144–149.

Rolin, C. (1993). Pectin. In Whistler, R.L., BeMiller, J.N. (Eds). *Industrial Gums: Polysaccharides and their Derivatives*, 3rd edn, San Diego: Academic Press, pp. 257–293.

Rollema, H.S. (1992). Casein association and micelle formation. In Fox, P.F. (Ed.). *Advanced Dairy Chemistry, Vol. 1: Proteins*, London: Elsevier Applied Science, pp. 111–140.

Rouimi, S., Schorsch, C., Valentini, C., Vaslin, S. (2005). Foam stability and interfacial properties of milk protein–surfactant systems. *Food Hydrocolloids*, **19**, 467–478.

Ruis, H.G.M., Venema, P., van der Linden, E. (2007). Relation between pH-induced stickiness and gelation behaviour of sodium caseinate aggregates as determined by light scattering and rheology. *Food Hydrocolloids*, **21**, 545–554.

Sagis, L.M.C., Veerman, C., van der Linden, E. (2004). Mesoscopic properties of semi-flexible amyloid fibrils. *Langmuir*, **20**, 924–927.

Salvetat, J.P., Bonard, J. M., Thomson, N.H., Kulik, A.J., Forro, L., Benoit, W., Zuppiroli, L. (1999). Mechanical properties of carbon nanotubes. *Applied Physics A*, **69**, 255–260.

Sanchez, C.C., Rodriguez Patino, J.M. (2005). Interfacial, foaming and emulsifying characteristics of sodium caseinate as influenced by protein concentration in solution. *Food Hydrocolloids*, **19**, 407–416.

Schmidt, D.G. (1982). Association of caseins and casein micelle structure. In Fox, P.F. (Ed.). *Developments in Dairy Chemistry*. London: Applied Science, pp. 61–86.

Schmidt, D.G., Payens, T.A. (1972). The evaluation of positive and negative contributions to the second virial coefficient of some milk proteins. *Journal of Colloid and Interface Science*, **39**, 655–662.

Semenova, M.G. (2007). Thermodynamic analysis of the impact of molecular interactions on the functionality of food biopolymers in solution and in colloidal systems. *Food Hydrocolloids*, **21**, 23–45.

Semenova, M.G., Plashchina, I.G., Braudo, E.E., Tolstoguzov, V.B. (1988). Structure formation in sodium κ-carrageenan solutions. *Carbohydrate Polymers*, **9**, 133–145.

Semenova, M.G., Antipova, A.S., Belyakova, L.E., Dickinson, E., Brown, R., Pelan, E., Norton, I. (1999). Effect of pectinate on properties of oil-in-water emulsions stabilized by α_{s1}-casein and β-casein. In Dickinson, E., Rodriguez Patino, J.M. (Eds). *Food Emulsions and Foams: Interfaces, Interactions and Stability*. Cambridge, UK: Royal Society of Chemistry, pp. 163–175.

Semenova, M.G., Myasoedova, M.S., Antipova, A.S. (2001a). Effect of starch components and derivatives on the surface behaviour of a mixture of protein and small-molecule surfactants. In Dickinson, E., Miller, R. (Eds). *Food Colloids: Fundamentals of Formulation*. Cambridge, UK: Royal Society of Chemistry, pp. 233–241.

Semenova, M.G., Chen, J., Dickinson, E., Murray, B.S., Whittle, M. (2001b). Sticking of protein-coated particles in a shear field. *Colloids and Surfaces B: Biointerfaces*, **22**, 237–244.

Semenova, M.G., Antipova, A.S., Belyakova, L.E. (2002). Food protein interactions in sugar solutions. *Current Opinion in Colloid and Interface Science*, **7**, 438–444.

Semenova, M.G., Il'in, M.M., Belyakova, L.E., Antipova, A.S. (2003). Protein + small-molecule surfactant mixtures: thermodynamics of interactions and functionality. In Dickinson, E., van Vliet, T. (Eds). *Food Colloids, Biopolymers and Materials*, Cambridge, UK: Royal Society of Chemistry, pp. 377–387.

Semenova, M.G., Belyakova, L.E., Dickinson, E., Eliot, C., Polikarpov, Yu.N. (2005). Caseinate interactions in solution and in emulsions: effect of temperature, pH and calcium ions. In Dickinson, E. (Ed.). *Food Colloids: Interactions, Microstructure and Processing*. Cambridge, UK: Royal Society of Chemistry, pp. 209–217.

Semenova, M.G., Belyakova, L.E., Polikarpov, Yu.N., Il'in, M.M., Istarova, T.A., Anokhina, M.S., Tsapkina, E.N. (2006). Thermodynamic analysis of the impact of the surfactant–protein interactions on the molecular parameters and surface behaviour of food proteins. *Biomacromolecules*, **7**, 101–113.

Semenova, M.G., Belyakova, L.E., Polikarpov, Yu.N., Stankovic, I., Antipova, A.S., Anokhina, M.S. (2007). Analysis of light scattering data on the sodium caseinate assembly as a response to the interactions with likely charged anionic surfactant. *Food Hydrocolloids*, **21**, 704–715.

Semenova, M.G., Belyakova, L.E., Polikarpov, Yu.N., Antipova, A.S., Anokhina, M.S. (2008). Utilization of sodium caseinate nanoparticles as molecular nanocontainers for delivery of bioactive lipids to food systems: relationship to the retention and controlled release of phospholipids in the simulated digestion conditions. In Williams, P.A., Phillips, G.O. (Eds). *Gums and Stabilisers for the Food Industry 14*, Cambridge, UK: Royal Society of Chemistry, pp. 326–333.

Semenova, M.G., Belyakova, L.E., Polikarpov, Yu.N., Antipova, A.S., Dickinson, E. (2009). Light scattering study of sodium caseinate + dextran sulfate in aqueous solution: relationship to emulsion stability. *Food Hydrocolloids*, **23**, 629–639.

Semo, E., Kesselman, E., Danino, D., Livney, Y.D. (2007). Casein micelle as a natural nano-capsular vehicle for nutraceuticals. *Food Hydrocolloids*, **21**, 936–942.

Shahidi, F., Han, X.Q. (1993). Encapsulation of food ingredients. *Critical Reviews in Food Science and Nutrition*, **33**, 501–547.

Swaisgood, H.E. (2003). Chemistry of the caseins. In Fox, P.F., McSweeney, P.L.H. (Eds). *Advanced Dairy Chemistry, Vol. 1 — Proteins*, 3rd edn, New York: Kluwer Academic/Plenum, part A, pp. 139–202.

Tanford, C. (1961). *Physical Chemistry of Macromolecules*, New York: Wiley.

Thongngam, M., McClements, D.J. (2005). Influence of pH, ionic strength, and temperature on self-association and interactions of sodium dodecyl sulfate in the absence and presence of chitosan. *Langmuir*, **21**, 79–86.

Thurn, A., Burchard, W., Niki, R. (1987a). Structure of casein micelles. I. Small-angle neutron scattering and light scattering from β- and κ-casein. *Colloid and Polymer Science*, **265**, 653–666.

Thurn, A., Burchard, W., Niki, R. (1987b). Structure of casein micelles. II. α_{s1}-Casein. *Colloid and Polymer Science*, **265**, 897–902.

Tsoga, A., Richardson, R.K., Morris, E.R. (2004a). Role of cosolutes in gelation of high-methoxy pectin. Part 1. Comparison of sugars and polyols. *Food Hydrocolloids*, **18**, 907–919.

Tsoga, A., Richardson, R.K., Morris, E.R. (2004b). Role of cosolutes in gelation of high-methoxy pectin. Part 2. Anomalous behaviour of fructose: calorimetric evidence of site-binding. *Food Hydrocolloids*, **18**, 921–932.

Tzannis, S.T., Prestrelski, S.J. (1999). Moisture effects on protein–excipient interactions in spray-dried powders: nature of destabilizing effects of sucrose. *Journal of Pharmaceutical Sciences*, **88**, 360–370.

Uricanu, V.I., Duits, M.H.G., Mellema, J. (2004). Hierarchical networks of casein proteins: an elasticity study based on atomic force microscopy. *Langmuir*, **20**, 5079–5090.

van der Linden, E. (2006). Innovations with protein nano-fibres. *World's Poultry Science Journal*, **62**, 439–442.

van der Linden, E., Sagis, L.M.C. (2001) Isotropic-force percolation in protein gels. *Langmuir*, **17**, 5821–5824.

Vasilescu, M., Angelescu, D., Almgren, M., Valstar, A. (1999). Interactions of globular proteins with surfactants studied with fluorescence probe methods. *Langmuir*, **15**, 2635–2643.

Veerman, C. (2004) Properties of fibrillar protein assemblies and their percolating networks. PhD Thesis, Wageningen University, the Netherlands.

Veerman, C., Ruis., H., Sagis, L.M.C., van der Linden E. (2002). Effect of electrostatic interactions on the percolation concentration of fibrillar β-lactoglobulin gels. *Biomacromolecules*, **3**, 869–873.

Veerman, C., Sagis, L.M. C., van der Linden, E. (2003a). Gels at extremely low weight fractions formed by irreversible self-assembly of proteins. *Macromolecular Bioscience*, **3**, 243–247.

Veerman, C., Sagis, L.M.C., Heck, J., van der Linden, E. (2003b). Mesostructure of fibrillar bovine serum albumin gels. *International Journal of Biological Macromolecules*, **31**, 139–146.

Vetier, N., Desobry-Banon, S., Ould Eleya, M.M., Hardy, J. (1997). Effect of temperature and acidification rate on the fractal dimension of acidified casein aggregates. *Journal of Dairy Science*, **80**, 3161–3166.

Vreeman, H.J., Brinkhuis, J.A., van der Spek, C.A. (1981). Some association properties of bovine SH-κ-casein. *Biophysical Chemistry*, **14**, 185–193.

Walstra P., Jenness, R. (1984). *Dairy Chemistry and Physics*, New York: Wiley.

Wang, J., Wang, L. (2000). Structures and properties of commercial maltodextrins from corn, potato, and rice starches. *Starch/Starke*, **52**, 296–304.

Wangsakan, A., Chinachoti, P., McClements, D.J. (2001). Maltodextrin–anionic surfactant interactions: isothermal titration calorimetry and surface tension study. *Journal of Agricultural and Food Chemistry*, **49**, 5039–5045.

Wangsakan, A., Chinachoti, P., McClements, D.J. (2003). Effect of different dextrose equivalent of maltodextrin on the interactions with anionic surfactant in an isothermal titration calorimetry study. *Journal of Agricultural and Food Chemistry*, **51**, 7810–7814.

Weijers., M., Sagis., L.M.C., Veerman., C., Sperber, B., van der Linden, E. (2002a). Rheology and structure of ovalbumin gels at low pH and low ionic strength. *Food Hydrocolloids*, **16**, 269–276.

Weijers, M., Nicolai, T., Visschers, R.W. (2002b). Light scattering study of heat-induced aggregation and gelation of ovalbumin. *Macromolecules*, **35**, 4753–4762.

Whittle, M., Murray, B.S., Chen, J., Dickinson, E. (2000). Simulation and experiments on colloidal particle capture in a shear field. *Langmuir*, **16**, 9784–9791.

Ye, A., Singh, H. (2000a). Interfacial composition and stability of sodium caseinate emulsions as influenced by calcium ions. *Food Hydrocolloids*, **15**, 195–207.

Ye, A., Singh, H. (2000b). Influence of calcium chloride addition on the properties of emulsions stabilized by whey protein concentrate. *Food Hydrocolloids*, **14**, 337–346.

CHAPTER SEVEN

EFFECTS OF INTERACTIONS BETWEEN DIFFERENT BIOPOLYMERS ON THE PROPERTIES OF FOOD COLLOIDS

Nowadays it is well established that the interactions between different macromolecular ingredients (*i.e.*, protein + protein, polysaccharide + polysaccharide, and protein + polysaccharide) are of great importance in determining the texture and shelf-life of multicomponent food colloids. These interactions affect the structure-forming properties of biopolymers in the bulk and at interfaces: thermodynamic activity, self-assembly, surface loading, thermodynamic compatibility / incompatibility, phase separation, complexation and rheological behaviour. Therefore, one may infer that a knowledge of the key physico-chemical features of such biopolymer–biopolymer interactions, and their impact on stability properties of food colloids, is essential in order to be able to understand and predict the functional properties of mixed biopolymers in product formulations.

Interactions between different kinds of food biopolymers may occur via physical bonding such as van der Waals, electrostatic, hydrophobic, hydrogen bonding and excluded volume effects, or by chemical bonding as in the case of Maillard-type protein–polysaccharide conjugates. The strength and character (net attractive or net repulsive) of the biopolymer–biopolymer interactions may also vary substantially, depending on the environmental conditions (pH, ionic strength and temperature).

A. IMPACT OF PHYSICAL INTERACTIONS BETWEEN BIOPOLYMERS ON STRUCTURE AND STABILITY OF COLLOIDAL SYSTEMS

1. *Thermodynamically Unfavourable Interactions between Biopolymers in the Bulk*

1.1. *The Nature of Thermodynamically Unfavourable Interactions*

Thermodynamically unfavourable interactions are ubiquitous in mixed biopolymer systems. As described in chapters 3 and 5, they arise mainly from excluded volume effects — the physical volume of one biopolymer molecule is inaccessible to other biopolymer molecules — and also from electrostatic repulsion between like-charged groups on different molecules (Ogston, 1970; Nagasawa and Takahashi, 1972; Tanford, 1961).

We can compare experimentally measured values of the cross second virial coefficient (A_{ij}^{*}) with theoretical ones ($A_{ij}^{*\mathrm{exc}}$) based on the contribution from the excluded volume, as calculated from the size and conformational features of the interacting biopolymers (see Table 5.1). The analysis of this comparison allows us to postulate that any increase in the strength of the thermodynamically unfavourable interactions between biopolymers can be largely attributable to four main factors (Semenova, 1996; Semenova and Savilova, 1998):

(i) an increase in molecular size;
(ii) a lowering of the degree of accessibility of the space occupied by one of the biopolymers for the others due to enhanced biopolymer flexibility;
(iii) a growth in the stiffness of the interacting biopolymer molecules leading to an increase in their size and consequently their excluded volume; and
(iv) an increase in the like net charge on the interacting biopolymers.

1.2. *The Phenomenon of Segregative Phase Separation — 'Simple Coacervation'*

The presence of thermodynamically unfavourable interactions between biopolymers (thermodynamic incompatibility) ($A_{ij} > 0$) generally leads to segregative phase separation in an aqueous medium containing increasing concentrations of the mixed biopolymers (Bungenberg de Jong, 1949; Albertsson, 1971; Tolstoguzov, 1991, 2000a,b, 2003; Piculell *et al.*, 1995; Grinberg and Tolstoguzov, 1997; Polyakov *et al.*, 1997; Semenova, 1996, 2007; Semenova *et al.*, 1990, 1999a; Semenova and Savilova, 1998; Wasserman *et al.*, 1997). The system typically separates into two distinct solvent-rich phases, each containing predominantly one of the two biopolymer components. Instead of two completely separated bulk phases, one often observes a water-in-water emulsion system (see Figures 7.2 and 7.3). The peculiarities of such an emulsion are the co-solubility of the two biopolymers in the coexisting phases and the very low interfacial tension (10^{-5}–10^{-6} N m^{-1}) (Ding *et al.*, 2002; Guido *et al.*, 2002; van Puyvelde *et al.*, 2002). The temperature-independent interfacial tension increases with the total concentration of polymers in the aqueous medium (in the absence of polymer gelation). In other words, the tension increases with the length of the tie-lines (Ding *et al.*, 2002).

The phenomenon was called 'simple coacervation' by Bungenberg de Jong (1949) in order to distinguish it from 'complex coacervation' where both polymers are concentrated in the same solvent-depleted phase. The phenomenon of 'simple coacervation' in aqueous food biopolymer systems has attracted considerable interest for many years. This is because of the perception of the potential of these phase-separated biopolymer solutions for the manufacture of zero/low-fat food products, and also because of the strong influence of the microstructure of such mixtures on the texture and the flavour release properties of these 'structured water' products (Norton and Frith, 2001).

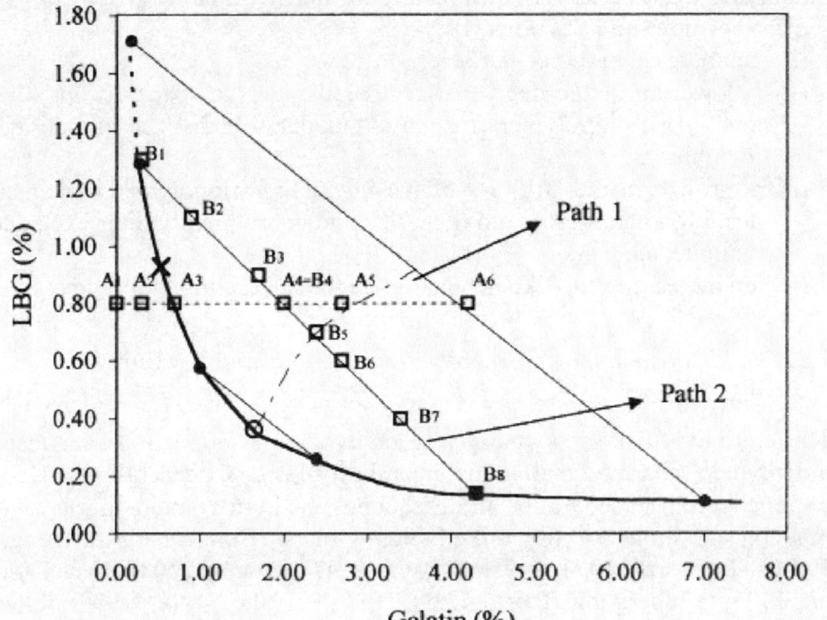

Figure 7.1 Phase diagram of the water + gelatin + locust bean gum (LBG) system at pH = 5.0, ionic strength = 0.002 M and T = 40 °C: (—) binodal; (•—•) tie lines; (- — -) rectilinear diameter; (•) composition of coexisting phases; (○) critical point; (×) threshold point. The various composition points (A_m, B_n) relate to the sets of images presented in Figures 7.2 and 7.3. Reproduced from Alves *et al.* (2001) with permission.

Nowadays it is established that confocal microscopy observation can be a more sensitive method to assess the phase state of mixed biopolymer systems than the traditional centrifugation or viscometric methods (Alves *et al.*, 1999, 2001; Vega *et al.*, 2005). Indeed, microscopy can demonstrate that a system may be already phase-separated at compositions well below the apparent binodal line (as determined by these other methods). The report of Alves *et al.* (2001) demonstrates the relationship between specific compositional points in the phase diagram (Figure 7.1) and the observed microstructure (Figures 7.2 and 7.3) for water + gelatin + locust bean gum (LBG). The white areas in Figures 7.2 and 7.3 correspond to gelatin-rich regions, and the dark areas correspond to regions rich in locust bean gum. Thus, as the gelatin content is increased along 'Path 1' (A_3, A_5 and A_6) in the phase diagram (Figure 7.1), the system goes from a water-in-water emulsion of gelatin droplets in a LBG continuous phase (Figure 7.2a) to a system where the gelatin phase becomes the continuous phase containing irregular inclusions of LBG (Figure 7.2c). At point A_5, located close to the phase inversion point, a bicontinuous structure composed of gelatin-rich and LBG-rich regions is clearly shown (Figure 7.2b). Along 'Path 2' in the phase diagram, the same kind of evolution is observed, with the system moving from the ultrastructure of an emulsion containing fewer and fewer spherical gelatin inclusions (B_3, Figure 7.3a; B_4, Figure 7.3b) to that of a gelatin-rich continuous phase containing LBG inclusions which are better and better defined (B_6, Figure 7.3d; B_7, Figure 7.3e). In between, a bicontinuous microstructure is observed (B_5, Figure 7.3c).

It has been demonstrated in many experimental studies (Grinberg and Tolstoguzov, 1997; Polyakov *et al.*, 1997; Tolstoguzov, 1991, 2000a,b, 2003; Closs *et al.*, 1999; Wang and Qvist, 2000; Alves *et al.*, 2001; Vega *et al.*, 2005) that simple coacervation is a ubiquitous phenomenon in non-dilute solutions of protein + polysaccharide, protein + protein or polysaccharide + polysaccharide at moderate or high ionic strength. As many as eighty different biopolymer + biopolymer + water systems were investigated by Tolstoguzov and his team of co-workers (Tolstoguzov, 1986, 1991, 2000a,b), and the vast majority of these were observed to exhibit phase separation due to thermodynamic incompatibility between the biopolymers.

Theoretical treatments of simple coacervation based on the statistical thermodynamics of polymer solutions have been set out by Scott (1949), Tanford (1961), Zeman and Patterson (1972), and Hsu and Prausnitz (1974). These treatments have shown that the main molecular factors affecting the phenomenon are:

(i) the values of the molar weights of the individual biopolymers;
(ii) the weight/molar ratio of the biopolymers in the mixed solution;
(ii) the strength of the unlike interactions between the biopolymers; and
(iv) the difference in thermodynamic affinity of the biopolymers for the solvent — described in the language of Flory–Huggins theory as the $\Delta\chi$-effect, where $\Delta\chi = \chi_{1i} - \chi_{1j}$. (Note that it is significant that the definition of $\Delta\chi$ is formally similar to $\Delta A_2 = A_{ii} - A_{jj}$, involving the second virial coefficients (see equation (3.31)).

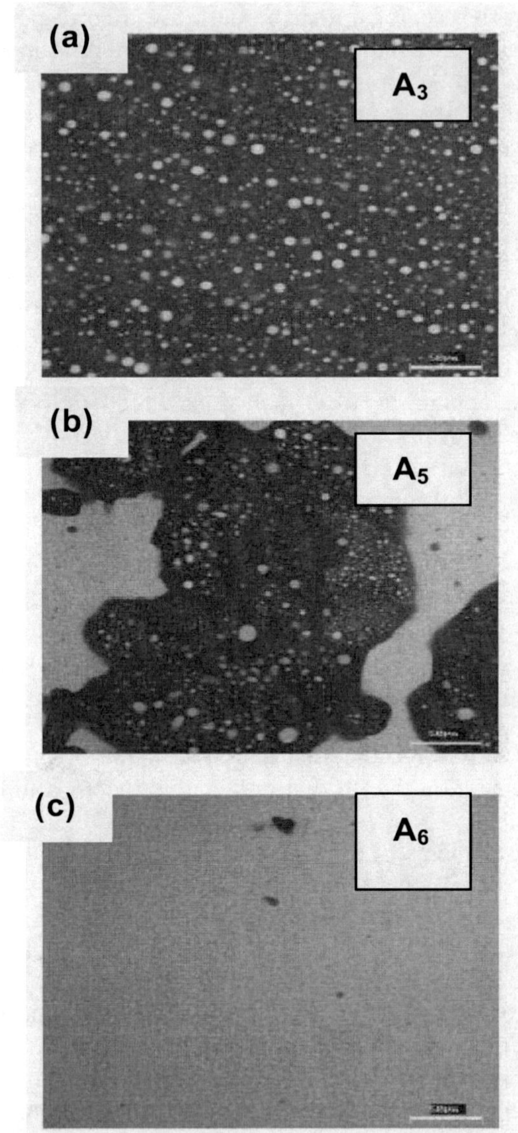

Figure 7.2 Evolution of the microstructure in the water + gelatin + LBG system (Figure 7.1) for compositions of 0.8 % LBG and increasing gelatin concentration, at pH = 5.0, ionic strength = 0.002 M and T = 40 °C. 'Path 1': (a) point A_3; (b) point A_5; (c) point A_6. Each scale bar represents 50 μm. Reproduced from Alves *et al.* (2001) with permission.

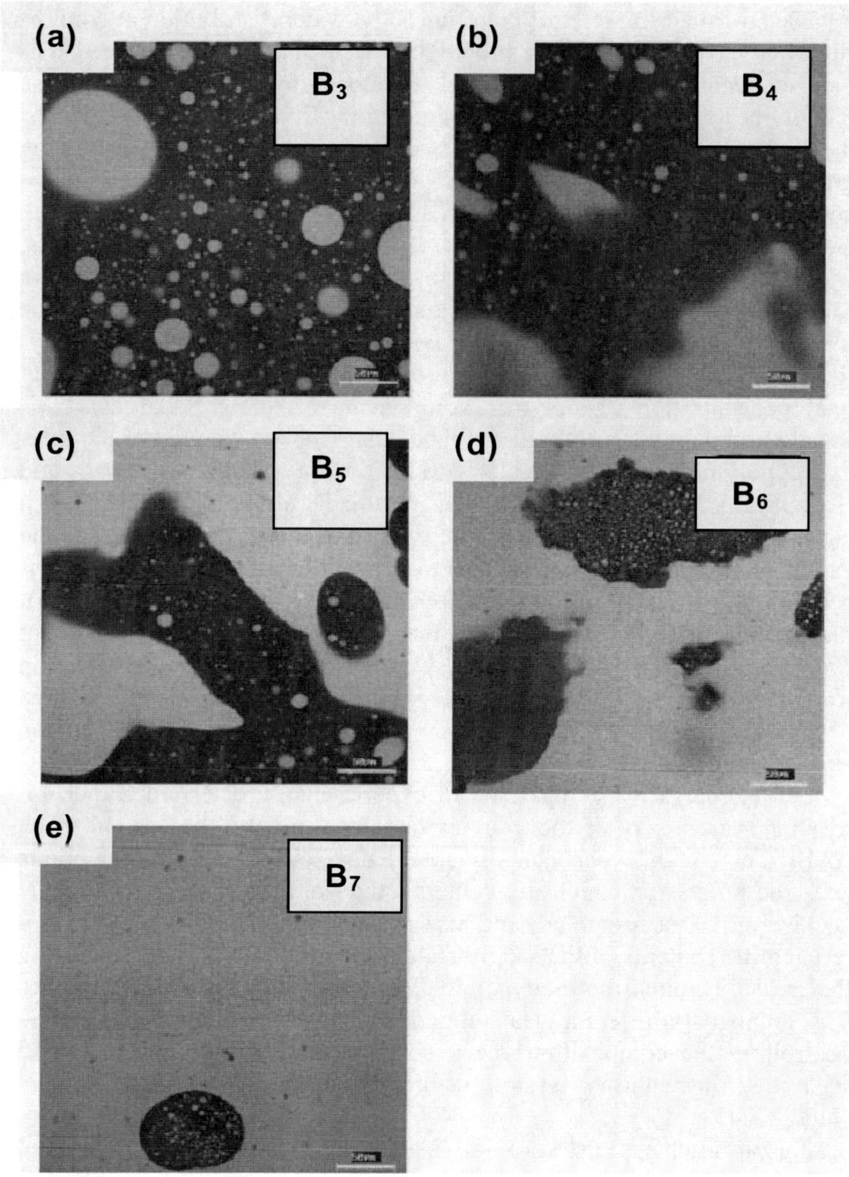

Figure 7.3 Evolution of the microstructure in the water + gelatin + LBG system (Figure 7.1) for compositions located along a tie-line, at pH = 5.0, ionic strength = 0.002 M and T = 40 °C. 'Path 2': (a) point B_3; (b) point B_4; (c) point B_5; (d) point B_6; (e) point B_7. Each scale bar represents 50 μm. Reproduced from Alves *et al.* (2001) with permission.

The so-called '$\Delta\chi$-effect' determines the water distribution at equilibrium between the coexisting biopolymer-rich phases. The phase that is more concentrated in polymer, and therefore the one that contains the biopolymer with the lower thermodynamic affinity for the solvent, has the higher density; so this phase becomes the lower phase following gravitational separation towards the bulk equilibrium state. For a mixed aqueous system of protein + polysaccharide, it is the protein component, being generally the less hydrophilic of the two biopolymer components, which tends to concentrate in the lower aqueous phase at equilibrium. Moreover, the $\Delta\chi$-effect results in a shift of the critical point along the binodal towards the concentration axis of the less hydrophilic biopolymer. The threshold of phase separation, corresponding to the minimum total concentration of polymer components on the binodal curve, is mainly defined by the ratio of the molecular (molar) weights of the mixing biopolymers (Albertsson, 1971; Semenova *et al.*, 1990; Quiroga and Bergenståhl, 2008) (see, as examples, Figure 7.1 above, and Figure 3.3 in chapter 3 for more details). The theory also predicts that changes in the values of χ_{1i} and χ_{1j} (the $\Delta\chi$-effect) may result in an increase in the concentration range of immiscibility, even coming to dominate the effects of the polymer$_i$–polymer$_j$ interaction; that is, phase separation may occur even for $\chi_{1j} \leq 0$. It should be noted, however, that this latter situation leads to a reshaping of the phase diagram, so that the immiscibility area becomes encircled into a closed loop (Zeman and Patterson, 1972; Hsu and Prausnitz, 1974).

Let us consider a comparison of experimentally determined thermodynamic parameters of the pair interactions (namely, the second virial coefficients) with experimentally found positions of the critical points (*i.e.*, the points for which the contents and volumes of the steady-state coexisting phases are equal) and also the binodals (*i.e.*, the lines characterizing the contents of the steady-state coexisting phases, and separating the areas of limited thermodynamic compatibility and phase separation). This comparison offers a clear view of the basic thermodynamic factors controlling the composition space corresponding to segregative phase separation in the ternary system of biopolymer$_i$ + biopolymer$_j$ + solvent (Figure 7.4).

For purposes of illustration in what follows, we consider the cases of various specific biopolymer mixtures demonstrating the roles of different thermodynamic parameters in determining the tendency towards phase separation. The deciding role of a greater positive value of A_{ij}^* is indicated for mixtures in which there is a small difference in thermodynamic affinities of the two biopolymers for the solvent, $\Delta A_2 = |A_{ii}^* - A_{jj}^*|$. A pair of mixed systems with similar values of $|\Delta A_2|$ but different values of A_{ij}^* are mixtures (3) and (1) in Figure 7.4:

mixture (3): legumin + dextran (Tsapkina *et al.*, 1992)
$|\Delta A_2| = 0.15 \times 10^5$ cm^3/mol
$A_{ij}^* = 0.31 \times 10^5$ cm^3/mol

mixture (1): legumin + ovalbumin (Wasserman *et al.*, 1997)
$|\Delta A_2| = 0.15 \times 10^5$ cm^3/mol
$A_{ij}^* = 0.07 \times 10^5$ cm^3/mol

Another pair of systems of this type are mixtures (5) and (6):

mixture (5): α_{s1}-casein + pectin (Semenova *et al.*, 1999a)
$|\Delta A_2| = 5.2 \times 10^5$ cm^3/mol
$A_{ij}^* = 0.0008 \times 10^5$ cm^3/mol

mixture (6): β-casein + pectin (Semenova *et al.*, 1999a)
$|\Delta A_2| = 5.2 \times 10^5$ cm^3/mol
$A_{ij}^* = 0.11 \times 10^5$ cm^3/mol

In contrast, the determining role of a greater value of $|\Delta A_2|$ makes itself evident when there are practically equal values of A_{ij}^*, as for mixtures (4) and (3) in Figure 7.4:

mixture (4): legumin + pectin (Semenova *et al.*, 1990)
$A_{ij}^* = 0.27 \times 10^5$ cm^3/mol
$|\Delta A_2| = 1.63 \times 10^5$ cm^3/mol

mixture (3). legumin + dextran (Tsapkina *et al.*, 1992)
$A_{ij}^* = 0.31 \times 10^5$ cm^3/mol
$|\Delta A_2| = 0.15 \times 10^5$ cm^3/mol

Furthermore, the governing role of the tendency of biopolymers towards self-aggregation ($A_{ii}^* < 0$) in aqueous medium seems to manifest itself in the greater area of phase separation for mixture (2) (pH = 7.0) as compared with mixture (1) (the same biopolymers at pH = 7.8), despite lower values for both A_{ij}^* and $|\Delta A_2|$ at pH = 7.0 (Wasserman *et al.*, 1997):

mixture (2): legumin + ovalbumin (pH = 7.0)
$A_{ii}^* = -0.04 \times 10^5$ cm^3/mol, $A_{jj}^* = 0.007 \times 10^5$ cm^3/mol
$|\Delta A_2| = 0.047 \times 10^5$ cm^3/mol, $A_{ij}^* = 0.007 \times 10^5$ cm^3/mol

mixture (1): legumin + ovalbumin (pH = 7.8)
$A_{ii}^* = 0.17 \times 10^5$ cm^3/mol, $A_{jj}^* = 0.02 \times 10^5$ cm^3/mol
$|\Delta A_2| = 0.15 \times 10^5$ cm^3/mol, $A_{ij}^* = 0.07 \times 10^5$ cm^3/mol

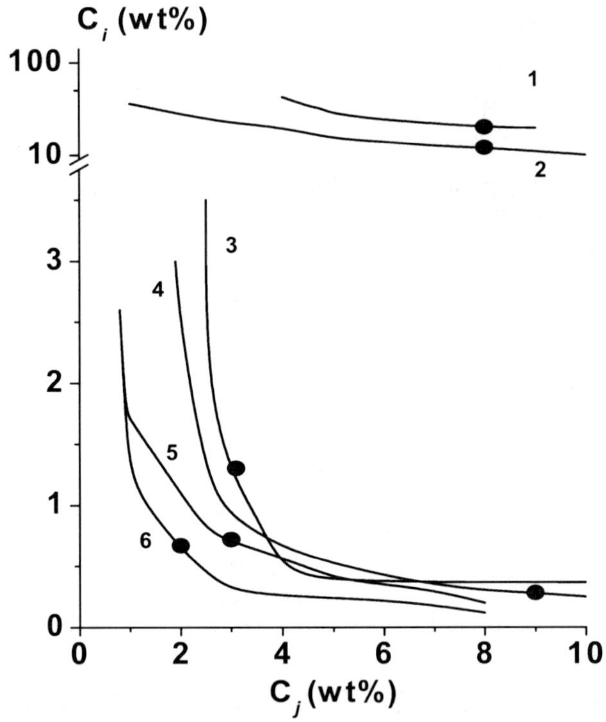

Figure 7.4 Phase behaviour of mixture of biopolymer i + biopolymer j at concentrations of C_i and C_j, respectively. Binodals and critical points (●) are shown for the following biopolymer mixtures in aqueous media: (1) legumin + ovalbumin (pH = 7.8, I = 0.1 M) (Wasserman *et al.*, 1997); (2) legumin + ovalbumin (pH = 7.0, I = 0.1 M) (Wasserman *et al.*, 1997); (3) legumin + dextran (pH = 7.8, I = 0.1 M) (Tsapkina *et al.*, 1992); (4) legumin + pectin (DE = 58%) (pH = 7.8, I = 0.3 M) (Semenova *et al.*, 1990); (5) α_{s1}-casein + pectin (DE = 76%) (pH = 7.0, I = 0.01 M) (Semenova *et al.*, 1999a); (6) β-casein + pectin (DE = 76%) (pH = 7.0, I = 0.01 M) (Semenova *et al.*, 1999a). Reproduced from Semenova (2007) with permission.

Biopolymers are, of course, polyelectrolytes. This means that electrostatic repulsion between them, as well as the contribution of counterions to the total free energy of the system, are to be included amongst the key factors affecting the character of the biopolymer interactions, and hence the stability of mixed biopolymer solutions with respect to phase separation (Antipova and Semenova, 1997; Grinberg and Tolstoguzov, 1997; Polyakov *et al.*, 1997; Semenova, 1996; Wasserman *et al.*, 1997). For example, the degree to which the net electrical charge on a protein can become involved in Coulombic repulsion depends on two considerations: (i) how far its isoelectric point p*I* differs from the solution pH, and (ii) how effectively the small ions of the solution can screen the intermolecular electrostatic interactions.

Actually, when the two biopolymers are like-charged, then either a lowering of the pH towards the p*I* of the protein or an increase in the ionic strength *I* is likely to reduce the strength of the thermodynamically unfavourable repulsive interactions, as in the case of legumin + ovalbumin (see Figure 7.4) (Wasserman *et al.*, 1997). Alternatively, such a change in the solution conditions could even have the effect of radically modifying the character of the biopolymer–biopolymer interaction, as in the case of α_{s1}-casein + pectin (Dickinson *et al.*, 1998; Semenova *et al.*, 1999a), where there is a change from slight repulsion ($A_{ij}^* = 0.0008 \times 10^5$ cm^3/mol) at pH = 7.0 and $I = 0.01$ M to strong attraction ($A_{ij}^* = -334.4 \times 10^5$ cm^3/mol) at pH = 5.5 and $I = 0.01$ M.

1.3. *Manipulation of Colloidal Systems via the Thermodynamically Unfavourable Interactions between Biopolymers*

For a colloidal system containing a mixture of different biopolymers, in particular a protein-stabilized emulsion containing a hydrocolloid thickening agent, it is evident that the presence of thermodynamically unfavourable interactions ($A_{ij}^* > 0$) between the biopolymers, which increases their chemical potentials (thermodynamic activity) in the bulk aqueous phase, has important consequences also for colloidal structure and stability (Antipova and Semenova, 1997; Antipova *et al.*, 1997; Dickinson and Semenova, 1992; Dickinson *et al.*, 1998; Pavlovskaya *et al.*, 1993; Tsapkina *et al.*, 1992; Semenova *et al.*, 1999a; Makri *et al.*, 2005; Vega *et al.*, 2005; Semenova, 2007).

1.3.1. *Adsorption Behaviour*

In line with the Gibbs adsorption equation (equation 3.33 in chapter 3), the presence of thermodynamically unfavourable interactions causes an increase in protein surface activity at the planar oil–water interface (or air–water interface). As illustrated in Figure 7.5 for the case of legumin adsorption at the *n*-decane–water interface (Antipova *et al.*, 1997), there is observed to be an increase in the rate of protein adsorption, and also in the value of the steady-state interfacial pressure π. (For the definition of this latter quantity, the reader is referred to the footnote on p. 96.)

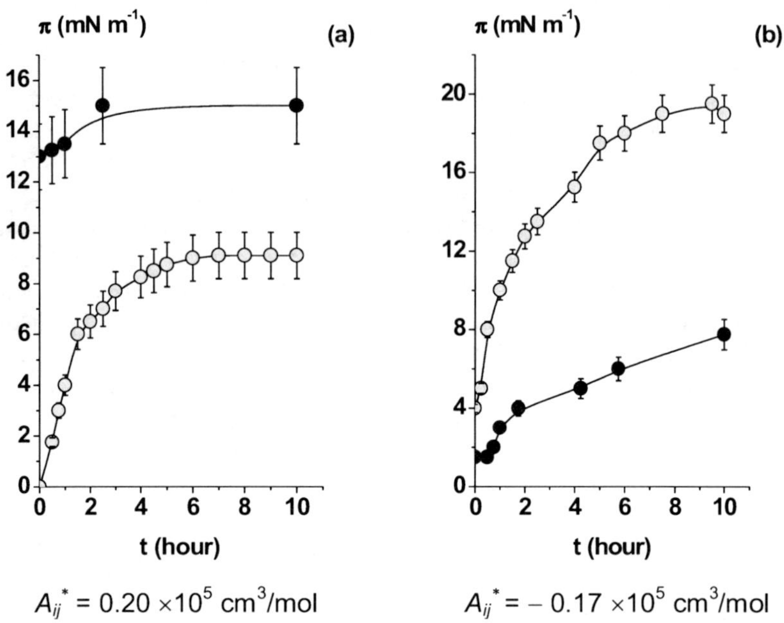

Figure 7.5 Effect of the character of the interactions between dextran and legumin on the time-dependent interfacial pressure π of the adsorbed layer of legumin at the planar *n*-decane–water interface: (○) 0.001 wt% legumin alone, and (●) 0.001 wt% legumin + 2 wt% dextran. (a) Thermodynamically unfavourable interaction: pH = 7.0, ionic strength = 0.01 M (dextran M_w = 48 kDa). (b) Thermodynamically favourable interaction: pH = 7.8, ionic strength = 0.01 M (dextran M_w = 270 kDa).

When a biopolymer mixture is either close to phase separation or lies in the composition space of liquid–liquid coexistence (see Figure 7.6a), the effect of thermodynamically unfavourable interactions is to induce biopolymer multilayer formation at the oil–water interface, as observed for the case of legumin + dextran (Dickinson and Semenova, 1992; Tsapkina *et al.*, 1992). Figure 7.6b shows that there are three concentration regions describing the protein adsorption onto the emulsion droplets. The first one ($C_{protein} < 0.6$ wt%) corresponds to incomplete saturation of the protein adsorption layer. The second concentration region (0.6 wt% < $C_{protein} < 6$ wt%) represents protein monolayer adsorption ($\Gamma \sim 2$ mg m^{-2}). And the third region ($C_{protein} > 6$ wt%) relates to formation of adsorbed protein multilayers on the emulsion droplets.

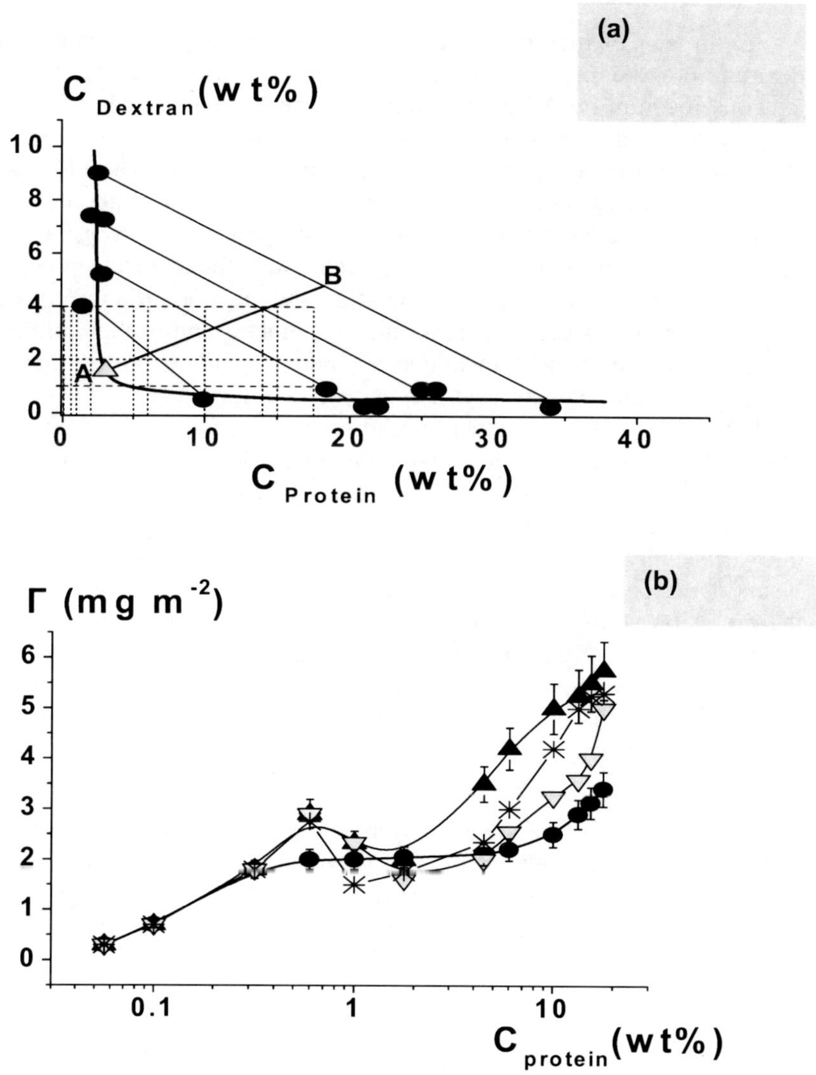

Figure 7.6 Effect of the phase state of mixed aqueous solutions of legumin + dextran (M_w = 270 kDa) (pH = 7.8, ionic strength = 0.1 M, 25 °C) on the protein adsorption Γ on emulsion droplets (30 vol% *n*-decane in water). (a) Phase diagram: (Δ) the critical point; A–B, rectilinear diameter; ●–●, tie lines. The grid shows the range of concentrations of protein and polysaccharide studied. (b) Concentration isotherm of protein adsorption: (●) legumin alone; (▲) legumin + 1 wt% dextran; (▽) legumin + 2 wt% dextran; (✳) legumin + 4 wt% dextran. The surface coverage Γ is plotted against the bulk protein concentration $C_{protein}$.

As a result of incorporation of the thermodynamically incompatible dextran into the protein solution, the net repulsive interaction in the aqueous medium produces an increase in the chemical potentials of the biopolymer components in the system. This leads, on the one hand, to a shortening of the concentration region of protein monolayer adsorption, and, on the other, to an increase in protein loading (Γ) on the emulsion droplets in the concentration region of protein multilayer adsorption. For the case of legumin + dextran, this effect was found to be more pronounced in the presence of ~ 1 % dextran, *i.e.*, for the protein + dextran compositions lying below the rectilinear diameter and in the vicinity of the lower branch of the binodal curve, where legumin forms the continuous phase of the water-in-water emulsion (see Figure 7.6a).

Table 7.1 shows that rather similar results were also found by Makri *et al.* (2005) for samples of coarse emulsions containing thermodynamically incompatible mixtures of legume seed protein + xanthan gum. The protein surface load was found to be enhanced in the presence of xanthan gum, especially at elevated ionic strengths. That is, there was observed to be an increase in the adsorption of legume seed proteins at the surface of the emulsion droplets which could be attributed to an increase in the thermodynamic activity of the proteins in the system in the presence of the incompatible polysaccharide (see Table 7.1). Associated with the greater extent of protein adsorption, the authors reported an enhancement in the emulsion stability.

Table 7.1 Effect of xanthan (XG) and NaCl on oil-in-water emulsions (50 vol% corn oil) made at pH = 7.0 with legume seed protein isolate (LSPI) as emulsifying agent: total protein adsorbed (Γ_T), average droplet size d_{32}, and amount of LSPI adsorbed per unit area of surface (Γ_S). Data from Makri *et al.* (2005) with permission.

Sample	d_{32} (μm)	Γ_T (mg)	Γ_S (mg/m^2)
1 wt% LSPI	10.5	13.6	0.79
1 wt% LSPI + 0.25 wt% XG	9.8	27.2	1.48
1 wt% LSPI + 0.25 wt% XG + 0.1 M NaCl	15.6	34.4	2.98
1 wt% LSPI + 0.25 wt% XG + 0.5 M NaCl	17.0	41.6	3.91

degree of flocculation; and, for 0.2 wt% XG, extensive flocculation was observed, leading to the formation of an emulsion gel.

A similar finding of accelerated creaming was found with emulsions stabilized by either skim milk powder (SPM) or sodium caseinate in the presence of incompatible locust bean gum (LBG) in systems formulated to resemble soft-serve ice-cream mixes (see Figure 7.7) (Vega *et al.*, 2005). In technological practice, in order to inhibit the creaming of such emulsions containing LBG, a small amount of κ-carrageenan is routinely added. It seems interesting to take note from Figure 7.7 that the added κ-carrageenan was found not to be as effective at inhibiting the formation of a cream layer in the sodium caseinate system as in the SMP system. The behaviour in the caseinate emulsions was attributed by the authors to κ-carrageenan self-association, rather than to direct interaction with the casein proteins, again in contrast to the behaviour in the SMP emulsions, where attractive interactions were expected (Vega *et al.*, 2005).

Figure 7.8 shows a set of images describing the microstructure of emulsions made with LBG and SMP or sodium caseinate in the presence or absence of κ-carrageenan, at the stage just after emulsion preparation, when no visual creaming had occurred (Vega *et al.*, 2005). All images reveal clear evidence of microscopic phase separation, with distinct protein domains (fluorescently stained, hence white) and protein-depleted domains (unstained, hence black). Fat globules can be resolved as dark spheres within the protein domains in all the images. (If the fat globules had resided within the protein-depleted phase, they would not have been seen by this staining technique). In the case of the emulsion made with sodium caseinate, the microstructure in the absence of κ-carrageenan (Figure 7.8c) was very different from that of the corresponding SMP emulsion (Figure 7.8a). In fact, the structure resembled more closely the SMP emulsion with κ-carrageenan added (Figure 7.8b), in that the phase-separated regions seemed to be more dispersed into smaller domains or a bicontinuous network, even though the emulsion readily phase separated on the macroscopic scale (see Figure 7.7). The SMP emulsion system showed a dramatic change in the size of the protein domains after the addition of 0.02 wt% κ-carrageenan, and it also demonstrated a much better stability against macroscopic serum separation (Vega and Goff, 2005), despite the evidence for microscopic phase separation (Figure 7.8b). In contrast, there was no real difference in the microscopic image following the addition of κ-carrageenan to the corresponding sodium caseinate emulsions (Figures 7.8c and 7.8d). And in the presence of 0.02 wt% κ-carrageenan, the sodium caseinate emulsions still showed visual serum separation.

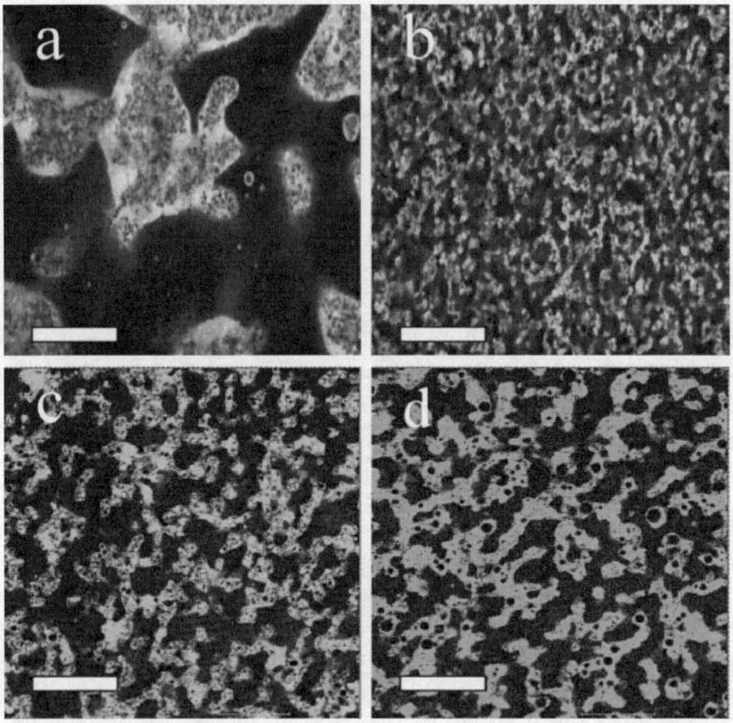

Figure 7.8 Confocal microscopy images for some freshly prepared emulsions containing 0.14 wt% locust bean gum, made with either skim milk powder (a, b) or sodium caseinate (c, d), and with no added κ-carrageenan (a, c) or 0.02 wt% κ-carrageenan (b, d). As the protein phase was fluorescently stained, it appears white. Scale bars correspond to 40 μm. Reproduced from Vega *et al.* (2005) with permission.

Whereas many protein-stabilized emulsions are close to Newtonian in their flow properties, after the addition of polysaccharide they become substantially shear-thinning. Sometimes the change in rheology of the emulsion can be directly attributed to the role of the polysaccharide acting as a hydrocolloid thickener. But when combined with no significant incorporation of the added polysaccharide into the protein adsorbed layer, and no significant viscous contribution to the aqueous phase from the non-adsorbed biopolymer, the behaviour is characteristic of depletion flocculation. Figure 7.9 shows illustrative data for the creaming stability and rheology of some casein-stabilized emulsions containing high-methoxy pectin (Semenova *et al.*, 1999a).

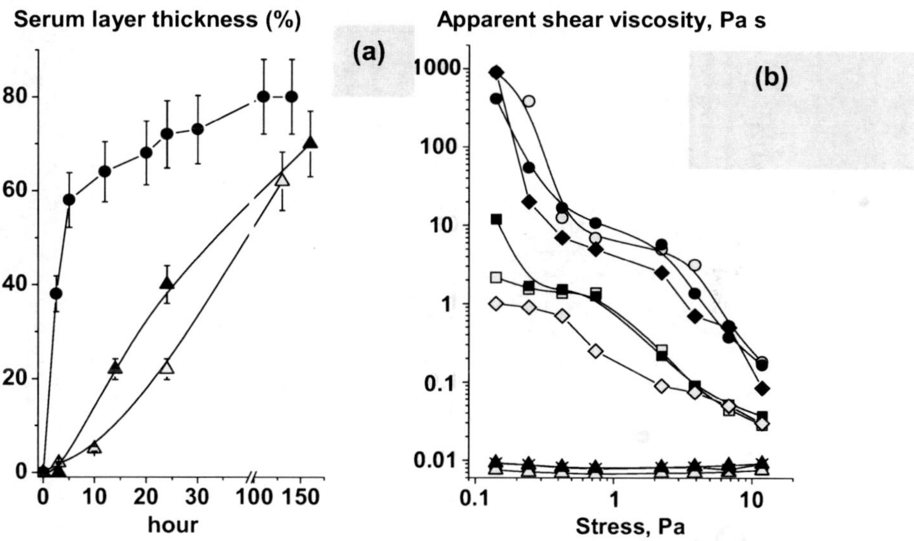

Figure 7.9 Effect of pectin (DE = 76%) on (a) creaming of protein-stabilized emulsions (11 vol% oil, 0.6 wt% protein, 0.28 wt% pectin, $I = 0.01$ M) containing (△) α_{s1}-casein (pH = 7), (▲) β-casein (pH = 7), and (●) α_{s1}-casein (pH = 5.5) and (b) steady-state shear viscometry of casein-stabilized emulsions (40 vol% oil, 2 wt% protein). Apparent shear viscosity at 22 °C is plotted against stress: pH = 7.0, $I = 0.01$ M, (△) α_{s1}-casein, (▲) β-casein, (□) α_{s1}-casein + 0.5 wt% pectin, (■) β-casein + 0.5 wt% pectin, (●) β-casein + 1.0 wt% pectin, (○) α_{s1}-casein + 1.0 wt% pectin; pH = 5.5, $I = 0.01$ M, (×) α_{s1}-casein, (◇) α_{s1}-casein + 0.5 wt% pectin, (◆) α_{s1}-casein + 1.0 wt% pectin. Reproduced from Semenova (2007) with permission.

We observe that a marked increase in the complex shear modulus G^* of emulsions stabilized by protein under conditions of thermodynamic incompatibility with pectin occurs at the polysaccharide concentrations in the phase diagram corresponding to the area of phase separation (Semenova *et al.*, 1999a). This trend can be seen by comparing the position of the binodals in Figure 7.10a with the compositions corresponding to the rise in G^* in Figure 7.10b. The effect is more pronounced for β-casein + pectin: this can be attributed to the thermodynamically more unfavourable interaction for this biopolymer combination ($A_{ij} = 0.46 \times 10^{-4}$ cm^3 mol g^{-2}) as compared with α_{s1}-casein + pectin ($A_{ij} = 0.008 \times 10^{-4}$ cm^3 mol g^{-2}) (Semenova *et al.*, 1999a).

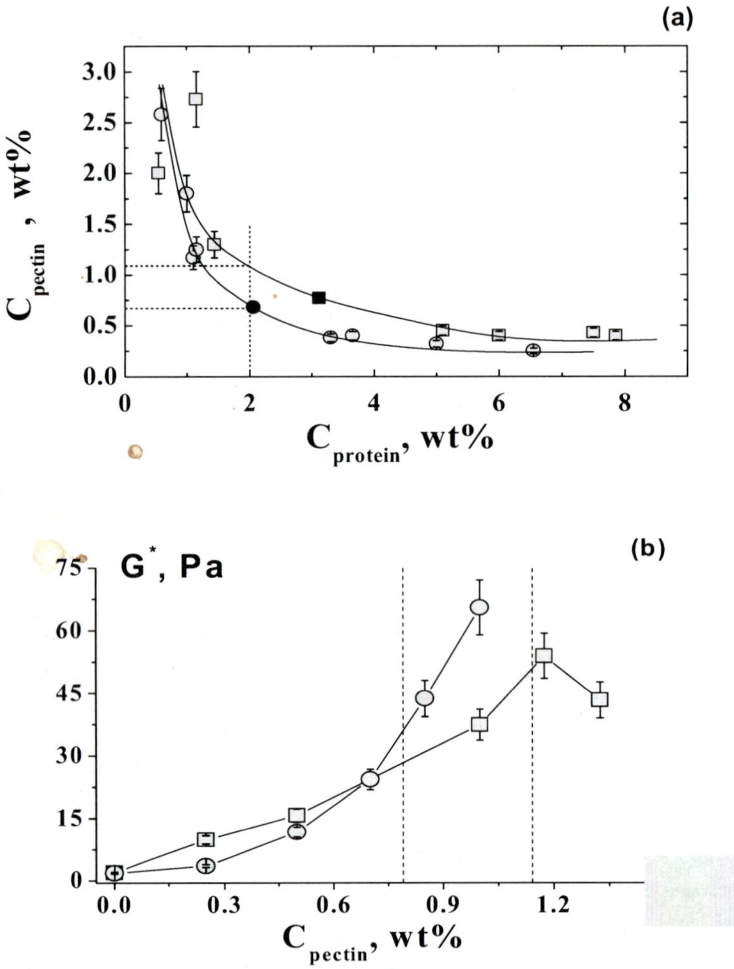

Figure 7.10 Effect of the thermodynamic incompatibility of α_{s1}/β-casein + high-methoxy pectin (pH = 7.0, I = 0.01 M) on phase diagram of the mixed solutions and elastic modulus of corresponding casein-stabilized emulsions (40 vol% oil, 2 wt% protein). (a) (○) Binodal line for β-casein + pectin solution with critical point (●); (□) binodal line for α_{s1}-casein + pectin solution with critical point (■). (b) Complex shear modulus G^* (1 Hz) is plotted against the pectin concentration: (○) β-casein; (□) α_{s1}-casein. Dotted lines indicate the range of pectin concentration for phase separation in the mixed solutions. The pectin was added to the protein solution before emulsion preparation. Data are taken from Semenova *et al.* (1999a).

1.3.5. *Gelation of Biopolymers in Mixed Aqueous Solutions*

Biopolymer gelation is sensitive to the presence of thermodynamically unfavourable interactions. This is mainly due to the role of the excluded volume effect. The gelation rate and gel rigidity tend to increase, and the minimum gelation concentration tends to decrease, because the excluded volume effect makes each biopolymer component behave as though it were at a higher effective concentration in the same volume (Piculell and Lindman, 1992; Tolstoguzov, 1991, 1995, 2003; Sciortino *et al.*, 1993; Manoj *et al.*, 1996; Beaulieu *et al.*, 2001; Lazaridou and Biliaderis, 2009). Additionally, when there is phase separation, the competition for hydration between the two biopolymer phases is an intrinsic part of the evolving gelation process (Beaulieu *et al.*, 2001).

In terms of the classification scheme of Morris (1986), the mixed gels formed from thermodynamically incompatible biopolymers are of the type commonly referred to as phase-separated networks. The ratio of the viscosities of the two coexisting phases has a controlling influence on the bulk rheological behaviour of these complex phase-separated biopolymer mixtures (Alves *et al.*, 2001). In general terms, however, the gel rheology is not easily related to the phase diagram or to the overall system composition. The difficulty in establishing this link is because the gel microstructure is highly sensitive to the specific chemical characteristics of the biopolymers in the system and to the processing conditions.

An example of how biopolymer interactions can influence the microstructure of a gel formed from an aqueous system of thermodynamically incompatible biopolymers (gelatin + oxidized starch) is given in Figure 7.11 (Firoozmand *et al.*, 2007). Pictures (a) and (b) are low-resolution confocal microscopy images of gels formed on cooling from solutions of 1 and 3 wt% gelatin, respectively, in the presence of 7 wt% oxidized starch. The light areas are the gelatin-rich features, and the dark areas are gelatin-depleted (starch-rich). At this level of resolution, the 3 wt% sample has a rather uniform distribution of fluorescently labelled gelatin. In contrast, the 1 wt% sample has the appearance of a fractal-type particle gel made from colloidal particles joined together into a stringy aggregated network. In fact, the particle gel microstructure of this biopolymer mixture resembles somewhat that of an acid-induced casein(ate) gel (Dickinson and Matia-Merino, 2002; Pugnaloni *et al.*, 2005) or maybe even a heat-set globular protein emulsion gel (Kerstens *et al.*, 2006). The particulate character of the aggregate/gel structure can be seen more clearly in the higher resolution images in Figure 7.11c–h for mixtures containing 0.1–3.0 wt% gelatin.

ref to micro. pictures

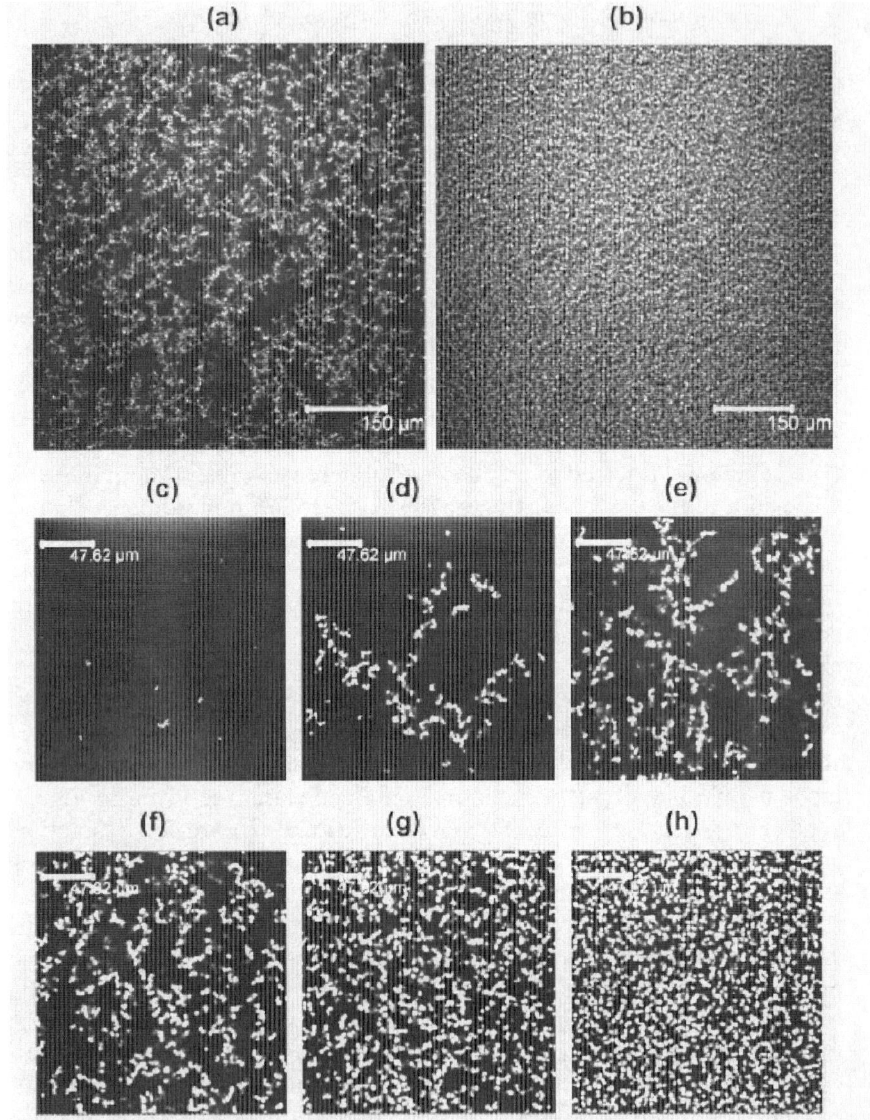

Figure 7.11 CLSM images of aqueous systems of gelatin + oxidized starch (7 wt% starch) after 24 hours at 24 °C. Low resolution: (a) 1 wt% gelatin; (b) 3 wt% gelatin. High resolution: (c) 0.1 wt% gelatin; (d) 0.5 wt% gelatin; (e) 1.0wt % gelatin; (f) 1.5 wt% gelatin; (g) 2.0 wt% gelatin; (h) 3.0 wt% gelatin. White regions are rich in gelatin. Reproduced from Firoozmand *et al*. (2007) with permission.

The proposed explanation for the formation of the fractal-type gel network shown in Figure 7.11 involves two distinct stages (Firoozmand *et al.*, 2007), as indicated below.

(i) The presence of the dissolved starch produces a thermodynamic driving force for gelatin self-association and local phase separation of gelatin-rich microgel particles on quenching the mixed biopolymer solution from 40 to 24 °C. The triggering of a phase transition through conformational ordering associated with the coil–helix transition was previously shown for gelatin + maltodextrin (Loren and Hermansson, 2000; Loren *et al.*, 2001; Williams *et al.*, 2001; Butler and Butler-Heppenstall, 2003).

(ii) Once formed, the sticky microgel particles become susceptible to association via a process of diffusion-limited cluster aggregation. The interparticle crosslinks are presumably the same sort of hydrogen-bonded triple-helix crosslinks that hold together the microgel particles themselves, and which also form the physical crosslinks in a normal cold-set pure gelatin gel.

When the gelatin content is ~ 1 wt%, there is a sufficient concentration of the microgel particles following the temperature quench for them to aggregate, irreversibly and randomly, into a particle gel network. This is the same type of fractal structure formation as generated in computer simulations of aggregation of hard spheres (Dickinson, 2000; Rottereau *et al.*, 2004) and in experiments on casein particle gelation (Walstra *et al.*, 1991; Dickinson, 2006a). A rigid gel is formed because, even though the overall protein content may be low, the local gelatin concentration in load bearing strands of the network is high (10–20 wt%) (Firoozmand *et al.*, 2007). An attractive feature of this transformation is its thermoreversibility at around body temperature, as visualized by CLSM (Figure 7.12). This behaviour offers an opportunity for designing new types of food gel architecture, while continuing to exploit gelatin's unique 'melt-in-the-mouth' property. A possible technical application, with potential for 'in-mouth'-triggered flavour release, could involve the encapsulation of lipophilic compounds or oil droplets within the particle gel network.

There is considerable evidence in the literature that the effect of phase separation on the viscoelastic properties of mixed biopolymer gels can be related to the strength of the adhesion forces between the dispersed phase (deformable filler particles) and the continuous phase (matrix). Depending on the system, these forces may lead to synergistic or antagonistic effects (Appelqvist and Debet, 1997; Grinberg and Tolstoguzov, 1997; Sanchez *et al.*, 1997; Tolstoguzov, 1997; Zasypkin *et al.*, 1997).

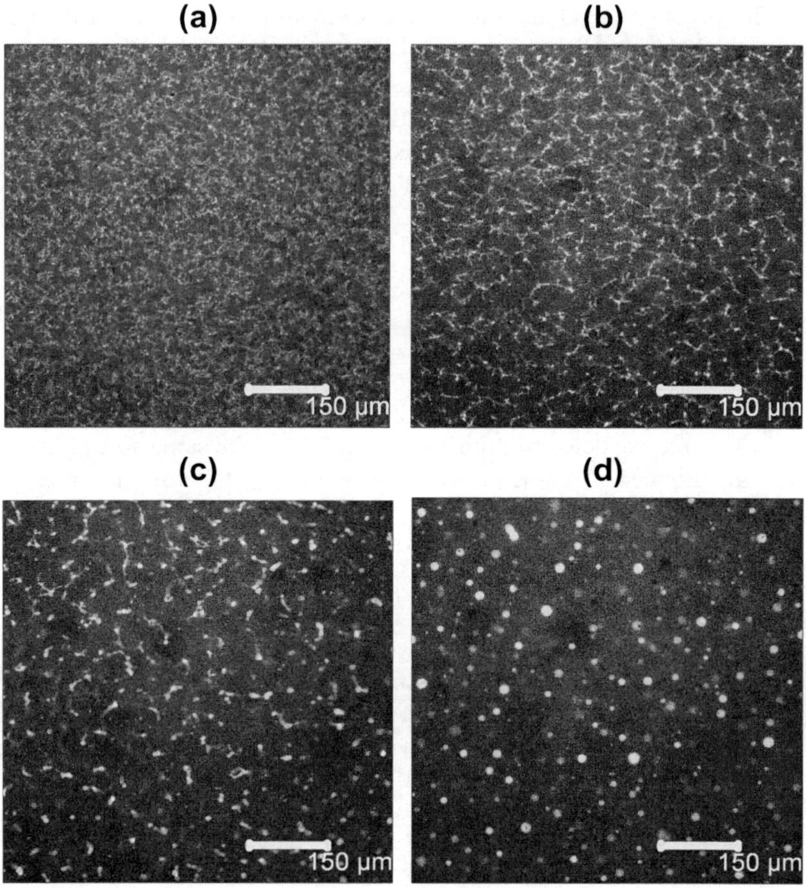

Figure 7.12 Sequence of CLSM images recorded during melting of a 24-hour-old fractal particle gel formed by quenching a 1 wt% gelatin + 7 wt% oxidized starch system from 40 to 24 °C: (a) immediately before heating; (b) network rearrangement and coarsening; (c) network break-up; (d) aggregate melting into polydisperse liquid droplets. Reproduced from Firoozmand *et al.* (2007) with permission.

There have been numerous systematic studies of model systems of mixed biopolymers describing the relationship between phase separation and gelation microstructure (Kasapis *et al.*, 1993; Alevisopoulos *et al.*, 1996; Alves *et al.*, 2000; Loren *et al.*, 1999; Bourriot *et al.*, 1999; Loren and Hermansson, 2000; Butler, 2002; Lazaridou and Biliaderis, 2009).

As a general principle, it is established that the gelation of one (or both) of the biopolymer components retards the process of phase separation and so leads to kinetically trapped microstructures (Bourriot *et al.*, 1999; Butler and Butler-Heppenstall, 2003; Loren and Hermansson, 2000).

A study of kinetic trapping of microstructure was reported recently for thermodynamically incompatible mixtures of oat β-glucans + sodium caseinate (or pullulan) (Lazaridou and Biliaderis, 2009). The authors found that the process of phase separation coexisted with (and competed with) chain aggregation / gelation phenomena. Thus, the phase separation could drive and accelerate the formation of β-glucan-enriched network structures. At the same time, the process of gelation appeared to induce and promote phase separation on the *micro*scopic scale, although bulk phase separation was seemingly 'arrested' by the gelation. The kinetic entrapment of the components within a highly viscous medium was found to influence the composition and volume fraction of the individual phases, resulting in an incomplete phase separation. For these β-glucan mixtures, Lazaridou and Biliaderis (2009) showed that the enhanced viscosity or gelation of the system influences the phase separation kinetics by restricting the mobility of the biopolymer components, thus leading to non-equilibrium situations and kinetically trapped states. Because this phenomenon hinders macroscopic phase separation, it is referred to as 'arrested' phase separation.[1]

The thermal gelation of β-lactoglobulin in the presence of non-gelling amylopectin was investigated by Olsson *et al.* (2002). Dynamic measurements of the protein's aggregation and gel formation were carried out in the vicinity of p*I* in the presence of the thermodynamically incompatible polysaccharide. The measurements indicated a faster formation of the protein network at a lower gelling temperature. The images in Figure 7.13 show that, whereas the pure β-lactoglobulin gel (images (a) and (d)) was composed of a dense network, the microstructure of this protein network became more open when amylopectin was introduced (images (b), (c), (f) and (e)). The increase in thermodynamic activity of the protein in the presence of amylopectin caused inhomogeneity in the protein structure formation, leading to larger pores within the network of protein clusters. This feature was more pronounced for the system with the low-viscosity amylopectin sample, because apparently this biopolymer did not hinder so much the formation of the thicker strands of protein clusters in the aqueous medium (see Figure 7.13c,f) (Olsson *et al.*, 2002).

[1] There seems to be a sort of analogy here with the arrested phase separation of a protein-stabilized depletion-flocculated emulsion containing a thermodynamically incompatible hydrocolloid like xanthan gum (Moschakis *et al.*, 2005; Dickinson, 2006b).

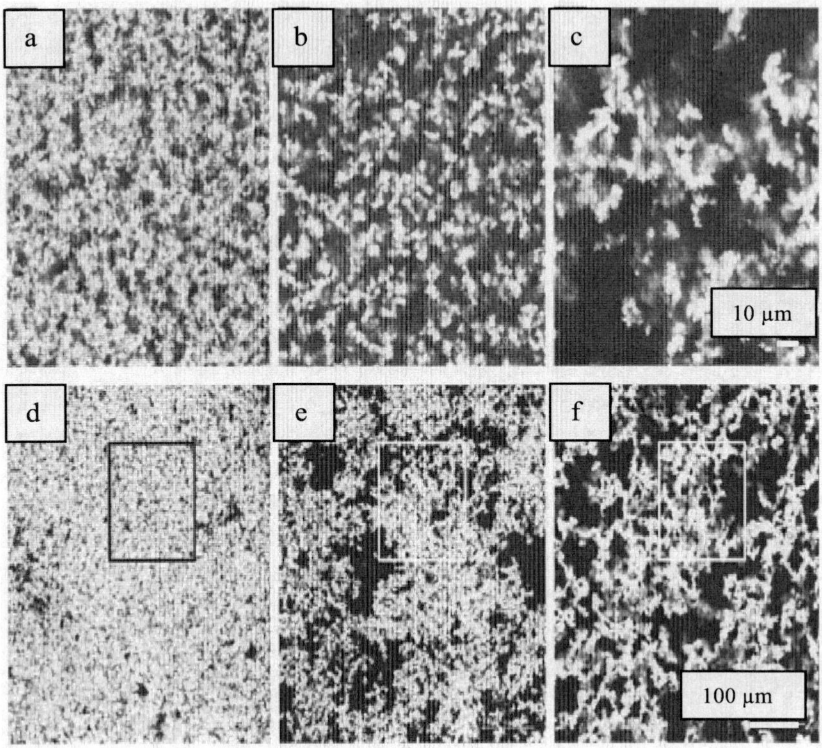

Figure 7.13 CLSM micrographs of heat-set gels of β-lactoglobulin + amylopectin: (a) 6 wt% β-lactoglobulin; (b) 6 wt% β-lactoglobulin + 0.75 wt% (high-viscosity) amylopectin; (c) 6 wt% β-lactoglobulin + 2.0 wt% (low-viscosity) amylopectin. Images (d), (e) and (f) are for the same gels as (a), (b) and (c), respectively, but at a lower magnification. Reproduced from Olsson *et al.* (2002) with permission.

In the study of Musampa *et al.* (2007), steady shear measurements were carried out at 60 °C on a thermodynamically incompatible mixture of pea protein + κ-carrageenan at pH = 7 (*i.e.*, above the p*I* of the protein) at two different ionic strengths (0.05 and 0.2 M). The rheological measurements showed a clear correlation with the phase diagram, in the sense that there was a marked increase in the viscosity of the mixtures on passing from the single-phase region to the biphasic region. Furthermore, some complementary CLSM measurements showed the formation of a network of pea protein aggregates in the presence of κ-carrageenan.

It was reported earlier by Capron *et al.* (1999) that the addition of κ-carrageenan reduces the gelation time of heated aqueous solutions of β-lactoglobulin. The result suggests that it is not the denaturation step that is affected by the added κ-carrageenan — rather it is that the subsequent aggregation of the unfolded protein molecules becomes faster. Bryant and McClements (2000), in an investigation of whey protein + xanthan gum solutions, suggested a possible explanation for this behaviour. The authors observed that the thermodynamically unfavourable interactions between protein and polysaccharide promotes phase separation with the denatured protein but not with the native protein. It would seem, therefore, that there is a significant strengthening of excluded volume interactions in this system as a result of the increase in effective molecular size following unfolding of the heat-treated whey protein (Bryant and McClements, 2000).

2. *Thermodynamically Favourable Interactions between Biopolymers in the Bulk*

2.1. *The Nature of Thermodynamically Favourable Interactions*

Thermodynamically favourable interaction (mutual attraction) between biopolymers is generally driven either by direct intermolecular forces (complexation) or by indirect entropy contributions from counterions or water molecules. In the case of mixtures of charged proteins and polysaccharides, electrostatic complexes have attracted considerable interest, in the past because of their stabilizing properties at interfaces in food colloids, and more recently because of the recognition of their potential for the (nano)encapsulation of flavour compounds and nutraceuticals (Dickinson, 2003, 2008a,b; Turgeon *et al.*, 2003, 2007; de Kruif *et al.*, 2004; McClements, 2005; de Vries and Cohen Stuart, 2006; Guzey and McClements, 2006; Damianou and Kiosseoglou, 2006; Benichou *et al.*, 2007; Jourdain *et al.*, 2008, 2009; McClements *et al.*, 2009).

In thermodynamic terms, the process of biopolymer complexation takes place spontaneously if the total free energy of the system decreases with the mixing of its components. The combination of the formation of direct contacts between associating biopolymers and the solvent reorganization in the solution together contribute to the favourable enthalpic contribution to the change in total free energy of the system. In turn, the favourable entropic contribution to complexation includes the effect of the release into bulk solution of counterions and water molecules. Any unfavourable entropic effect originates mainly from the configurational restrictions of the biopolymers following the association (Gilsenan *et al.*,

2003; Tolstoguzov, 2003; de Kruif *et al.*, 2004; de Vries and Cohen Stuart, 2006; Turgeon *et al.*, 2007; Dickinson, 2008a,b).

First and foremost, complexation is an intrinsic consequence of the electrostatic interactions between oppositely charged functional groups on the biopolymers. We find soluble complex formation for low charge densities, and precipitation (or complex coacervation) and/or gelation for high charge densities and equal molar ratios of the interacting biopolymers (Semenova *et al.*, 1991a; Dickinson, 1998a, 2003, 2008a; Ledward, 1994; Tolstoguzov, 1996; Dickinson and James, 2000; Gilsenan *et al.*, 2003; de Kruif *et al.*, 2004; Neirynck *et al.*, 2007; Sperber *et al.*, 2009). In particular, the maximum coacervation yield occurs for the case of a mixture of equal masses of the two biopolymers at the pH where they carry equal and opposite charges (Schmitt *et al.*, 1998). It has also been found that, in order to make a soluble acidic protein + anionic polysaccharide complex below the protein's isoelectric point, the amount of the polysaccharide present should be sufficient to ensure that the polysaccharide/protein ratio is much larger than that required at the electrical equivalence point (Neirynck *et al.*, 2007).

Of course, the strength of attractive protein–polysaccharide electrostatic interactions depends on pH and ionic strength, and also on other environmental factors such as temperature, high-pressure treatment, and calcium ion concentration (Schmitt *et al.*, 1998; Dickinson and James, 2000; Gilsenan *et al.*, 2003; de Kruif *et al.*, 2004; Turgeon *et al.*, 2007; McClements, 2006; Ye, 2008). Hence, it is found that strong attractive electrostatic complexes are typically formed with mixtures of positively charged protein (pH $<$ p*I*) with a negatively charged polysaccharide. Weaker reversible complexes can be formed between a highly charged sulfated polysaccharide, like κ- or ι-carrageenans or dextran sulfate, and a protein carrying nearly zero overall charge (pH $\approx$ p*I*) or even a net negative charge (pH $>$ p*I*) (Galazka *et al.*, 1997, 1999; Dickinson, 1998a, 2008a,b; Alexander and Dalgleish, 2007; Jourdain *et al.*, 2008). Complexation of the latter type has been traditionally attributed to the interaction of the anionic polysaccharide with positively charged 'patches' on the protein's surface (Hattori *et al.*, 2000, 2001; Park *et al.*, 1992). For example, at neutral pH (7.0), when the globular protein bovine serum albumin carries an overall net negative charge, there is a clearly defined patch (size ~ 2–3 nm) of positive charge on the protein's globular surface (Hattori *et al.*, 2001).

The presence of casein–polysaccharide interactions is commonly invoked to explain the mechanistic stabilizing role of food hydrocolloids in dairy colloids (Dickinson, 1998a). Thus, for example, under conditions where the casein micelles and κ-carrageenan both carry a net negative charge, an attractive interaction between them is revealed by an increase in the mean size of the casein micelles in the presence of this polysaccharide (Spagnuolo *et al.*, 2005). It was suggested many years ago (Snoeren *et al.*, 1975) that such interactions could occur via localized electrostatic attractions between the sulfated groups on κ-carrageenan and the very short positively charged region of κ-casein. More recently, for this particular pair of biopolymer ingredients, direct evidence for the formation of casein micelle / polysaccharide networks has been demonstrated by electron microscopy (Martin *et al.*, 2006; Spagnuolo *et al.*, 2005).

An alternative explanation for complex formation with strong polyelectrolytes on the 'wrong side' of the protein p*I* involves the concept of 'protein charge reversal' (Biesheuvel and Cohen Stuart, 2004; Dickinson, 2008a). An intrinsic property of a protein, expressing the ability for charge regulation, is its electrical capacitance. It appears that globular food proteins can display a high effective capacitance around p*I*, leading to an additional attractive energy of several kT between the protein and some polyelectrolyte molecules as a result of these charge-induced charge interactions (da Silva *et al.*, 2006; Jönsson *et al.*, 2007; Dickinson, 2008a).

According to Ledward (1994), the strength of complexation between a protein and a polysaccharide depends on three general factors:

(i) the distribution of ionisable groups on the surface of the protein,
(ii) the ease of unfolding of the protein's native structure, and
(iii) the flexibility and charge distribution of the polysaccharide.

The extent of reversibility of complexation depends mainly on the aqueous environment and the mixing conditions (Tolstoguzov, 1986; Galazka *et al.*, 1999). The tendency towards non-equilibrium formation of an insoluble coacervate is enhanced at low ionic strength and at pH values significantly below the p*I* of the protein.

A simple practical application of protein–polysaccharide complexation is in the protection of proteins against the loss of solubility caused by certain processing operations. Loss of solubility may occur as a result of pH lowering towards the p*I*, or by heating and/or high-pressure processing in the case of the globular proteins. For instance, the acid-induced precipitation of casein(ate) is readily inhibited by its complexation with charged polysaccharide. The mechanism of protein protection against aggregation and precipitation probably mainly involves the blocking of potentially hydrophobic binding sites on the protein as a consequence of the close proximity of the bulky polysaccharide (Imeson *et al.*, 1977; Ledward, 1979; Galazka *et al.*, 1997, 1999; Dickinson, 1998a, 2008a; Jourdain *et al.*, 2008).

2.2. *Physico-Chemical Features of 'Complex Coacervation'*

The term 'complex coacervation' is defined (Bungenberg de Jong, 1949) as the spontaneous phase separation of a solution of protein + polysaccharide into a solvent-rich phase and a solvent-depleted *liquid* phase. This phase separation is driven by attractive electrostatic interactions between the macromolecules, resulting in the primary formation of electrostatic protein–polysaccharide soluble complexes. The rearrangement of such complexes leads to the formation of soluble and insoluble liquid droplets, also called coacervates, which ultimately coalesce and sediment to form the coacervate phase (Schmitt *et al.*, 1999, 2000; de Kruif *et al.*, 2004). The solvent present in the coacervate phase largely constitutes the residual solvation liquid, whereas the supernatant is a very dilute polyelectrolyte solution. The coacervate phase possesses the microstructure and rheology of a heterogeneous viscous material (Singh *et al.*, 2007a).

Traditional colloid science textbooks give as the classical example of a complex coacervate the system gelatin + gum arabic under pH conditions where the two biopolymers are oppositely charged (Bungenberg de Jong, 1949). Although the topic was largely neglected for much of the second half of the last century, there is now much renewed interest in the phenomenon of coacervation because of its potential in developing new functional food ingredients (Schmitt *et al.*, 1998; Turgeon *et al.*, 2003, 2007; de Kruif *et al.*, 2004; Singh *et al.*, 2007a,b; Dickinson, 2008a). In particular, food biopolymer coacervates have attracted significant attention for use as possible fat replacers or meat analogues (Bakker *et al.*, 1994), for microcapsules and matrices for encapsulation of flavours or drugs (Burgess, 1994; Luzzi, 1970; McClements *et al.*, 2009), and for composite biomaterials in edible films and packaging (Kester and Fennema, 1986).

Complex coacervation has been observed for various aqueous solutions of gelatin or globular proteins mixed with carboxylated polysaccharides (gum arabic, pectin, xanthan, *etc.*) (Girard *et al.*, 2003; Burgess *et al.*, 1991; Burgess and Singh, 1993; Lii *et al.*, 2002; Weinbreck *et al.*, 2004c). For the case of β-lactoglobulin + gum arabic (Schmitt *et al.*, 1999), Figure 7.14 shows the ternary phase diagram (water / β-lactoglobulin / gum arabic) containing the characteristic 'drop-shaped' two-phase region located towards the water-rich corner. The thermodynamic behaviour in Figure 7.14 is identified by the partition of the tie lines on each side of the water corner bisector, as is generally the case for complex coacervation. Confinement of the two-phase region into the solvent corner is the signature of the existence of an equilibrium between a solvent-rich phase and a biopolymer-rich phase (Schmitt *et al.*, 1999).

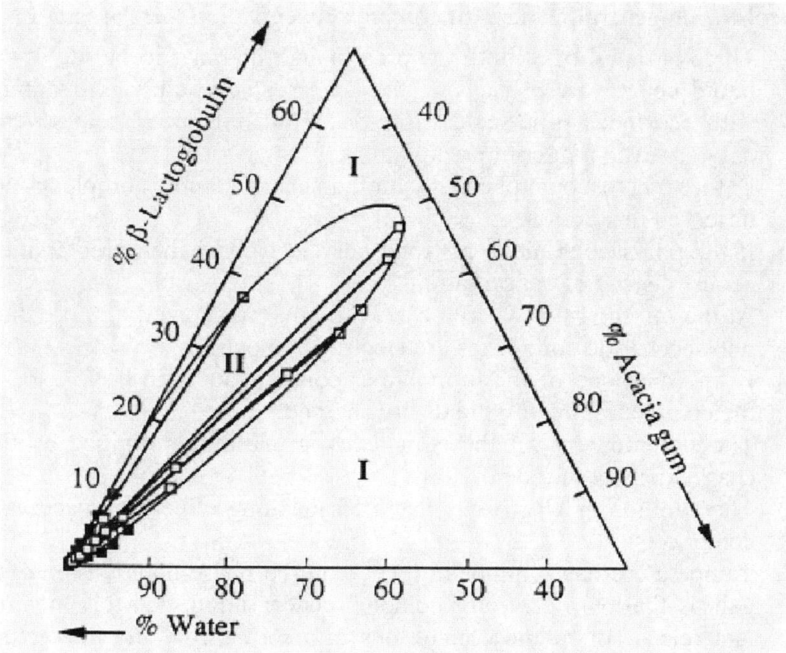

Figure 7.14 Complete phase diagram of the water + β-lactoglobulin + gum arabic system at 20 °C and pH = 4.2. Features indicated are: □, tie-lines; ■, binodal points; I, one-phase region; II, two-phase region. Reproduced from Schmitt *et al.* (1999) with permission.

The electrostatic interaction between whey protein(s) and gum arabic leads to the formation of a liquid coacervate in the pH window between 2.5 and 4.8 (Schmitt *et al.*,1999; Weinbreck *et al.*, 2003a, 2004c; Weinbreck and de Kruif, 2003). Denser or more open structures of the whey protein / gum arabic coacervates can be obtained by changing the solution compositional parameters — such as the pH, the protein / polysaccharide ratio, or the ionic strength. Diffusion and barrier properties are sensitive to coacervate structure, and these are the properties that are most relevant when using coacervates in encapsulation applications (Weinbreck *et al.*, 2004c). Mixtures of proteins with sulfated polysaccharides (*e.g.*, carrageenans, dextran sulfate) tend to produce more solid-like phases (*i.e.*, co-precipitates) rather than liquid-like coacervates (Gurov *et al.*, 1988; Semenova *et al.*, 1991a; Galazka *et al.*, 1999; Haug *et al.*, 2003; Weinbreck *et al.*, 2003b, 2004a).

In a critical and definitive review of this area (de Kruif *et al.*, 2004), the following general features of complex coacervation were asserted:

(1) There is mobility of both the protein and the polysaccharide in the liquid coacervate phase, *i.e.*, the coacervate structure is dynamic, with enormous time-scale variations from milliseconds to several days (or even longer in precipitates).
(2) Before macroscopic phase separation occurs, soluble complexes are observed in solution.
(3) If the polysaccharide is a strong polyelectrolyte, then precipitation occurs instead of coacervation.
(4) At the mixing ratio where coacervation is maximized, the protein–polysaccharide complexes are electrically neutral.
(5) Even when one of the biopolymer components is in great excess, the complexes are only modestly charged.
(6) The stoichiometry of the complexes is mainly determined by the charge density of the polyions.
(7) The addition of electrolyte has a dissociating effect on coacervate complexes.
(8) Temperature has a minor influence on the phase diagram; entropy gain is the main driver of complex coacervation, due to concomitant release of bound counterions into solution by the interacting charged biopolymers, which increases the solution entropy.
(9) The biopolymer conformation is a key property determining the ability of the coacervate to encapsulate oil droplets.
(10) For the production of good capsules, it is desirable to promote complex coacervation, rather than precipitation.

Based on analysis of the available experimental data, de Kruif *et al.* (2004) have suggested that at low ionic strength the lowering of the free energy is due to the enthalpic effects, while at higher ionic strength (> 10 mM) it is the entropic effects that largely contribute. Such a conclusion is in line with the observed temperature independence of complex coacervation (Kaibara *et al.*, 2000; Weinbreck *et al.*, 2004b). This entropy gain is due to the 'liberation' of salt ions as a result of electrostatic attractions between the opposite charges of the biopolymers (Overbeek and Voorn, 1957) and the delocalization of the polyelectrolyte in the concentrated coacervate phase (de Kruif *et al.*, 2004). Indeed, adding a large amount of salt suppresses the primary complexation (Weinbreck *et al.*, 2004a).

There is evidence that random mixing of partially charge-neutralized hydrated polyelectrolyte complexes inside the coacervate phase imparts higher configurational entropy to these less stiff polyelectrolyte molecules as compared to those in the pre-coacervation phase (Kaibara *et al.*,

2000; Mohanty and Bohidar, 2003; Gupta and Bohidar, 2005). It was suggested by de Kruif *et al.* (2004) that entropy gain is the main driver in complex coacervation based on isothermal calorimetry measurements showing the temperature independence of the transition. Girard *et al.* (2003) inferred from the binding isotherms of pectin to β-lactoglobulin that two kinds of complexes were formed. The first ones were soluble intrapolymeric complexes, whose formation was driven by enthalpy gain (binding stoichiometry of 6–8). The second ones arose from aggregation of the first kind, and were mostly controlled by the entropy change (binding stoichiometry in the range 15–16.5).

It has also been suggested that the structure of electrostatic complexes can be influenced by a coil-to-helix transition in either of the complexing polyelectrolytes (de Kruif *et al.*, 2004). Thus, for example, it was found (Burova *et al.*, 2007) that β-casein forms stable electrostatic complexes (precipitates) with κ- or ι-carrageenan. These nanocomposites are soluble over a notably wider pH range than the β-casein alone. In addition, it was found that β-casein reduces the helicity of the carrageenan molecules as a consequence of its preferential interaction with the disordered parts of the polysaccharide chains. The dissociation of these nanocomposites and recovery of the helical structure of the carrageenans could be achieved by increasing the ionic strength.

Returning again to the case of whey protein/gum arabic coacervates, it was found (Weinbreck *et al.*, 2004b) that the pH plays a major role in determining the viscosity of the coacervate phase, because the dynamical properties of the coacervate are directly related to the strength of the electrostatic interactions. This effect was investigated by comparing the viscosity of mixtures of biopolymers at the same concentration as in the coacervate phase, but at pH = 7 where no interaction took place. The results showed that the coacervate phase was highly viscous in nature, as compared to the reference biopolymer mixture, which was more elastic. In addition, the coacervate phase was only slightly shear-thinning at low shear-rates, but it exhibited some hysteresis and thixotropic behaviour. That is, with time, the initial viscosity was completely recovered, showing the structural reorganization of the system. A maximum viscosity was found at pH = 4.0, which is the pH at which the concentrations of whey protein and gum arabic in the coacervate are greatest, and the pH for which the electrostatic interaction is strongest. On the strength of the data obtained, it could be inferred that the coacervate phase is a liquid phase with a dynamic structure, reversible upon deformation, and from a rheological point of view one behaving less like a concentrated visco-elastic polymer solution and more like a viscous particle dispersion (de Kruif *et al.*, 2004).

For the system gelatin + agar, Singh *et al.* (2007b) used a combination of techniques — static and dynamic light scattering, viscometry, and electrophoretic measurements — to characterize the nanostructure of the supernatant in thermodynamic equilibrium with the coacervate phase. The supernatant could be defined as a dilute liquid phase, rich in soluble complexes and biopolymer nanoparticles (Mohanty *et al.*, 2004; Singh *et al.*, 2007b). These complexes were demonstrated to be asymmetric intermolecular aggregates held together by polarization-induced electrostatic interactions. The viscosity data for pure gelatin and agar solutions were compared in order to estimate the level of the mixed interactions (Singh *et al.*, 2007b). The agar solutions exhibited a viscosity (1.88 mPa s) that was independent of pH, whereas the viscosity of the gelatin solutions showed a strong pH dependency. For non-interacting mixed solutions (pH > p$I \approx$ 9), the viscosity was constant (1.23 mPa s), and much lower than that of the pure agar solution. The addition of gelatin was therefore responsible for reducing the viscosity of the agar solution by as much as 30 %, thereby functioning as a 'thinner'. For pH < pI, as a result of the intermolecular attractive interactions, the viscosity fell by a further 25 %. This could be attributed to the binding of gelatin to the agar, leading to a reduction in its molecular stiffness (due to partial charge neutralization) and its molecular spatial extension (Singh *et al.*, 2007b).

In technological practice, the phenomenon of complex coacervation has been widely used for microencapsulation of various substances (de Kruif *et al.*, 2004; Xing *et al.*, 2004; Mayya *et al.*, 2003). For efficient encapsulation, it is well established that biopolymer complexes should be in the coacervate form rather than the precipitate form. It seems that the appropriate steric conformation of the molecules forming the complex is a key consideration for effective encapsulation, and that adding a surfactant increases the encapsulation yield. Thus Mayya *et al.* (2003) have found that an added ionic surfactant initially interacts with the oppositely charged polymer, forming a primary layer which lowers the tension at the oil–water interface; this is followed by a secondary adsorbed layer of polyelectrolyte–polyelectrolyte complex. According to the work of Xing *et al.* (2004), the use of a surfactant with a rigid structure, and one which can form hydrogen bonds, is especially favourable for the formation of stable microcapsules.

The trend in microencapsulation research nowadays is to replace the traditional gelatin + gum arabic system by other biopolymers. The motivation is to design capsules with improved properties, and also to avoid the use of gelatin for health, ethical or religious reasons (de Kruif *et al.*, 2004). For example, gelatin has been replaced by plant proteins in mixtures with gum arabic (Ducel *et al.*, 2004). The physicochemical conditions (pH and biopolymer ratio) under which a coacervate was formed between gliadin or pea globulin and gum arabic (or carboxymethyl cellulose) were investigated using ζ-potential measurements and turbidimetry. The optimum coacervation pH was found to be rather low (pH ~ 3) compared with that for more familiar proteins like gelatin or whey proteins. Nevertheless, effective encapsulation of oil was achieved using these plant protein systems, and coacervate morphology could be determined.

The complexation of chitosan with alginate has attracted considerable attention recently. Thus Zhang *et al.* (2004) encapsulated bovine serum albumin (BSA) with an efficiency of 80 % in microspheres of carboxymethyl-chitosan + alginate hardened with $CaCl_2$. The BSA was found to be quickly released at pH = 7.2 (*i.e.*, the pH of the intestinal fluid). The capsules were found to swell more at neutral pH than at pH = 1.0 (the pH of the gastric juice). Due to strong electrostatic interactions between the chitosan (a weak polybase) and alginate, the BSA was hardly released at all at pH = 1.0. It was observed that the chitosan concentration had a greater influence on capsule formation than the alginate concentration. Also it was found that a longer hardening time led to a more tightly structured gel network, thereby reducing the extent of protein release.

Peniche *et al.* (2004) successfully encapsulated up to 65 % of shark liver oil (rich in polyunsaturated fatty acids) in chitosan/alginate capsules in order to mask the oil's unpleasant taste. Here again it was found that the chitosan coating allowed a greater degree of control of capsule permeability. The capsules could be degraded by enzymes such as lipase or pancreatin. They were initially resistant to the acid environment of the stomach, although after 4 hours under intestinal conditions (pH = 7.4) the capsule walls were finally disrupted.

2.3. *Thermodynamically Favourable Interactions Involving Neutral Polysaccharides*

In the case of neutral polysaccharides, complexation with other biopolymers tends to have its main molecular origin in the hydrogen bonding of appropriate functional groups (Semenova *et al.*, 1999a,b). In addition, specific thiol–disulfide interchange reactions may contribute to mutual biopolymer attraction (Gezimati *et al.*, 1997). Furthermore, an indirect contribution from the translational entropy of counterions is a characteristic feature of the thermodynamically favourable interaction between a charged biopolymer and a non-ionic one. The magnitude of this indirect contribution increases with the charge on the biopolymer (Antipova and Semenova, 1997; Khokhlov and Nyrkova, 1992; Piculell and Lindman, 1992; Semenova, 1996).

Let us consider the specific case of a mixture of legumin and neutral dextran (Semenova, 1996; Antipova *et al.*, 1997). Either a change in pH away from the p*I* or a decrease in the ionic strength is considered favourable for an increase in translational entropy of the counterions, causing a transformation in the character of the protein–polysaccharide interaction from thermodynamically unfavourable ($A_{ij}^* = 0.31 \times 10^5$ cm^3/mol at pH = 7.8 and ionic strength = 0.1 M) to thermodynamically favourable ($A_{ij}^* = -0.17 \times 10^5$ cm^3/mol at pH = 7.8 and ionic strength = 0.01 M). In terms of the phase diagram, a dramatic increase in the area of composition space corresponding to compatibility of the biopolymers was observed (Semenova, 1996). That is, the coordinates of the experimentally determined critical point were found to be 3.1 wt% legumin and 1.3 wt% dextran at pH = 7.8 and ionic strength = 0.1 M. But when the ionic strength was reduced to 0.01 M, it was found that thermodynamic compatibility of legumin + dextran was established over the whole concentration range studied, *i.e.*, up to 10 wt% for each component in the mixed solution (20 wt% total concentration of biopolymers).

Finally, we note here that the dehydration of biopolymers associated with their attractive non-ionic interactions (dipole–dipole, hydrophobic) can increase the contribution of the mixing entropy. This in turn can lead to a tendency towards enhancement of the thermodynamically favourable interactions between them (Appelqvist and Debet, 1997; Cai and Arntfield, 1997; Semenova *et al.*, 1991b).

2.4. *Manipulation of Colloidal Systems via the Thermodynamically Favourable Interactions between Biopolymers*

In considering the impact of thermodynamically favourable interactions between biopolymers on the formation and stabilization of food colloids, a number of regular trends can be identified. One of the most important aspects is the effect of complexation on interfacial properties, including rates of adsorption and surface rheological behaviour.

2.4.1. *Adsorption Behaviour*

The presence of a thermodynamically favourable interaction between protein and polysaccharide is commonly associated with a marked decrease in protein surface activity at the air–water or oil–water interface (see Figures 7.5b and 7.15). There is a slower decay in the surface tension for complexes in comparison with the pure protein, and also higher values of the tension in the steady state. Data establishing these trends have been reported for the following biopolymer pairs in aqueous media: legumin + dextran and legumin + maltodextrin (Antipova and Semenova,

1997; Antipova *et al.*, 1997; Pavlovskaya *et al.*, 1993); legumin + ι-carrageenan (Galazka *et al.*, 2000); β-lactoglobulin + pectin (Ganzevles *et al.*, 2006); bovine serum albumin + ι-carrageenan (Dickinson and Pawlowsky, 1997); bovine serum albumin + κ-carrageenan (Dickinson and Pawlowsky, 1998); and sodium caseinate + dextran sulfate (Jourdain *et al.*, 2009).

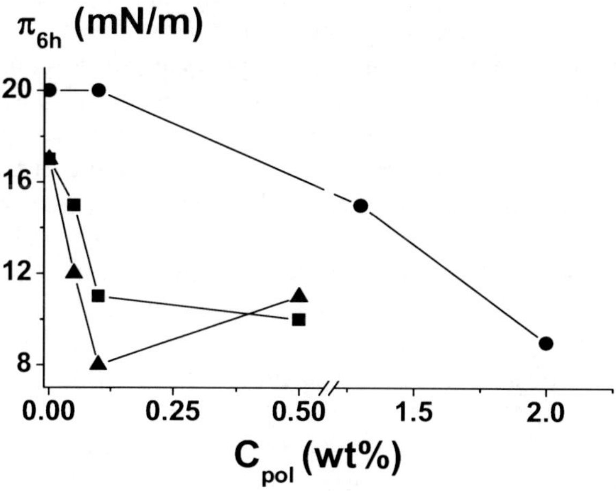

Figure 7.15 Effect of thermodynamically favourable interactions between biopolymers on protein surface activity at the planar oil–water or air–water interface. The surface pressure π reached after 6 hours is plotted against the polysaccharide concentration: (●), legumin (0.001 wt%) + dextran (M_w = 270 kDa) at *n*-decane–water surface at pH = 7.8 and ionic strength = 0.01 M, ($A_{ij}^* = -0.2 \times 10^5$ cm^3 mol^{-1}) (Pavlovskaya *et al.*, 1993); (■), legumin (0.001 wt%) + maltodextrin (MD6, M_w = 102 kDa) at air–water surface at pH = 7.2 and ionic strength = 0.05 M ($A_{ij}^* = -0.02 \times 10^5$ cm^3 mol^{-1}) (Belyakova *et al.*, 1999); (▲), legumin (0.001 wt%) + maltodextrin (MD10, M_w = 45 kDa) at air–water surface at pH = 7.2 and ionic strength = 0.05 M ($A_{ij}^* = -0.08 \times 10^5$ cm^3 mol^{-1}) (Belyakova *et al.*, 1999).

On the basis of available data, it would appear that there are several possible reasons that may account for the observed decrease in surface activity of proteins, depending on the strengths of their thermodynamically favourable interactions with different polysaccharides. In the case of a rather weak interaction, which does not lead to the formation of a stable complex between protein and polysaccharide, the decrease in the surface activity of protein is evidently determined by the corresponding decrease in thermodynamic activity (chemical potential) of the protein in the bulk of the mixed solution. For rather strong associative interactions, however, which lead to stable complex formation between protein and polysaccharide, the reduction in surface activity of protein is associated with the higher overall hydrophilicity of the complex. This increased hydrophilicity can be attributed to the attachment of hydrophilic molecules of polysaccharide to the protein, and also to blocking of potentially surface-active hydrophobic patches on the protein molecule due to the close proximity of the bulky polysaccharide.

The expected greater size of protein–polysaccharide complexes can reduce the diffusion rate of the adsorbing species towards the interface. This effect is especially important for small monomeric proteins. In addition, Ganzevles and co-workers (2006) have suggested that the diffusion of protein in the complexes may not solely be responsible for the slow surface tension decay. Rather, the gradual dissociation (and subsequent adsorption) of protein from complexes, when they are in close proximity to the interface, could also contribute to the behaviour.

As a specific example, we consider here the case of rather strong electrostatic complexes of sodium caseinate nanoparticles (0.5 wt%) with the highly charged polysaccharide dextran sulfate (1 wt%) at pH = 6.0 in 20 mM imidazole buffer solution. There was observed to be an increase in the steady-state value of the surface tension from 15.6 mN m^{-1} for the protein alone to 17 mN m^{-1} for the complex (Jourdain *et al.*, 2009). This reduction in surface activity could be attributed mainly to the fact that the complexes carry a larger total negative charge (Table 7.2) and are more hydrophilic than the pure protein (Figure 7.16a) (Semenova *et al.*, 2009). The greater hydrophilicity of the complex particles is reflected in the increasing values of their second virial coefficients with increasing dextran sulfate concentration (Figure 7.16a). Furthermore, the higher net charge of the complexes does not favour further complex adsorption once some complex particles have become adsorbed at the interface.

It was also demonstrated that the reduced surface activity of the protein did not affect the protein loading significantly (see Table 3.1 in chapter 3) (Semenova *et al.*, 1999a,b), or at the very least only slightly reduced it (Jourdain *et al.*, 2008, 2009). That is, for sodium caseinate (0.5 wt%) + dextran sulfate (1 wt%), the lower surface activity of the complexes, in comparison with the protein alone, led to a protein surface concentration in emulsions of 1.7 mg m^{-2} for the complex, as compared to 2.3 mg m^{-2} for the protein alone (Jourdain *et al.*, 2008, 2009).

Table 7.2 Effect of the presence of an anionic polysaccharide on the measured zeta potential (ζ) of emulsion droplets stabilized by proteins under experimental conditions corresponding to protein–polysaccharide complexation. In all cases the complexes were formed in the bulk aqueous medium before emulsification.

System	ζ (mV)		Reference(s)
	Protein Alone	Complex	
sodium caseinate + dextran sulfate (0.1 wt%) at pH = 6.0, 20 mM imidazole buffer	−32	−42	Jourdain *et al.* (2008)
sodium caseinate + dextran sulfate (1 wt%) at pH = 6.0, 20 mM imidazole buffer	−32	−55	Jourdain *et al.* (2008)
α_{s1}-casein + pectin (DE 76%) (10^{-4} wt%) at pH = 5.5, ionic strength = 0.01 M, ($A_{ij} = -7.6 \times 10^{-4}$ cm^3 mol g^{-2})	−15	−28	Dickinson *et al.* (1998); Semenova *et al.* (1999a)
0.5 wt% β-lactoglobulin + 0.1 wt% ι-carrageenan at pH = 6.0, without added salt	−44	−54	Gu *et al.* (2005b)
acidified milk drink + 0.5 wt% pectin (DE 72%) at pH = 4.0	+24	−19	Sejersen *et al.* (2007)

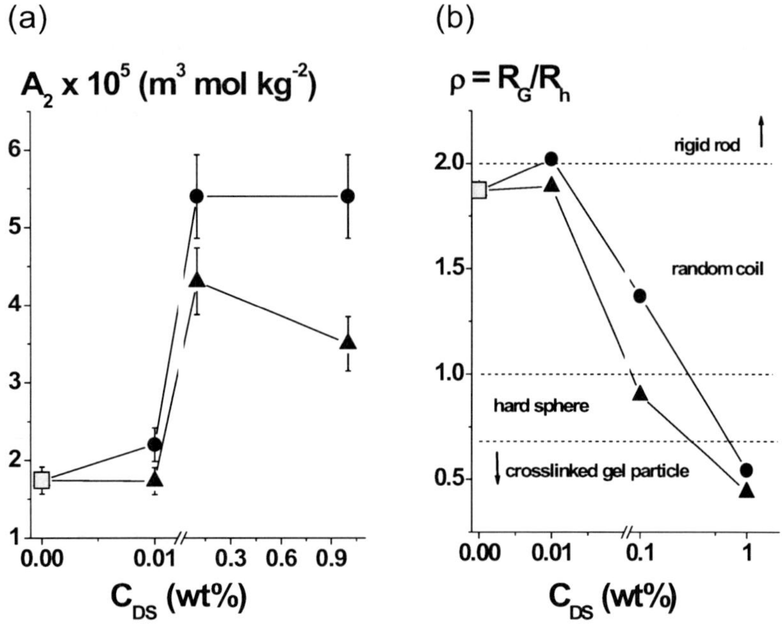

Figure 7.16 Dependence on the polysaccharide concentration C_{DS} of (a) the second virial coefficient A_2 and (b) the structure-sensitive parameter ρ of complexes of sodium caseinate + dextran sulfate: ●, complexes prepared in bulk solution; ▲, complexes prepared at the interface in a protein-stabilized foam; □, sodium caseinate alone. Reproduced from Semenova *et al.* (2009) with permission.

2.4.2. *Emulsifying Properties and Emulsion Droplet Size*

One might expect that complexation with polysaccharide could lead to a change in the emulsifying properties of protein due to the associative adsorption of polysaccharide onto the surface of the droplets together with the protein. Zeta potential measurements are indicative of a higher surface charge density on emulsion droplets prepared with complexes that those prepared with the protein alone (see Table 7.2). The formation of a more highly charged layer would be expected to confer a greater degree of electrostatic stabilization, and the formation of a thicker layer would be expected to provide additional steric stabilization during and after emulsion preparation (Dickinson and Eriksson, 1991; Dickinson *et al.*, 1998; Semenova, *et al.*, 1999a; Jourdain *et al.*, 2008, 2009). These expectations are indeed realized in practice. For instance, the associative interactions between β-lactoglobulin and various polysaccharides have been found to produce biopolymer complexes with impressive emulsifying properties (Dickinson and Galazka, 1991; Nagasawa *et al.*, 1996).

Using a combination of ultrasound measurements and diffusing wave spectroscopy (DWS), Gancz *et al.* (2006) have investigated emulsions (10 vol%) containing high-methoxyl pectin at pH = 3.5. They reported that sequential additions of the polysaccharide demonstrate four different stages in their interactions with the emulsion droplets, which were stabilized primarily with whey protein isolate. At a very low pectin concentration (< 0.02 wt%) no change could be detected. At a pectin content of 0.02–0.04 wt%, the presence of some new long-range interactions was indicated. Between 0.045 and 0.06 wt% pectin, there was a dramatic change in the ultrasound velocity and attenuation, and also in the DWS characteristics. This was attributed to a change in hydration layer properties and in the surface charge distribution on the oil droplets. Finally, at high concentrations of pectin (> 0.06 wt%), it was reported that bridging flocculation occurred, leading to changes in the effective droplet size and size distribution (Gancz *et al.*, 2006).

In the study of Neirynck *et al.* (2007), the electrophoretic mobility data indicated that whey protein-stabilized emulsion droplets became gradually more negatively charged with pectin addition at pH = 5.5. This change was not only reflected in a smaller average droplet size, but also in a significant improvement in the creaming stability of the emulsions.

2.4.3. *Surface Shear Viscosity*

Surface shear rheology at the oil–water interface is a sensitive probe of protein–polysaccharide interactions. In particular, there is considerable experimental evidence for a general increase in surface shear viscosity of protein adsorbed layers as a result of interfacial complexation with polysaccharides (Dickinson *et al.*, 1998; Dickinson and Euston, 1991; Dickinson and Galazka, 1992; Semenova *et al.*, 1999a; Jourdain *et al.*, 2009). One such example is the case of α_{s1}-casein + pectin at pH = 5.5 and ionic strength = 0.01 M (A_{ij}^{*} = – 334 × 10^5 cm^3/mol): the interfacial viscosity after 24 hours was found to be five times larger in the presence of pectin (*i.e.*, values of 820 ± 80 and 160 ± 20 mN m^{-1} with and without pectin, respectively) (Semenova *et al.*, 1999a).

2.4.4. *Stability of Emulsions with Respect to Flocculation*

Evidence for the flocculation of emulsion droplets is commonly derived from a combination of rheological and creaming stability experiments. For instance, a marked increase in both the viscoelasticity of emulsions of moderately high oil volume fraction (40 vol%) and the rapid serum separation at moderately low oil volume fraction (11 vol%) with addition of polysaccharide was found for emulsions containing α_{s1}-casein + high-methoxy pectin at pH = 5.5 and ionic strength = 0.01 M ($A_{ij}^{*} = -334 \times 10^5$ cm^3/mol) (see Figure 7.9a,b). This behaviour is attributable to droplet flocculation by a bridging mechanism (Dickinson *et al.*, 1998; Semenova *et al.*, 1999a), with the adsorbed molecules of both protein and polysaccharide playing a role in the biopolymer bridges between emulsion droplets covered by the mixed protein + polysaccharide adsorbed layers (Dickinson, 1998a).

Experiments on interactions of polysaccharides with casein micelles show similar trends to those with casein-coated droplets. For example, Maroziene and de Kruif (2000) demonstrated the pH-reversible adsorption of pectin molecules onto casein micelles at pH = 5.3, with bridging flocculation of casein micelles observed at low polysaccharide concentrations. In turn, Tromp *et al.* (2004) have found that complexes of casein micelles with adsorbed high-methoxy pectin (DE = 72.2%) form a self-supporting network which can provide colloidal stability in acidified milk drinks. It was inferred that non-adsorbed pectin in the serum was linked to this network owing to the absence of mobility of all the pectin in the micellar casein dispersion. Hence it seems that the presence of non-adsorbed pectin is not needed to maintain stability of an acid milk drink system. It was stated by Tromp *et al.* (2004) that the adsorption of pectin was irreversible in practical terms, *i.e.*, the polysaccharide did not desorb under the influence of thermal motion.

It has generally been observed that bridging flocculation is a feature of emulsions containing rather low concentrations of adsorbing biopolymer. However, a substantial increase in the biopolymer concentration tends to enhance steric and/or electrostatic stabilization of protein-coated droplets, hence leading to restabilization of the emulsion system (Dickinson and Pawlowsky, 1997; Dickinson, 1998a; Dickinson *et al.*, 1997; Jourdain *et al.*, 2008, 2009). Figure 7.17 shows mean droplet-size data from static light scattering demonstrating the restabilization of BSA-stabilized droplets with the addition of ι-carrageenan at contents beyond ~ 0.1 wt% (Dickinson and Pawlowsky, 1997). In this type of emulsion, the extent of flocculation (and restabilization) depends on the density of charged sulfate groups on the polymer — and therefore, by inference, on the strength of the protein–polysaccharide complex (Dickinson and Pawlowsky, 1997, 1998). That is to say, the associative interaction of the globular protein with κ-carrageenan is weaker than the interaction with ι-carrageenan owing to the greater proportion of anionic sulfate groups in the latter polysaccharide.

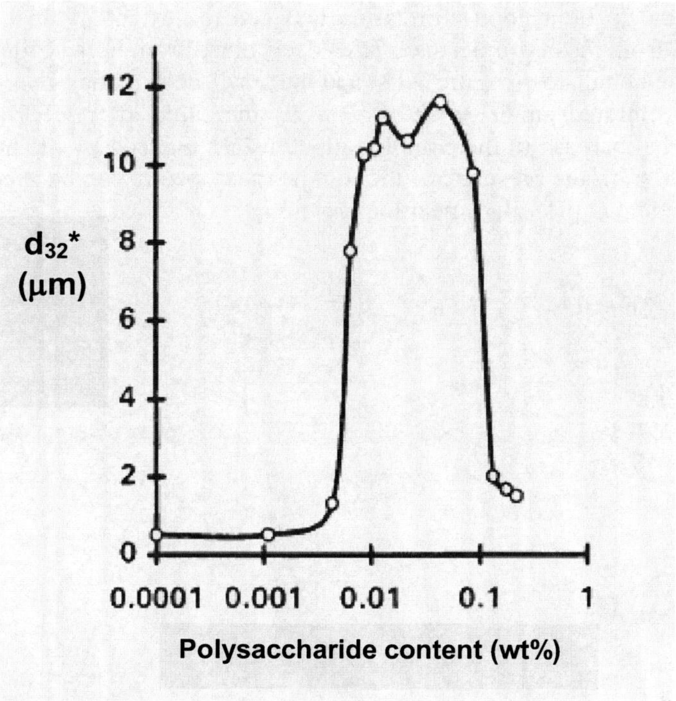

Figure 7.17 Influence of ι-carrageenan on the state of flocculation of BSA-stabilized emulsions (20 vol% oil, 1.7 wt% protein, pH = 6, ionic strength = 0.005 M) stored at 25 °C for 40 hours. The average droplet size measured by static light scattering (Malvern Mastersizer), d_{32}^*, is plotted against the polysaccharide concentration c_h added to the freshly made emulsion. Reproduced from Dickinson and Pawlowsky (1997) with permission.

Protein–polysaccharide interactions in food colloids can be affected by processing operations such as high-pressure treatment (HPT). Oil-in-water emulsion systems containing β-lactoglobulin + low-methoxy pectin were investigated by Dickinson and James (2000). It was suggested by the authors that, during sufficiently intense HPT, a globular protein such as β-lactoglobulin undergoes unfolding which can provide further opportunities for associative biopolymer–biopolymer attractive interactions. As a result of intensification of β-lactoglobulin–pectin attractive interactions induced by HPT, there was observed to be some oil droplet flocculation induced immediately following emulsification, which was attributed to a bridging mechanism. Figure 7.18 shows that this flocculation was more pronounced at pH < 6.0, where the electrostatic repulsion between like charges on the protein and polysaccharide is reduced. There was found to be a good correlation between the extent of flocculation inferred from mean droplet sizes (d_{43}) determined from light scattering of the diluted emulsion (Figure 7.18) and the small-deformation rheology of the concentrated emulsion (20 vol% oil) measured after HPT (Figure 7.19). The increase in the complex modulus G^* induced by the bridging flocculation in the presence of the low-methoxy pectin can be seen to be most evident at pH values near the protein pI.

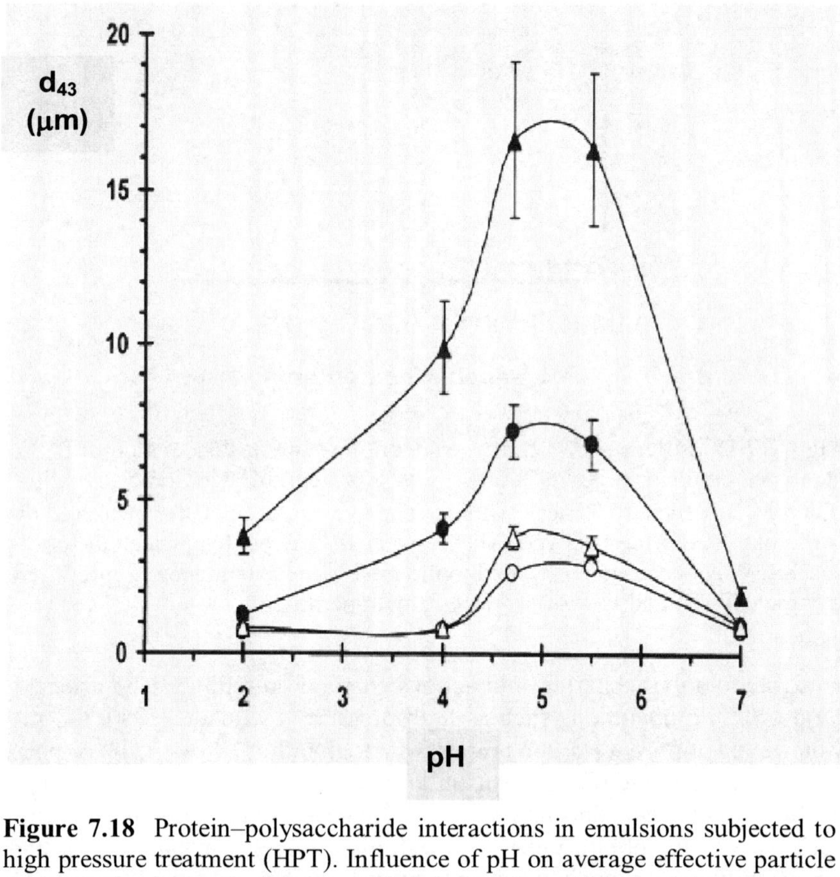

Figure 7.18 Protein–polysaccharide interactions in emulsions subjected to high pressure treatment (HPT). Influence of pH on average effective particle diameter d_{43} determined by static light scattering (Malvern Mastersizer) in emulsions (20 vol% soybean oil, 0.5 wt% β-lactoglobulin) prepared with untreated protein (open symbols) and high-pressure-treated (800 MPa for 30 min; filled symbols) protein in the absence (○,●) and presence (△, ▲) of 0.5 wt% pectin. Reproduced from Dickinson and James (2000) with permission.

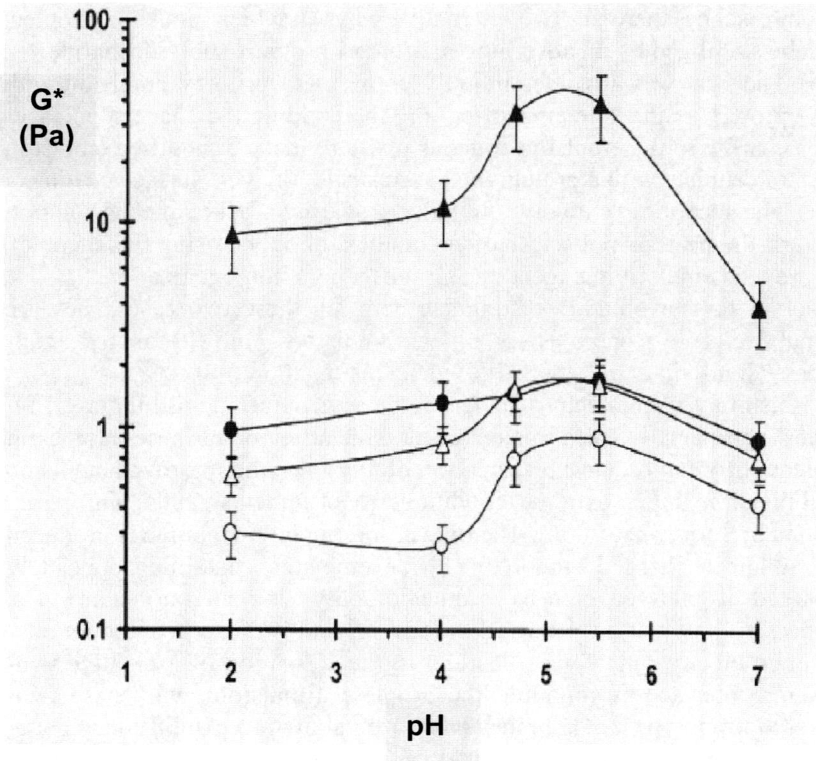

Figure 7.19 Influence of pH on the complex shear modulus G^* (at 1 Hz) of emulsions (20 vol% soybean oil, 0.5 wt% β-lactoglobulin) prepared with untreated (open symbols) and high pressure treated (800 MPa for 30 min; filled symbols) protein in the absence (○,●) and presence (△, ▲) of 0.5 wt% pectin. Reproduced from Dickinson and James (2000) with permission.

2.4.5. *Structure of Adsorbed Layers of Interacting Biopolymers — Effect of the Order of Addition to the Interface*

Adsorbed layers of mixed biopolymers are potentially non-equilibrium systems in terms of their structure and composition. Therefore one has to be aware that the impact of thermodynamically favourable interactions between biopolymers on the formation and stabilization of food colloids is dependent, not only on the total system composition, but also on the experimental procedure whereby the two interacting biopolymers are brought to the interface (McClements, 2004; Jourdain *et al.*, 2008, 2009; Dickinson, 2008a).

In essence, there are two alternative ways in which emulsion droplets can be stabilized by an adsorbing mixture of protein + polysaccharide:

(i) The first way involves initially preparing a primary emulsion with protein as the sole emulsifier, and then adding the charged polysaccharide to the emulsion aqueous phase to make a modified emulsion of droplets with a protein–polysaccharide 'bilayer' surface coating.

(ii) The second way involves initially preparing a bulk aqueous solution of the protein–polysaccharide complex, and then using this complex as the emulsifying agent during subsequent homogenization.

In order to conveniently distinguish between these two cases, they are simply referred to here as (i) 'bilayer emulsions' and (ii) 'mixed emulsions' (Jourdain *et al.*, 2008, 2009; Dickinson, 2008a).

The bilayer approach — under the more general title of the 'layer-by-layer' approach — has attracted recent wide attention because of its great potential for nanoscale encapsulation of nutrients and improved emulsion stabilization under extreme conditions (McClements, 2004; Gu *et al.*, 2005a,b; Surh *et al.*, 2006). The bilayer interfacial membrane is produced by adding a charged biopolymer to an emulsion containing oppositely charged droplets covered by another biopolymer (or a surfactant). The added biopolymer is attracted to the originally adsorbed biopolymer through electrostatic forces, leading to the formation of a putative two-layered interface surrounding the droplets. Emulsions with these two-layered biopolymer membranes have been shown to exhibit much better stability against environmental stresses (such as thermal processing, freeze–thaw cycling, lipid oxidation, and high salt content) as compared with those having single-layered biopolymer membranes (McClements, 2004; Gu *et al.*, 2005a,b; Surh *et al.*, 2006).

Figure 7.20 shows some data from the McClements group comparing properties of primary emulsions (β-lactoglobulin-stabilized) and secondary emulsions (β-lactoglobulin / ι-carrageenan-stabilized) at pH = 6.0 in the presence and absence of NaCl (Gu *et al.*, 2005b). Firstly, it is evident from Figure 7.20a that the ζ-potential values of droplets in the secondary emulsions are substantially more negative than those for the primary emulsions. This is consistent with adsorption of anionic ι-carrageenan on the surface of the negatively charged β-lactoglobulin-coated droplets. When the NaCl concentration was increased from 0 to 100 mM, the magnitude of the ζ-potential for the primary emulsion dropped to 68 % of its initial value, but that for the secondary emulsion fell only to 95 % of its initial value (Gu *et al.*, 2005b). On increasing the salt content from 0 to 500 mM, the average particle diameter measured for the primary emulsions increased considerably, particularly for NaCl concentrations beyond 100 mM (Figure 7.20b).

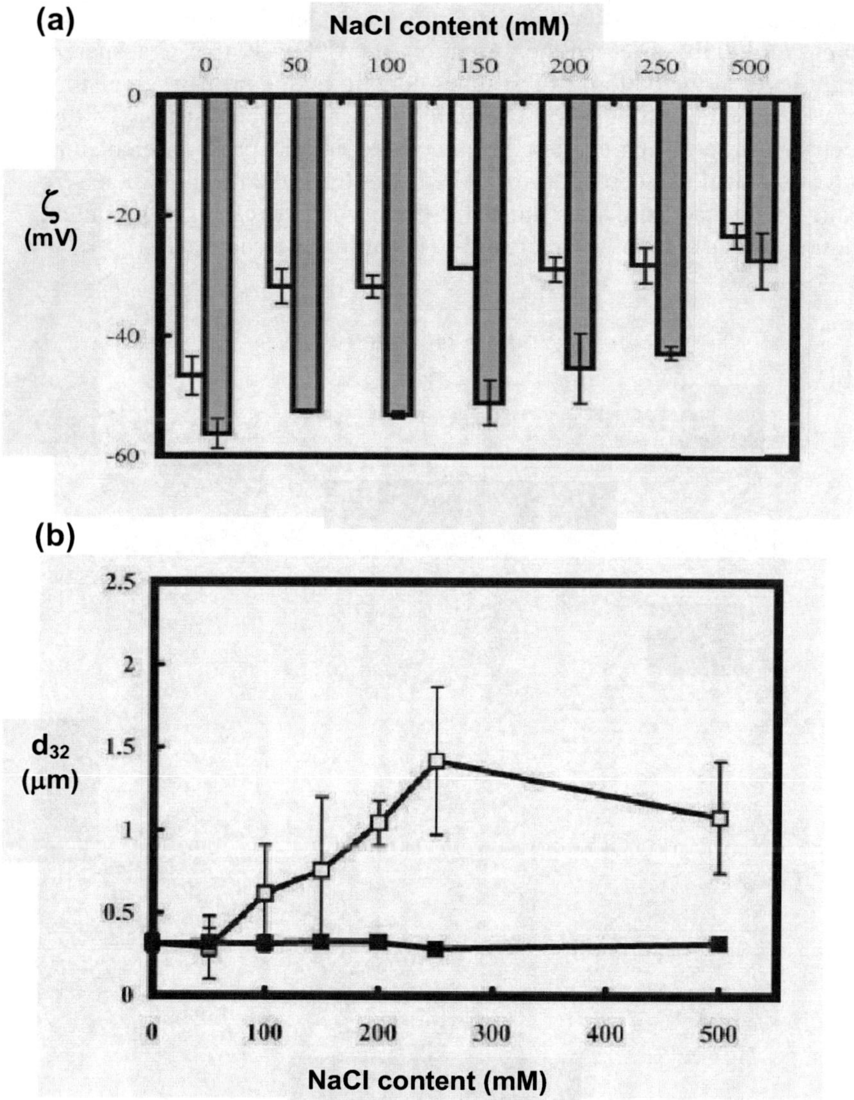

Figure 7.20 Influence of salt content on properties of bilayer emulsions based on β-lactoglobulin + ι-carrageenan at pH = 6.0: (a) zeta potential ζ, and (b) mean particle diameter d_{32}. The primary emulsion (open symbols) contained 5 wt% oil and 0.5 wt% protein; the secondary emulsions (filled symbols) contained an additional 0.1 wt% polysaccharide. Reproduced from Gu *et al*. (2005b) with permission.

In addition to reporting the data for the droplet sizes (Figure 7.20b), Gu *et al.* (2005b) observed the formation of a relatively thick transparent serum layer at the bottom of test tubes containing the primary emulsions for a salt content of ≥ 0.1 M. The stability measurements suggested that extensive droplet aggregation had occurred in the primary emulsions above a critical salt content (~ 0.1 M). In contrast, the mean particle size of the secondary emulsions did not change with increasing salt content (Figure 7.20b), and no visual creaming instability was detected.

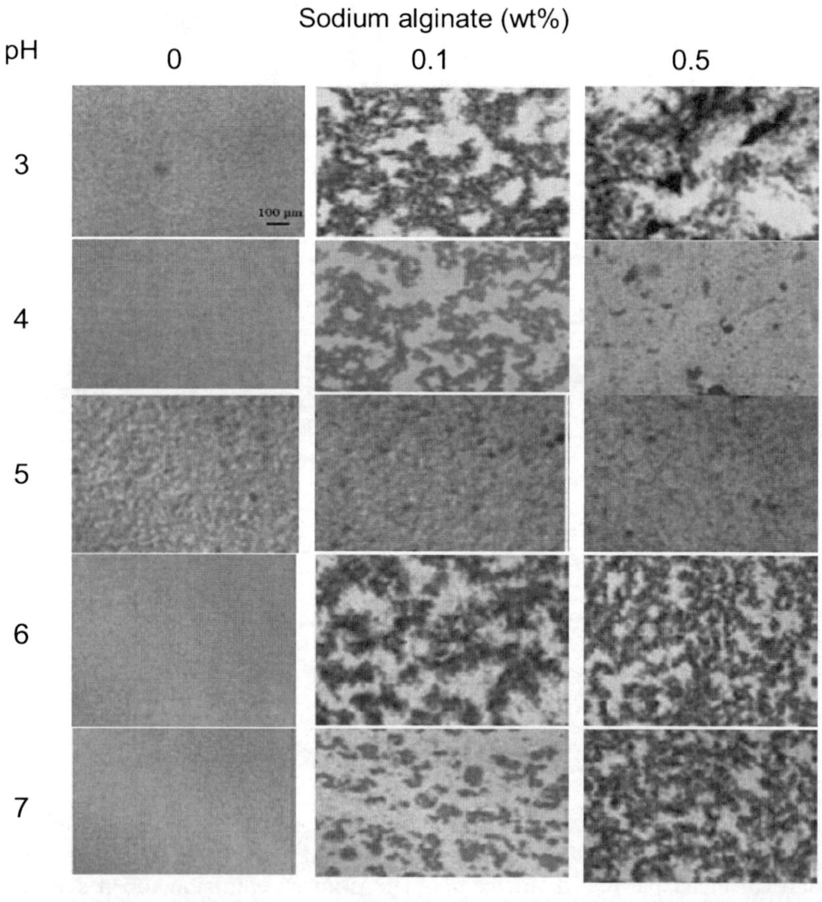

Figure 7.21 Effect of pH on microstructure of primary emulsions (no added polysaccharide) and secondary emulsions (0.1 or 0.5 wt% sodium alginate) based on 5 wt% oil, 0.45 wt% β-lactoglobulin, and 5 mM phosphate buffer. Reproduced from Pongsawatmanit *et al.* (2006) with permission.

Despite its obvious conceptual appeal, there is unfortunately a major practical problem in trying to exploit the sequential adsorption approach. The problem concerns the tendency of such emulsions to become extensively flocculated during preparation, even under apparently favourable conditions where the protein surface should be completely saturated with the polysaccharide according to normal thermodynamic considerations (McClements, 2005; Guzey and McClements, 2006; Dickinson, 2007, 2008a). Two different mechanisms are involved: bridging flocculation, when the polysaccharide content is so low that droplet collisions occur faster than the rate of polysaccharide saturation of the protein-coated droplet surfaces (Dickinson and Pawlowsky, 1997; McClements, 2005; Jourdain *et al.*, 2008); and depletion flocculation, when the concentration of the non-adsorbed polymer exceeds a certain critical value (Dickinson, 2003; McClements, 2005; Surh *et al.*, 2006). As an illustration of the problem, Figure 7.21 compares some micrographs of primary emulsions (β-lactoglobulin-stabilized) and secondary emulsions (β-lactoglobulin–alginate-stabilized) over a range of pH values. This set of images clearly indicates that there is extensive droplet flocculation in these emulsions on both sides of the isoelectric point in the presence of polysaccharide (Pongsawatmanit *et al.*, 2006).

Especially troublesome is bridging flocculation. It is therefore much more convenient to prepare emulsions with protein and polysaccharide components both present together in the aqueous medium before homogenization (Dickinson *et al.*, 1998; Garti *et al.*, 1999; Dickinson, 2008a). Moreover, in a direct comparison between the two techniques (Jourdain *et al.*, 2008), it has been demonstrated that the experimentally more straightforward mixed emulsion approach can actually produce a better level of stability than the bilayer approach.

Using light-scattering measurements, an attempt was made recently (Semenova *et al.*, 2009) to understand the differences between properties of 'normal' sodium caseinate–dextran sulfate complexes, as formed in aqueous solution by simple mixing, and equivalent complexes formed initially at the interface in an aqueous foamed system ('interface' complexes). In this investigation at pH = 6.0 (20 mM imidazole buffer), the assumption was made that the complexes formed at the air–water interface of a foam would resemble those residing at the oil–water interface in an emulsion. In the first place, as a result of sodium caseinate–dextran sulfate interactions, there was observed to be a decrease in weight-average molar weight, M_w, of the complexes with increasing polysaccharide concentration. This implies some dissociation of the original sodium caseinate nanoparticles as a consequence of interaction with the highly charged polysaccharide, involving strong electrostatic repulsion between the sulfate groups on the complexed molecules of dextran sulfate. As a result of such interactions, the derived values of the second virial coefficient, A_2, were found to become more positive with the addition of the highly negatively charged dextran sulfate molecules to the sodium caseinate particles (Figure 7.16a). This implies a significant increase in the thermodynamic affinity of the biopolymer particles for the aqueous medium (*i.e.*, an enhanced hydrophilicity) due to enrichment of aggregated caseinate nanoparticles with the electrostatically bound polysaccharide. However, although the positive values of A_2 for dextran sulfate concentrations of 0.1 wt% and 1 wt% were found to be substantially larger than the value determined for the original sodium caseinate particles ($A_2 = 1.7 \times 10^{-5}$ m^3 mol kg^{-2}), the values were still two orders of magnitude lower than that for pure dextran sulfate ($A_2 = 264 \times 10^{-5}$ m^3 mol kg^{-2}). This suggests that, whilst there is some incorporation of additional negative charge from dextran sulfate molecules associating with the original negatively charged protein particle, there is also extensive neutralization of the net negative charge on the dextran sulfate molecules as a result of the strong attraction between individual opposite charges on the protein and polysaccharide during complex formation.

There were found to be three important differences between the 'normal' complexes made in the bulk solution and the 'interface' complexes made from the foam (Semenova *et al.*, 2009).

(1) The M_w value obtained for 'interface' complexes at the maximum studied concentration of dextran sulfate (1 wt%) was smaller than that obtained for 'normal' complexes.

This result is indicative of a greater extent of dissociation of the original micelle-like protein particles when the interaction first occurs between polysaccharide and protein already adsorbed at the air–water interface. In an adsorbed casein(ate) layer, there is strong interaction of hydrophobic regions of the protein molecules with the fluid interface (Dickinson, 1995; 1998b, 1999, 2003). This may lead to greater availability of positively charged residues on the protein for close electrostatic interaction with sulfate groups on the anionic polysaccharide. Hence the fluid interface may act as a template for formation of smaller composite sodium caseinate–dextran sulfate particles, which are able to maintain their characteristic sizes and nanostructure following desorption.

(2) The larger thermodynamic affinity for the aqueous medium for the case of 'normal' complexes as compared to 'interface' complexes (Figure 7.16a) correlates well with ζ-potential values of oil droplets in 'mixed' and 'bilayer' emulsions (Table 7.3).

It seems that there is probably greater availability of positively charged residues on the adsorbed protein for electrostatic interaction with sulfate groups of the anionic polysaccharide. This could lead to a greater extent of neutralization of dextran sulfate as a result of complex formation, and consequently to a lower thermodynamic affinity of the complexes for the aqueous medium and a lower value of the ζ-potential for emulsion droplets in 'bilayer' emulsions.

(3) Derived values of the structure parameter ρ for the complex particles formed at the air–water interface were smaller than those for complexes formed in the bulk aqueous solution (see Figure 7.16b and Table 7.3).

This result seems to be in conformity with the inference above concerning the greater availability of positively charged protein residues available for attraction to negatively charged groups on dextran sulfate when the two biopolymers come together in the protein adsorbed layer at the air–water interface. Consequently there is dissociation of the original protein particles and enhanced electrostatic bonding in the interior of the complex particles, resulting in formation of gel-like complex particles with 1 wt% of dextran sulfate.

Therefore, two contributory factors may provide an explanation for more effective electrostatic / steric stabilization of the so-called 'mixed' emulsions in comparison with the sequentially assembled biopolymer interfaces of the 'bilayer' emulsions: firstly, a greater hydrophilicity of the adsorbed protein–polysaccharide complexes, caused by the larger net negative charge, and, secondly, a more bulky architecture of the 'normal' complexes as compared to the 'interface' complexes.

Evidence for a substantial dependence of the surface properties of protein + polysaccharide layers on the sequence of adsorption of the biopolymer components has also been found for mixtures of β-lactoglobulin + low-methoxy pectin adsorbing under acidic conditions at macroscopic air–water and oil–water interfaces (Ganzevles *et al.*, 2007). Based on neutron reflectivity analysis, a significantly different structure was determined for the two-component interfacial layer prepared by sequential adsorption as compared to the layer formed by simultaneous adsorption (Ganzevles *et al.*, 2008). The direct co-adsorption of the protein–polysaccharide complexes to the interface led to rather uniform mixed layers. But for complex adsorption at the previously formed protein monolayer, the postulated structure was interpreted as being more like a bilayer, with the density of packing in the secondary layer dependent on whether the protein–polysaccharide complex was negatively charged or neutral.

Table 7.3 Relationship between molecular parameters (A_2, ρ) of sodium caseinate (0.5 wt%) + dextran sulfate complexes at pH = 6.0 formed in the bulk and at the interface of a protein foam, and the corresponding properties (d_{43}, ζ) of the 'bilayer' and 'mixed' emulsions (20 vol% oil, 0.5 wt% sodium caseinate) containing 0.1 or 1.0 wt% dextran sulfate (Jourdain *et al.*, 2008; Semenova *et al.*, 2009).

Method of preparation	d_{43} (μm)	ζ-potential (mV)	$A_2 \times 10^5$ (m^3 mol kg^{-2})	$\rho = R_G/R_h$
0.1 wt% dextran sulfate				
'bilayer emulsion'	8.3	−39.4 ± 2.5	4.3 ± 0.4	0.9
'mixed emulsion'	1.1	−41.8 ± 2.9	5.4 ± 0.5	1.4

'bilayer emulsion'

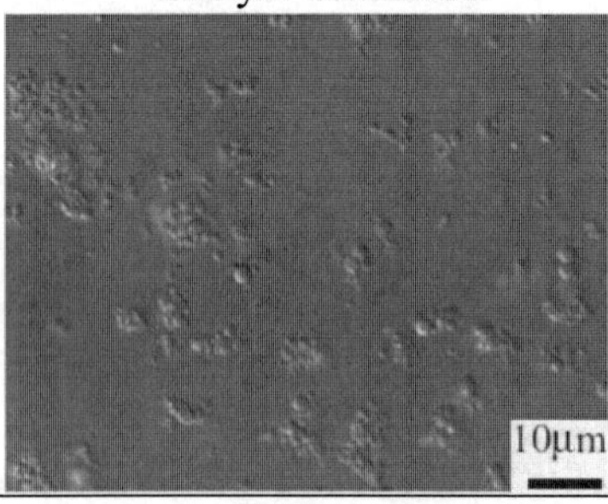

'mixed emulsion'

Method of preparation	d_{43} (μm)	ζ-potential (mV)	$A_2 \times 10^5$ (m^3 mol kg^{-2})	$\rho = R_G/R_h$
1 wt% dextran sulfate				
'bilayer emulsion'	1.8	−50.0 ± 2.5	3.5 ± 0.4	0.4
'mixed emulsion'	1.1	−55.2 ± 2.9	5.4 ± 0.5	0.5

'bilayer emulsion'

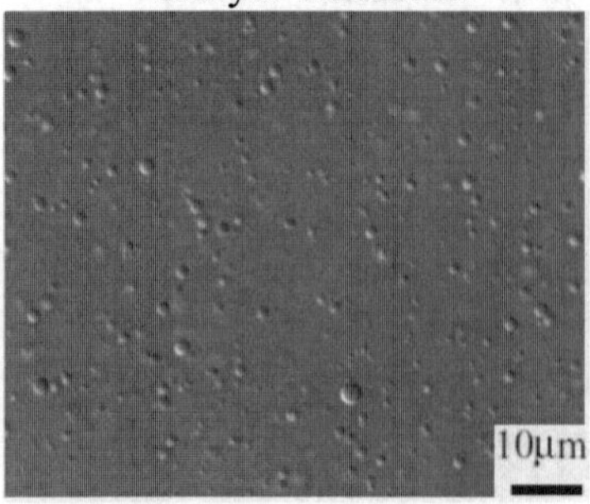

'mixed emulsion'

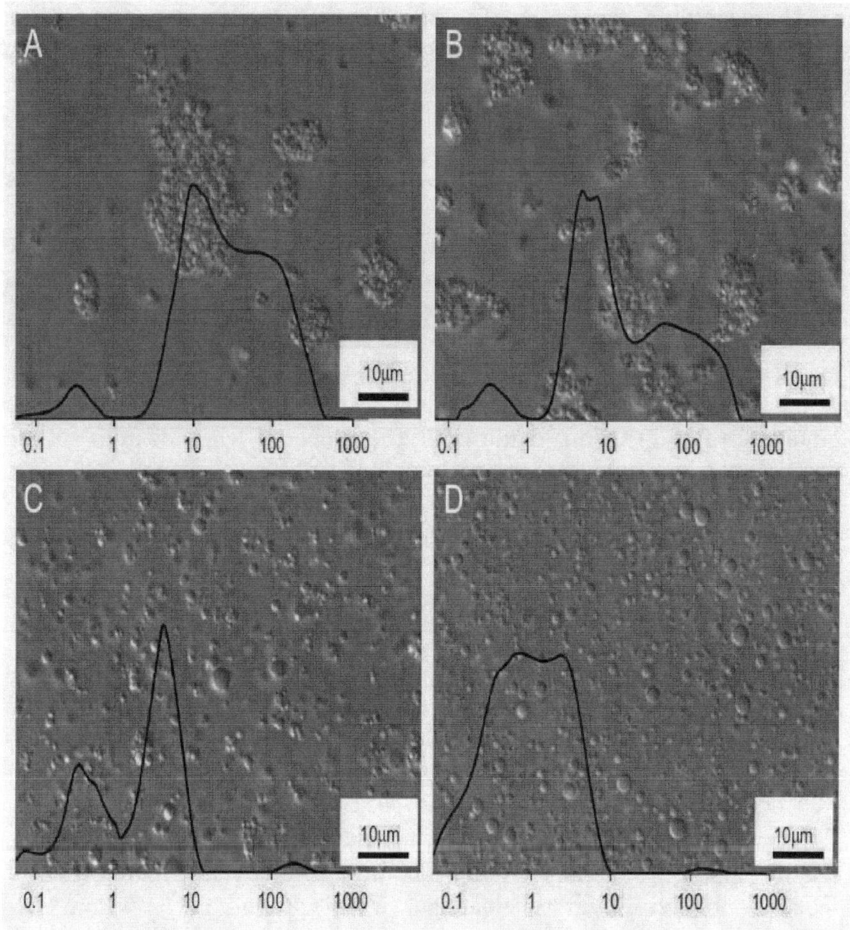

Figure 7.22 Microstructure of acidified 'mixed' emulsions (20 vol% oil, 0.5 wt% sodium caseinate) containing different concentrations of dextran sulfate (DS). Samples were prepared at pH = 6 in 20 mM imidazole buffer and acidified to pH = 2 by addition of HCl. Emulsions were diluted 1 : 10 in 20 mM imidazole buffer before visualization by differential interference contrast microscopy: (A) no added DS; (B) 0.1 wt% DS; (C) 0.5 wt% DS; (D) 1 wt% DS. Particle-size distributions of the diluted emulsions determined by light-scattering (Mastersizer) are superimposed on the micrographs, with horizontal axial labels indicating the particle diameter (in µm). Reproduced with permission from Jourdain *et al.* (2008).

Using soluble complexes of sodium caseinate + dextran sulfate, some impressively stable 'mixed' emulsions could be prepared by single-step homogenization under acidic conditions (pH = 2) (Jourdain *et al.*, 2008). Without any added dextran sulfate, the average droplet size was found to be slightly greater at pH = 2 than at pH = 6 (d_{43} = 2.0 μm as compared to d_{43} = 1.2 μm), presumably due to enhanced flocculation. The d_{43} values for the emulsions prepared at pH = 2 with 0.5 or 1 wt% dextran sulfate (*i.e.*, corresponding to polysaccharide contents sufficient to saturate the surface of protein-coated droplets) were similar to those for emulsions made at pH = 6, where no flocculation was observed (see Figure 7.22).

It was reported by Jourdain *et al.* (2008) that it was impossible to make a proper emulsion with sodium caseinate alone at pH = 4 because of the very low protein solubility close to p*I*. Nevertheless, a fine stable emulsion ($d_{43} \approx 1.0$ μm) could easily be prepared when dextran sulfate was present in a modest amount. In contrast, the monodisperse 'bilayer' emulsion could not be prepared at low pH because the primary caseinate-stabilized emulsion was already either partially aggregated (pH = 2) or completely insoluble (pH = 4) under these conditions.

2.4.6. *Gelation in Mixed Biopolymer Systems*

The existence of thermodynamically favourable interactions between two biopolymers influences their gel-forming ability in aqueous media. As an example, let us refer here to the effect of low-methoxyl amidated (LMA) pectin (DE = 30 %) on the gelation ability of sodium caseinate (Matia-Merino, 2003; Matia-Merino *et al.*, 2004; Dickinson, 2008a,b).

The presence of LMA pectin during the acidification with glucono-δ-lactone caused a dramatic drop in the value of the elastic modulus G' of the acid-induced sodium caseinate gel (2 wt% protein, pH ≈ 4) from G' = 80 Pa without pectin to $G' \to 0$ with 0.8 wt% pectin. Without the added polysaccharide, the protein gel had the character of a soft solid with a yoghurt-type appearance and texture. Its fractal-type microstructure contained casein particles, chains, strands and clusters (white areas in Figure 7.23a), which immobilized the water in pores of size up to 30 μm. The dramatic reduction in elasticity of the protein gel with increasing pectin content could be directly correlated with the loss in aggregation capacity of the protein under acidification, due to the rather strong complexation with pectin as the pH approaches the p*I* of the protein (Figure 7.23b–d) (Matia-Merino *et al.*, 2004; Dickinson, 2008a,b). The interruption of the formation of interconnected casein strands and clusters in the presence of the hydrocolloid is attributed to the blocking of hydrophobic sites on the caseinate particles by the much more hydrophilic molecules of pectin.

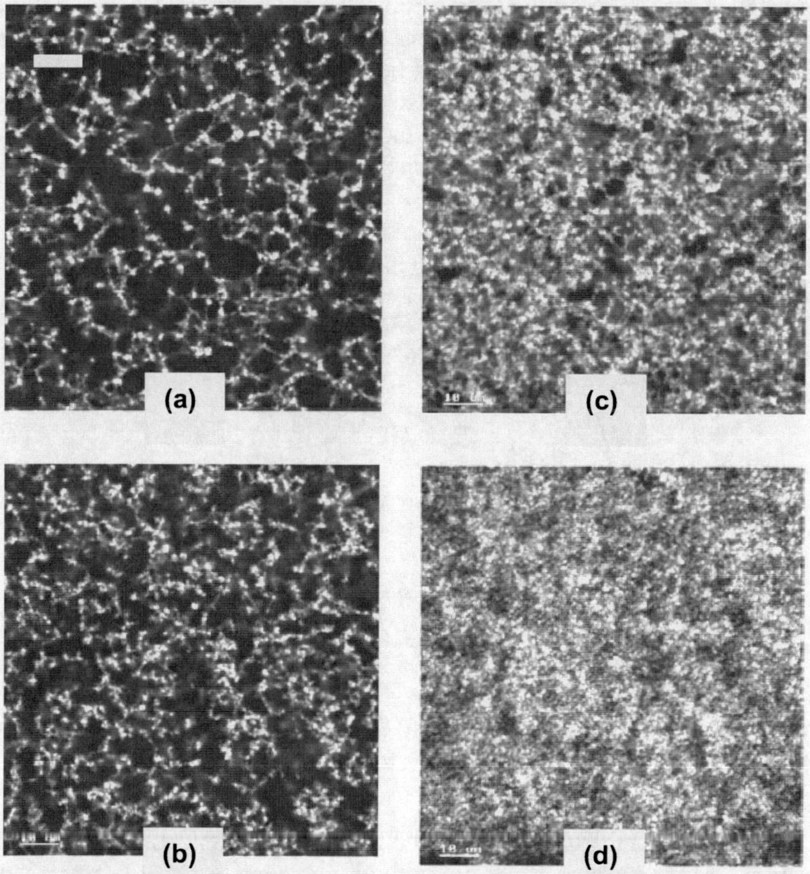

Figure 7.23 Influence of low-methoxyl amidated pectin on microstructure of caseinate gels acidified with glucono-δ-lactone. The images show the gels (2 wt%) after 6 hours of acidification to pH = 4.0 for different added polysaccharide concentrations (wt%): (a) 0; (b) 0.1; (c) 0.4; (d) 0.8. Scale bar = 10 μm. Reproduced from (Matia-Merino *et al.*, 2004) with permission.

2.5. *Application of Biopolymer Complexes for Encapsulation*

In principle, an emulsion contained within an emulsion droplet, as shown schematically in Figure 7.24, can be regarded as an excellent reservoir for 'active matter'. Benichou *et al.* (2004) have suggested that complexes of proteins with polysaccharides can be employed to improve the steric stabilization of such a double ('duplex') emulsion by forming a thick multilayered coating on the outer droplet surface. The motivation behind this suggestion is the recognition that the relatively large size of double (multiple) emulsion droplets — and hence their intrinsic thermodynamic instability — is a significant drawback of multiple emulsion technology. Moreover, use of conventional low-molecular-weight emulsifiers has not really provided an adequate solution (Dickinson and McClements, 1995).

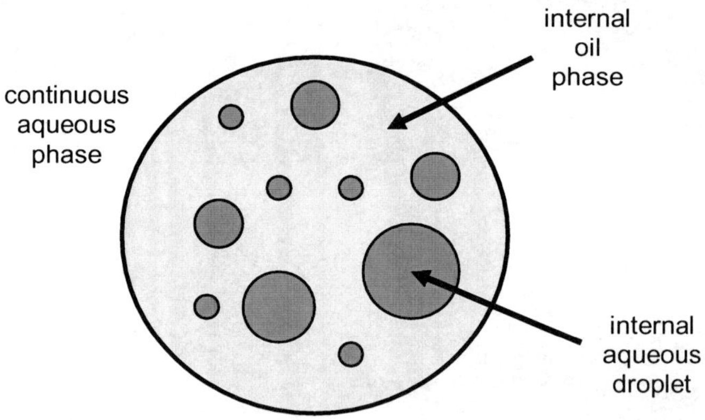

Figure 7.24 Schematic presentation of a multiple (double) emulsion droplet of the most common water-in-oil-in-water type.

In addition to the necessary protection of the contents of the emulsion droplets, effective encapsulation technology requires that the release of the 'active matter' be controlled at a specified rate. Benichou *et al.* (2004) have demonstrated that a mixture of whey protein isolate (WPI) and xanthan gum can be successfully used for the controlled release of vitamin B_1 entrapped within the inner aqueous phase of a multiple emulsion. The release profile, as a function of the pH of the external aqueous phase, is plotted in Figure 7.25. We can observe that the external interface appears more effectively 'sealed' against release of the entrapped vitamin at pH = 2 than at pH = 4 or 7. It was reported that an increase in the protein-to-polysaccharide ratio reduced the release rate at pH = 3.5 (Benichou *et al.*, 2004). More broadly, the authors suggest that compatible blends of biopolymers (hydrocolloids and proteins) should be considered excellent amphiphilic candidates to serve as 'release controllers' and 'stability enhancers' in future formulations of double emulsions. So perhaps mixed compatible biopolymers will at last allow researchers to realise the (as yet) unfulfilled promise of double emulsions for food-related applications — namely, to make colloid-size double emulsion droplets of good long-term stability[2] using food-grade ingredients, and then to be able to trigger and control release of encapsulated compounds at will.

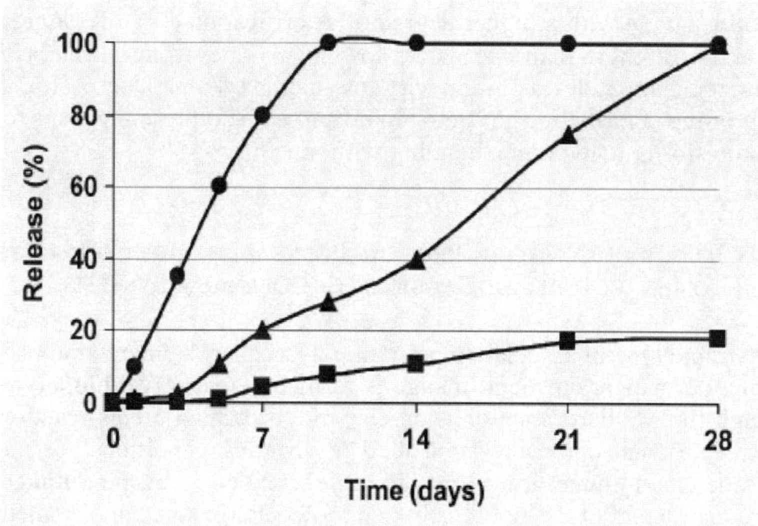

Figure 7.25 Effect of the external aqueous phase pH on the release profile of vitamin B_1 from multiple emulsions stabilized with WPI/xanthan gum as the external (secondary) emulsifier: (●) pH = 7, (▲) pH = 4, (■) pH = 2. Reproduced from Benichou *et al.* (2004) with permission.

The thermodynamically favourable interaction between biopolymers and mucin (or mucus) has received considerable attention in pharmaceutical formulation science as a platform for controlled delivery (Sriamornsak and Wattanakorn, 2008). The advantage of using biopolymers with muco-adhesive ability for delivery of nutraceuticals is the potential to prolong the residence time within a specified region of the body by creating intimate contact with the absorbing membrane (Gu *et al.*, 1988). It has been suggested that the extent of rheological synergism between biopolymer and mucin can be used as an *in vitro* parameter to determine the muco-adhesive properties of a material (Bonacucina *et al.*, 2004). In

[2] In principle, the long-term stability can be greatly improved by replacing the internal (macro)emulsion by a *nano*emulsion or, better still, a thermodynamically stable *micro*-emulsion. (We note that, in this context, the prefix 'micro' denotes an entity that is smaller than 'nano'!)

particular, Sriamornsak and Wattanakorn (2008) established that pectins show rheological synergism when mixed with mucin: there is an increase in the dynamic moduli and a decrease in the loss tangent. The extent of this synergism could be correlated with the degree of physical entanglement of the gel network — depending on pectin type, pectin concentration, mucin concentration, and the dispersion medium. It was found that a pectin sample with a higher degree of esterification (DE) has a greater interaction strength than one with a lower DE. The strongest interaction at low ionic strength was found with low-methoxyl amidated pectin, as a result of the contributions of both amide and carboxylic acid groups to its structure-forming hydrogen bonding with mucin.

B. Impact of Covalent Protein–Polysaccharide Conjugates on Structure and Stability of Colloidal Systems

An extreme type of associative protein–polysaccharide interaction occurs when a permanent covalent linkage is formed between two biopolymers through the Maillard reaction (non-enzymic browning). This reaction is typically carried out under controlled dry-heating conditions. Terminal and side-chain amine groups on the protein molecule become linked to the reducing end of the polysaccharide molecule (or to a simple sugar or oligosaccharide). The product of the Maillard reaction is therefore not a single composite macromolecular species. As a consequence of the large number of potentially reactive groups on each protein molecule, such conjugates are necessarily somewhat polydisperse with respect to both size and composition (Choi *et al.*, 2005).

It has been demonstrated in many studies that Maillard-type protein–polysaccharide conjugates can have excellent emulsifying and steric stabilizing properties, especially under solution conditions where the protein alone is poorly soluble (*i.e.*, at a pH in the vicinity of p*I*, or at a high ionic strength), or after thermal processing (Kato *et al.*, 1990; Dickinson and Galazka, 1991; Dickinson and Semenova, 1992; Dickinson, 1993, 2003, 2008a,b, 2009; Shepherd *et al.*, 2000; Akhtar and Dickinson, 2003, 2007; Wooster and Augustin, 2006; Fechner *et al.*, 2007). For example, whereas sodium caseinate loses its emulsifying properties when it starts to precipitate on acidification below pH ~ 5.5, the conjugate with dextran or maltodextrin exhibits good solubility and is an effective emulsifier under acidic conditions (Shepherd *et al.*, 2000; Fechner *et al.*, 2007). Figure 7.26 shows some solubility data for sodium caseinate following glycosylation with dextran.

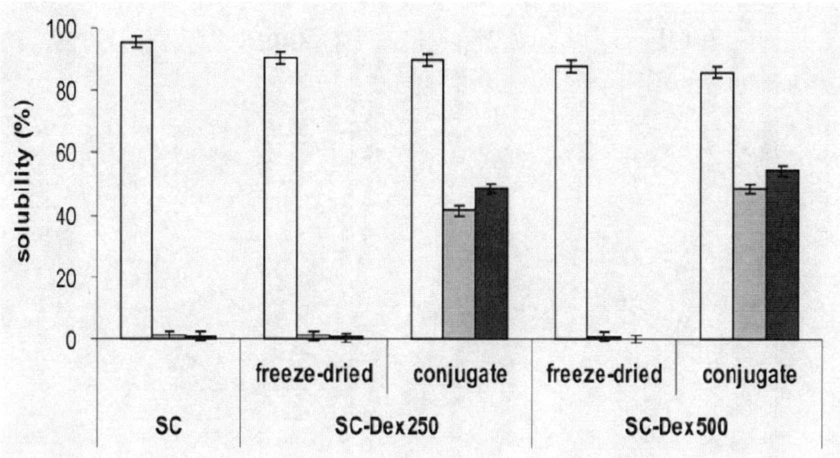

Figure 7.26 Solubility of sodium caseinate (SC) before and after glycosylation with dextran (Dex) (M_w = 250 kDa or 500 kDa): (white column) pH = 7; (grey column) pH = 4.0; (black column) pH = 4.5. The labels 'freeze-dried' and 'conjugate' relate to the non-treated and heat-treated samples, respectively. Reproduced from Fechner *et al.* (2007) with permission.

For a range of food protein systems, it has been established that the emulsifying capacity of the conjugate and the long-term stability of the resulting emulsion are sensitive to the polysaccharide molecular weight and the molar ratio of polysaccharide to protein (R). It would seem that, for each protein–polysaccharide combination, there is a critical value of R that confers optimum stabilization (Dickinson and Semenova, 1992). This is illustrated in Figure 7.27a for the case of conjugates of legumin with two dextran polymers of different molecular weight.

A significant degree of sensitivity of the functional properties of the conjugates to protein / polysaccharide ratio and polysaccharide molecular weight might be expected for various reasons.

(i) Each polysaccharide molecule has only one reducing group capable of reacting with amine groups on the protein. So one would expect the extent of covalent linkage formation to increase with increasing proportion of polysaccharide and with decreasing molecular weight.

(ii) It is important that binding of polysaccharide to the protein does not cause any substantial steric shielding of hydrophobic groups on the protein which might significantly reduce its surface activity. Such an effect would presumably be more likely to occur for a greater proportion of polysaccharide and for a higher molecular weight.

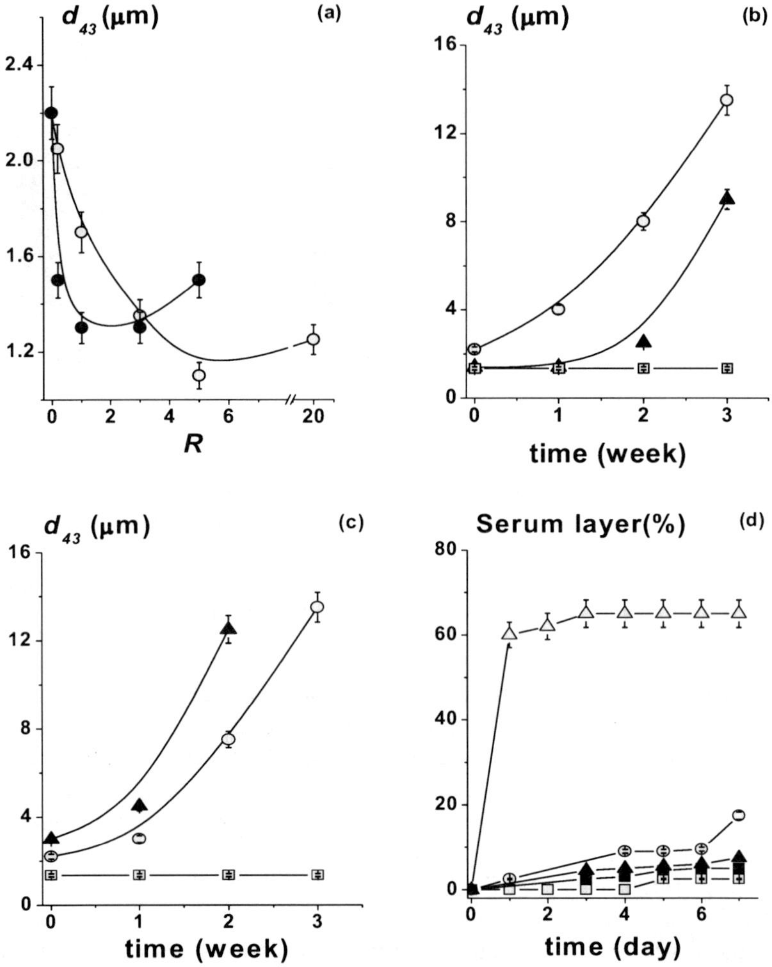

Figure 7.27 Effect of Maillard conjugate formation between legumin and two dextrans of different molecular weights (Dex1, $M_w = 4 \times 10^4$ Da; Dex2, $M_w = 5 \times 10^5$ Da) on properties of emulsions prepared with conjugates (10 vol% oil, pH = 8.0, ionic strength = 0.1 M, 0.5 wt% legumin): (a) average droplet size d_{43} plotted against polysaccharide/protein molar ratio R: (○) Dex1, (●) Dex2; (b) at $R = 3$, d_{43} plotted against emulsion storage time: (○) legumin alone, (▲) mixture of legumin + Dex1, and (□) legumin–Dex1 conjugate; (c) as for (b) except with Dex2; (d) at $R = 3$, time-dependent serum layer formation: (○) legumin alone, (□)legumin–Dex2 conjugate, (△) mixture of legumin + Dex2, (■) legumin–Dex1 conjugate, and (▲) mixture of legumin + Dex1. Data taken from Dickinson and Semenova (1992).

(iii) It is important that there is no excess of unreacted polysaccharide remaining after the dry-heating treatment. This is because any non-adsorbed polysaccharide present in the resulting emulsion aqueous phase could destabilize through depletion flocculation (Dickinson and Semenova, 1992; Dickinson, 2008a,b). This kind of instability phenomenon is rather a common one found in mixtures of thermodynamically incompatible biopolymers (as discussed in chapter 3). For the case of legumin + dextran, the effect was found to be more pronounced for the polysaccharide of higher molecular weight (Dex2) (see Figures 7.27b,c,d).

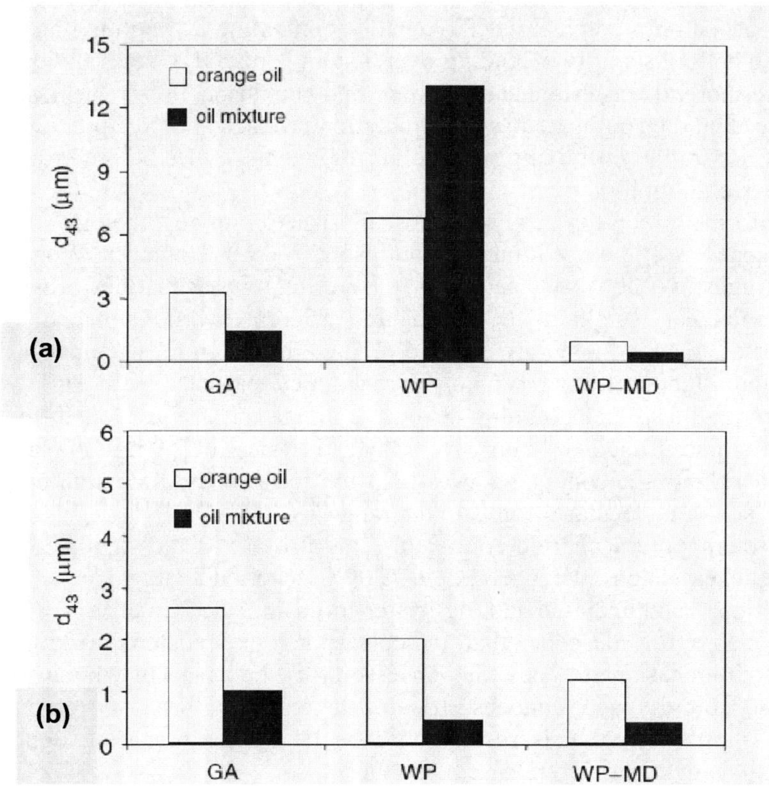

Figure 7.28 Average droplet size d_{43} of O/W emulsions (20 vol% oil, 2 wt% emulsifier) containing orange oil (□) or triglyceride oil (■), made with gum arabic (GA), whey protein (WP) or a whey protein–maltodextrin conjugate (WP–MP) as the emulsifier, after storage for 40 days at ambient temperature: (a) pH = 3.2, (b) pH = 7.0. Reproduced from Akhtar and Dickinson (2007) with permission.

A highly effective and rather inexpensive combination of ingredients for preparing Maillard-type conjugate emulsifiers for food use is whey protein isolate + maltodextrin. Under neutral (pH = 7.0) or acidic (pH = 3.2) conditions, it was demonstrated that a whey protein–maltodextrin conjugate made from maltodextrin of DE = 19 (8.7 kDa) can produce smaller and more stable emulsion droplets than whey protein alone or gum arabic (Akhtar and Dickinson, 2007; Dickinson, 2008a,b). Figure 7.28 shows that the emulsifying performance of the conjugate is especially impressive under acidic conditions using either triglyceride oil or orange oil as the dispersed phase. It was also demonstrated that this same conjugate emulsifier could be used to prepare low-pH beverage emulsions based on a commercial flavour oil (containing a weighting agent) having colloid stability extending over a storage period of several weeks. Furthermore, it was established (Akhtar and Dickinson, 2007; Dickinson, 2008b) that the conjugate-based formulation produced no precipitation or phase separation on mixing with colouring agents, either before or after extensive dilution.

Information on the detailed nanoscale structure of adsorbed layers of conjugates is still very limited (Dickinson, 2008a). However, Wooster and Augustin (2006) have shown that, by using polysaccharides of varying molecular weight in the conjugate, different hydrodynamic thicknesses of adsorbed layer are reached. That is, based on measurements of dynamic light-scattering on polystyrene latex particles with surfaces covered with β-lactoglobulin–dextran conjugates, it was found that, whereas pure β-lactoglobulin gave a layer thickness of 2.9 nm, the layer thickness with the conjugates was 8–23 nm for a series of dextran polymers having molecular weights in the range 18–440 kDa.

Protein–polysaccharide conjugates can also act as the stabilizers of multiple emulsions. Fechner *et al.* (2007) reported that, under acidic conditions, conjugate-containing water-in-oil-in-water emulsions were more stable to coalescence than the corresponding emulsions made with just sodium caseinate. They also observed that the extent of vitamin B_{12} release from the inner aqueous phase of the conjugate-based system was significantly lower. This result could be useful for preparing double emulsions with variable release behaviour.

For reasons that are probably unrelated to their technical performance, these covalent protein–polysaccharide conjugates have not yet been used commercially in food systems. But it seems that it is only a matter of time before the impressive potential of these highly functional ingredients becomes exploited on a commercial scale in various food applications — not just for emulsification, but also for foaming, gelation, water-holding, and encapsulation.

Concluding Remarks

We take the view that the recently improved understanding of the nature and strength of interactions between the basic biopolymer components of food colloids is now revealing the molecular mechanisms underlying the formation and breakdown of their colloidal structure. This understanding seems beneficial for prediction, regulation, and optimization of food colloid processing conditions, as well as for the improvement of the texture and shelf-life of products over a wide range of pH and ionic strength conditions. Moreover, this knowledge base should allow the elimination of undesirable food ingredients, and the introduction of alternative healthier ingredients based on biopolymer mixtures with functionality carefully designed, controlled and improved via their molecular interactions. In particular, we believe that the protein–polysaccharide electrostatic and covalent complexes have much to offer in the formulation of 'intelligent' nanoscale encapsulation systems for food applications.

BIBLIOGRAPHY

Akhtar, M., Dickinson, E. (2003). Emulsifying properties of whey protein–dextran conjugates at low pH and different salt concentrations *Colloids and Surfaces B: Biointerfaces*, **31**, 125–132.

Akhtar, M., Dickinson, E. (2007). Whey protein–maltodextrin conjugates as emulsifying agents: an alternative to gum arabic. *Food Hydrocolloids*, **21**, 607–616.

Albertsson, P.-Å. (1971). *Partition of Cell Particles and Macromolecules*, 2nd edn, Stockholm: Almqvist & Wiksell.

Alevisopoulos, S., Kasapis, S., Abeysekera, R. (1996). Formation of kinetically trapped gels in the maltodextrin–gelatin system. *Carbohydrate Research*, **293**, 79–99.

Alexander, M., Dalgleish, D.G. (2007). The interaction of casein micelles with κ-carrageenan studied by diffusing wave spectroscopy. *Food Hydrocolloids*, **21**, 128–136.

Alves, M.M., Antonov, Yu.A., Gonçalves, M.P. (1999). On the incompatibility of alkaline gelatin and locust bean gum in aqueous solution. *Food Hydrocolloids*, **13**, 77–80.

Alves, M.M., Antonov, Yu.A., Gonçalves, M.P. (2000). Phase equilibria and mechanical properties of gel-like water–gelatin–locust bean gum systems. *International Journal of Biological Macromolecules*, **27**, 41–47.

Alves, M.M., Garnier, C., Lefebvre, J., Gonçalves, M.P. (2001). Microstructure and flow behaviour of liquid water–gelatin–locust bean gum systems. *Food Hydrocolloids*, **15**, 117–125.

Antipova, A.S., Semenova, M.G. (1997). Effect of neutral carbohydrate structure in the set glucose/sucrose/maltodextrin/dextran on protein surface activity at the air/water interface. *Food Hydrocolloids*, **11**, 71–77.

Antipova, A.S., Semenova, M.G., Gauthier-Jacques, A. (1997). Effect of neutral carbohydrate structure on protein surface activity at air–water and oil–water interfaces. In Dickinson, E., Bergenståhl, B. (Eds). *Food Colloids: Proteins, Lipids and Polysaccharides*, Cambridge, UK: Royal Society of Chemistry, pp. 245–258.

Appelqvist, I., Debet, M. (1997). Starch–biopolymer interactions. *Food Reviews International*, **13**, 163–224.

Bakker, M.A.E., Konig, M.M.G., Visser, J. (1994). Fatty Ingredient. World Patent Application WO94/14334.

Beaulieu, M., Turgeon, S.L., Doublier, J.-L. (2001). Rheology, texture and microstructure of whey proteins/low methoxyl pectin mixed gels with added calcium. *International Dairy Journal*, **11**, 961–967.

Belyakova, L.E., Semenova, M.G., Antipova, A.S. (1999). Effect of small molecule surfactants on molecular parameters and thermodynamic properties of legumin in a bulk and at the air–water interface depending on a protein structure in an aqueous medium. *Colloids and Surfaces B: Biointerfaces*, **12**, 271–285.

Benichou, A., Aserin, A., Garti, N. (2004). Double emulsions stabilized with hybrids of natural polymers for entrapment and slow release of active matters. *Advances in Colloid and Interface Science*, **108–109**, 29–41.

Benichou, A., Aserin, A., Lutz, R., Garti, N. (2007). Formation and characterization of amphiphilic conjugates of whey protein isolate (WPI) / xanthan to improve surface activity. *Food Hydrocolloids*, **21**, 379–391.

Biesheuvel, P.M., Cohen Stuart, M.A. (2004). Electrostatic free energy of weakly charged macromolecules in solution and intermolecular complexes consisting of oppositely charged polymers. *Langmuir*, **20**, 2785–2791.

Bonacucina, G., Martelli, S., Palmieri, G.F. (2004). Rheological, mucoadhesive and release properties of Carbopol gels in hydrophilic cosolvents. *International Journal of Pharmaceutics*, **282**, 115–130.

Bourriot, S., Garnier, C., Doublier, J.-L. (1999). Micellar casein–κ-carrageenan mixtures. I. Phase separation and ultrastructure. *Carbohydrate Polymers*, **40**, 145–157.

Bryant, C.M., McClements, D.J. (2000). Influence of xanthan gum on physical characteristics of heat-denatured whey protein solutions and gels. *Food Hydrocolloids*, **14**, 383–390.

Bungenberg de Jong, H.G. (1949). In Kruyt, H.R. (Ed.). *Colloid Science*, Amsterdam: Elsevier, vol. 1, pp. 232–258.

Burgess, D.J. (1994). Complex coacervation: microcapsule formation. In Dubin, P., Bock, J., Davis, R.M., Schultz, D.N., Thies, C. (Eds). *Macromolecular Complexes in Chemistry and Biology*, Berlin: Springer-Verlag, pp. 285–300.

Burgess, D.J., Singh, O.N. (1993). Spontaneous formation of small sized albumin/acacia coacervate particles. *Journal of Pharmacy and Pharmacology*, **45**, 586–591.

Burgess, D.J., Kwok, K.K., Megremis, P.T. (1991). Characterization of albumin/acacia complex coacervation. *Journal of Pharmacy and Pharmacology*, **43**, 232–236.

Burova, T.V., Grinberg, N.V., Grinberg, V.Ya., Usov, A.I., Tolstoguzov, V.B., de Kruif, C.G. (2007). Conformational changes in ι- and κ-carrageenans induced by complexing with β-casein. *Biomacromolecules*, **8**, 368–375.

Butler, M.F. (2002). Mechanism and kinetics of phase separation in a gelatin/maltodextrin mixture studied by small-angle light scattering. *Biomacromolecules*, **3**, 676–683.

Butler, M.F., Butler-Heppenstall, M. (2003). Phase separation in gelatin/dextran and gelatin/maltodextrin mixtures. *Food Hydrocolloids*, **17**, 815–830.

Cai, R., Arntfield, S. (1997). Thermal gelation in relation to binding of bovine serum abumin–polysaccharides systems. *Journal of Food Science*, **62**, 1129–1134.

Capron, I., Taco, N., Durand, D. (1999). Heat induced aggregation and gelation of β-lactoglobulin in the presence of κ-carrageenan. *Food Hydrocolloids*, **13**, 1–5.

Choi, S.J., Kim, H.J., Park, K.H., Moon, T.W. (2005). Molecular characteristics of ovalbumin–dextran conjugates formed through the Maillard reaction. *Food Chemistry*, **92**, 93–99.

Closs, C.B., Conde-Petit, B., Roberts, I.D., Tolstoguzov, V.B., Escher, F. (1999). Phase separation and rheology of aqueous starch/galactomannan systems. *Carbohydrate Polymers*, **39**, 67–77.

Damianou, K., Kiosseoglou, V. (2006). Stability of emulsions containing a whey protein concentrate obtained from milk serum through carboxymethylcellulose complexation. *Food Hydrocolloids*, **20**, 793–799.

da Silva, F.L.B., Lund, M., Jönsson. B., Akesson, T.J. (2006). On the complexation of proteins and polyelectrolytes. *Journal of Physical Chemistry B,* **110**, 4459–4464.

de Kruif, C.G., Weinbreck, F., de Vries, R. (2004). Complex coacervation of proteins and anionic polysaccharides. *Current Opinion in Colloid and Interface Science*, **9**, 340–349.

de Vries, R., Cohen Stuart, M.A. (2006). Theory and simulations of macroion complexation. *Current Opinion in Colloid and Interface Science*, **11**, 295–301.

Dickinson, E. (1993). Protein–polysaccharide interactions in food colloids. In Dickinson, E., Walstra, P. (Eds). *Food Colloids and Polymers: Stability and Mechanical Properties*, Cambridge, UK: Royal Society of Chemistry, pp. 77–93.

Dickinson, E. (1995). Mixed biopolymers at interfaces. In Harding, S.E., Hill, S.E., Mitchell, J.R. (Eds). *Biopolymer Mixtures*, Leicestershire, UK: Nottingham University Press, pp. 349–372.

Dickinson, E. (1998a). Stability and rheological implications of electrostatic milk protein–polysaccharide interactions. *Trends in Food Science and Technology*, **9**, 347–354.

Dickinson, E. (1998b). Proteins at interfaces and in emulsions: stability, rheology and interactions. *Journal of the Chemical Society, Faraday Transactions*, **94**, 1657–1669.

Dickinson, E. (1999). Adsorbed protein layers at fluid interfaces: interactions, structure and surface rheology. *Colloids and Surfaces B: Biointerfaces*, **15**, 161–176.

Dickinson, E. (2000). Structure and rheology of simulated gels formed from aggregated colloidal particles. *Journal of Colloid and Interface Science*, **225**, 2–15.

Dickinson, E. (2003). Hydrocolloids at interfaces and the influence on the properties of dispersed systems. *Food Hydrocolloids*, **17**, 25–39.

Dickinson, E. (2006a). Structure formation in casein-based gels, foams, and emulsions. *Colloids and Surfaces A: Physicochemical and Engineering Aspects*, **288**, 3–11.

Dickinson, E. (2006b). Colloid science of mixed ingredients. *Soft Matter*, **2**, 642–652.

Dickinson, E. (2007). Food colloids… how do interactions of ingredients control structure, stability and rheology? *Current Opinion in Colloid and Interface Science,* **12**, 155–157.

Dickinson, E. (2008a). Interfacial structure and stability of food emulsions as affected by protein–polysaccharide interactions. *Soft Matter*, **4**, 932–942.

Dickinson, E. (2008b). Emulsification and emulsion stabilization with protein–polysaccharide complexes. In Williams, P.A., Phillips, G.O. (Eds). *Gums and Stabilisers for the Food Industry 14*, Cambridge, UK: Royal Society of Chemistry, pp. 221–232.

Dickinson, E. (2009). Hydrocolloids as emulsifiers and emulsion stabilizers. *Food Hydrocolloids*, **23**, 1473–1482.

Dickinson, E., Eriksson, L. (1991). Particle flocculation by adsorbing polymers. *Advances in Colloid and Interface Science*, **34**, 1–29.

Dickinson, E., Euston, S.R. (1991). Stability of food emulsions containing both protein and polysaccharide. In Dickinson, E. (Ed.). *Food Polymers, Gels and Colloids*, Cambridge, UK: Royal Society of Chemistry, pp. 132–146.

Dickinson, E., Galazka, V.B. (1991). Emulsion stabilization by ionic and covalent complexes of β-lactoglobulin with polysaccharides. *Food Hydrocolloids*, **5**, 281–296.

Dickinson, E., Galazka, V.B. (1992). Emulsion stabilization by protein–polysaccharide complexes. In Phillips, G.O., Wedlock, D.J., Williams, P.A. (Eds). *Gums and Stabilisers for the Food Industry 6*, Oxford: IRL Press, pp. 351–362.

Dickinson, E., James, J.D. (2000). Influence of high-pressure treatment on β-lactoglobulin–pectin associations in emulsions and gels. *Food Hydrocolloids*, **14**, 365–376.

Dickinson, E., Matia-Merino, L. (2002). Effect of sugars on the rheological properties of acid caseinate-stabilized emulsion gels. *Food Hydrocolloids*, **16**, 321–331.

Dickinson, E., McClements, D.J. (1995). *Advances in Food Colloids*, Glasgow: Blackie, pp. 280–300.

Dickinson, E., Pawlowsky, K. (1997). Effect of ι-carrageenan on flocculation, creaming, and rheology of a protein-stabilized emulsion. *Journal of Agricultural and Food Chemistry*, **45**, 3799–3806.

Dickinson, E., Pawlowsky, K. (1998). Influence of κ-carrageenan on the properties of a protein-stabilized emulsion. *Food Hydrocolloids*, **12**, 417–423.

Dickinson, E., Semenova, M.G. (1992). Emulsifying behaviour of protein in the presence of polysaccharide under conditions of thermodynamic incompatibility. *Journal of the Chemical Society, Faraday Transactions*, **88**, 849–854.

Dickinson, E., Pinfield, V.J., Horne, D.S., Leermakers, F.A.M. (1997). Self-consistent field modelling of adsorbed casein: interaction between two protein-coated surfaces. *Journal of the Chemical Society, Faraday Transactions*, **93**, 1785–1790.

Dickinson, E., Semenova, M.G., Antipova, A.S., Pelan, E. (1998). Effect of high-methoxy pectin on properties of casein-stabilized emulsions. *Food Hydrocolloids*, **12**, 425–432.

Ding, P., Wolf, B., Frith, W.J., Clark, A.H., Norton, I.T., Pacek, A.W. (2002). Interfacial tension in phase-separated gelatin / dextran aqueous mixtures. *Journal of Colloid and Interface Science*, **253**, 367–376.

Ducel, V., Richard, J., Saulnier, P., Popineau, Y., Boury, F. (2004). Evidence and characterization of complex coacervates containing plant proteins: applications to the microencapsulation of oil droplets. *Colloids and Surfaces A: Physicochemical and Engineering Aspects*, **232**, 239–247.

Ercelebi, E.A., Ibanoğlu, E. (2007). Influence of hydrocolloids on phase separation and emulsion properties of whey protein isolate. *Journal of Food Engineering*, **80**, 454–459.

Fechner, A., Knoth, A., Scherze, I., Muschiolik, G. (2007). Stability and release properties of double-emulsions stabilised by caseinate–dextran conjugates. *Food Hydrocolloids*, **21**, 943–952.

Firoozmand, H., Murray, B.S., Dickinson, E. (2007). Fractal-type particle gel formed from gelatin + starch solution. *Langmuir*, **23**, 4646–4650.

Galazka, V.B., Ledward, D.A., Sumner, I.G., Dickinson, E. (1997). Influence of high pressure on bovine serum albumin and its complex with dextran sulfate. *Journal of Agricultural and Food Chemistry*, **45**, 3465–3471.

Galazka, V.B., Smith, D., Ledward, D.A., Dickinson, E. (1999). Complexes of bovine serum albumin with sulfated polysaccharides: effects of pH, ionic strength and high pressure treatment. *Food Chemistry*, **64**, 303–310.

Galazka, V.B., Dickinson, E., Ledward, D.A. (2000). Emulsifying properties of ovalbumin in mixtures with sulfated polysaccharides: effects of pH, ionic strength, heat and high-pressure treatment. *Journal of the Science of Food and Agriculture*, **80**, 1219–1229.

Gancz, K., Alexander, M., Corredig, M. (2006). *In situ* study of flocculation of whey protein-stabilized emulsions caused by addition of high-methoxyl pectin. *Food Hydrocolloids*, **20**, 293–298.

Ganzevles, R.A., Cohen Stuart, M.A., van Vliet, T., de Jongh, H.H.J. (2006). Use of polysaccharides to control protein adsorption to the air–water interface. *Food Hydrocolloids*, **20**, 872–878.

Ganzevles, R.A., van Vliet, T., Cohen Stuart, M.A., de Jongh, H.H.J. (2007). Manipulation of adsorption behaviour at liquid interfaces by changing protein–polysaccharide electrostatic interactions. In Dickinson, E., Leser, M.E. (Eds). *Food Colloids: Self-Assembly and Material Science*, Cambridge, UK: Royal Society of Chemistry, pp. 195–208.

Ganzevles, R.A., Fokkink, R., van Vliet, T., Cohen Stuart, M.A., de Jongh, H.H.J. (2008). Structure of mixed β-lactoglobulin–pectin adsorbed layers at air/water interfaces: a spectroscopic study. *Journal of Colloid and Interface Science*, **317**, 137–147.

Garti, N., Slavin, Y., Aserin, A. (1999). *Portulaca oleracea* gum and casein interactions and emulsion stability. *Food Hydrocolloids*, **13**, 127–138.

Gezimati, J., Creamer, L., Singh, H. (1997). Heat-induced interactions and gelation of mixtures of β-lactoglobulin and α-lactalbumin. *Journal of Agricultural and Food Chemistry*, **45**, 1130–1136.

Gilsenan, P.M., Richardson, R.K., Morris, E.R. (2003). Associative and segregative interactions between gelatin and low-methoxy pectin. Part 1. Associative interactions in the absence of Ca^{2+}. *Food Hydrocolloids*, **17**, 723–737.

Girard, M., Turgeon, S.L., Gauthier, S.F. (2003). Thermodynamic parameters of β-lactoglobulin–pectin complexes assessed by isothermal titration calorimetry. *Journal of Agricultural and Food Chemistry*, **51**, 4450–4455.

Grinberg, V.Ya., Tolstoguzov, V.B. (1997). Thermodynamic incompatibility of proteins and polysaccharides in solutions. *Food Hydrocolloids*, **11**, 145–158.

Gu, J.M., Robinson, J.R., Leung, S.S. (1988). Binding of acrylic polymers to mucin epithelial surfaces: structure–property relationships. *Critical Reviews in Therapeutic Drug Carrier Systems*, **5**, 21–67.

Gu, Y.S., Decker, E.A., McClements, D.J. (2005a). Influence of pH and carrageenan type on properties of β-lactoglobulin stabilized oil-in-water emulsions. *Food Hydrocolloids*, **19**, 83–91.

Gu, Y.S., Regnier, L., McClements, D.J. (2005b). Influence of environmental stresses on stability of oil-in-water emulsions containing droplets stabilized by β-lactoglobulin–ι-carrageenan membranes. *Journal of Colloid and Interface Sci*ence, **286**, 551–558.

Guido, S., Simeone, M., Alfani, A. (2002). Interfacial tension of aqueous mixtures of Na-caseinate and Na-alginate by drop deformation in shear flow. *Carbohydrate Polymers*, **48**, 143–152.

Gupta, A., Bohidar, H.B. (2005). Kinetics of phase separation in systems exhibiting simple coacervation. *Physical Review E*, **72**, 011507.

Gurov, A.N., Gurova, N.V., Leontiev, A.L., Tolstoguzov, V.B. (1988). Equilibrium and non-equilibrium complexes between bovine serum albumin and dextran sulfate: I. Complexing conditions and composition of non-equilibrium complexes. *Food Hydrocolloids*, **2**, 267–283.

Guzey, D., McClements D.J. (2006). Formation, stability and properties of multilayer emulsions for application in the food industry. *Advances in Colloid and Interface Science*, **128–130**, 227–248.

Hattori, T., Hallberg, R., Dubin, P.L. (2000). Roles of electrostatic interaction and polymer structure in the binding of β-lactoglobulin to anionic polyelectroytes: measurement of binding constants by frontal analysis continuous capillary electrophoresis. *Langmuir*, **16**, 9738–9743.

Hattori, T., Kimura, K., Seyrek, E., Dubin, P.L. (2001). Binding of bovine serum albumin to heparin determined by turbidimetric titration and frontal analysis continuous capillary electrophoresis. *Analytical Biochemistry*, **295**, 158–167.

Haug, I., Williams, M.A.K., Lundin, L., Smidsrød, O., Draget, K.I. (2003). Molecular interactions in, and rheological properties of, a mixed biopolymer system undergoing order/disorder transitions. *Food Hydrocolloids*, **17**, 439–444.

Hsu, C.C., Prausnitz, J.M. (1974). Thermodynamics of polymer compatibility in ternary systems. *Macromolecules*, **7**, 320–324.

Imeson, A.P., Ledward, D.A., Mitchell, J.R. (1977). On the nature of the interaction between some anionic polysaccharides and proteins. *Journal of Agricultural and Food Chemistry*, **28**, 661–667.

Jönsson, B., Lund, M., da Silva, F.L.B. (2007). Electrostatics in macromolecular solutions. In Dickinson, E., Leser, M.E. (Eds). *Food Colloids: Self-Assembly and Material Science*. Cambridge, UK: Royal Society of Chemistry, pp. 129–154.

Jourdain, L., Leser, M.E., Schmitt, C., Michel M., Dickinson, E. (2008). Stability of emulsions containing sodium caseinate and dextran sulfate: relationship to complexation in solution. *Food Hydrocolloids*, **22**, 647–659.

Jourdain, L.S., Schmitt, C., Leser, M.E., Murray, B.S., Dickinson, E. (2009). Mixed layers of sodium caseinate + dextran sulfate: influence of order of addition to oil–water interface. *Langmuir*, **25**, 10026–10037.

Kaibara, K., Okazaki, T., Bohidar H.B., Dubin, P.L. (2000). pH-induced coacervation in complexes of bovine serum albumin and cationic polyelectrolytes. *Biomacromolecules*, **1**, 100–107.

Kasapis, S., Morris, E.R., Norton, I.T., Gidley, M.J. (1993). Phase equilibria and gelation in gelatin/maltodextrin systems. Part II. Polymer incompatibility in solution. *Carbohydrate Polymers*, **21**, 249–259.

Kato, A., Sasaki, Y., Furuta, R., Kobayashi, K. (1990). Functional protein/polysaccharide conjugate prepared by controlled dry heating of ovalbumin/dextran mixtures. *Agricultural and Biological Chemistry*, **54**, 107–112.

Kerstens, S., Murray, B.S., Dickinson, E. (2006). Microstructure of β-lactoglobulin-stabilized emulsions containing non-ionic surfactant and excess free protein: influence of heating. *Journal of Colloid and Interface Science*, **296**, 332–341.

Kester, J.J., Fennema, O.R. (1986). Edible films and coatings: a review. *Food Technology*, **40**(12), 47–59.

Khokhlov, A.R., Nyrkova, I.A. (1992). Compatibility enhancement and microdomain structuring in weakly charged polyelectrolyte mixtures. *Macromolecules*, **25**, 1493–1502.

Lazaridou, A., Biliaderis, C.G. (2009). Concurrent phase separation and gelation in mixed oat β-glucans/sodium caseinate and oat β-glucans/pullulan aqueous dispersions. *Food Hydrocolloids*, **23**, 886–895.

Ledward, D.A. (1979). Protein–polysaccharide interactions. In Blanchard, J.M.V., Mitchell, J.R. (Eds). *Polysaccharides in Food*, London: Butterworth, pp. 205–217.

Ledward, D.A. (1994). Protein–polysaccharide interactions. In Hettiarachchy, N.S., Ziegler, G.R. (Eds). *Protein Functionality in Food Systems*, New York: Marcel Dekker, pp. 225–259.

Lii, C.Y., Liaw, S.C., Lai, V.M.F., Tomasik, T. (2002). Xanthan gum–gelatine complexes. *European Polymer Journal*, **38**, 1377–1381.

Loren, N., Hermansson, A.-M. (2000). Phase separation and gel formation in kinetically trapped gelatin/maltodextrin gels. *International Journal of Biological Macromolecules*, **27**, 249–262.

Loren, N., Langton, M., Hermansson, A.-M. (1999). Confocal laser scanning microscopy and image analysis of kinetically trapped phase-separated gelatin/maltodextrin gels. *Food Hydrocolloids*, **13**, 185–198.

Loren, N., Hermansson, A.-M., Williams, M.A.K., Lundin, L., Foster, T.J., Hubbard, C.D., Clark, A.H., Norton, I.T., Bergström, E.T., Goodall, D.M. (2001). Phase separation induced by conformational ordering of gelatin in gelatin/maltodextrin mixtures. *Macromolecules*, **34**, 289–297.

Luzzi, L.A. (1970). Microencapsulation. *Journal of Pharmaceutical Sciences*, **59**, 1367–1376.

Makri, E., Papalamprou, E., Doxastakis, G. (2005). Study of functional properties of seed storage proteins from indigenous European legume crops (lupin, pea, broad bean) in admixture with polysaccharides. *Food Hydrocolloids*, **19**, 583–594.

Manoj, P., Kasapis, S., Chronakis, I.S. (1996). Gelation and phase separation in maltodextrin–caseinate systems. *Food Hydrocolloids*, **10**, 407–420.

Maroziene, A., de Kruif, C.G. (2000). Interaction of pectin and casein micelles. *Food Hydrocolloids*, **14**, 391–394.

Martin, A., Goff, H.D., Smith, A., Dalgleish, D.G. (2006). Immobilization of casein micelles for probing their structure and interactions with polysaccharides using scanning electron microscopy (SEM). *Food Hydrocolloids*, **20**, 817–824.

Matia-Merino, L. (2003). PhD thesis, University of Leeds, UK.

Matia-Merino, L., Lau, K., Dickinson, E. (2004). Effects of low-methoxyl amidated pectin and ionic calcium on rheology and microstructure of acid-induced sodium caseinate gels. *Food Hydrocolloids*, **18**, 271–281.

Mayya, K.S., Bhattacharyya, A., Argillier, J.F. (2003). Micro-encapsulation by complex coacervation: influence of surfactant. *Polymer International*, **52**, 644–647.

McClements, D.J. (2004). *Food Emulsions: Principles, Practice and Techniques*, 2nd edn, Boca Raton, FL: CRC Press.

McClements, D.J. (2005). Theoretical analysis of factors affecting formation and stability of multilayered colloidal dispersions. *Langmuir*, **21**, 9777–9785.

McClements, D. J. (2006). Non-covalent interactions between proteins and polysaccharides. *Biotechnology Advances*, **24**, 621–625.

McClements, D.J., Decker, E.A., Park, Y., Weiss, J. (2009). Structural design principles for delivery of bioactive components in neutraceuticals and functional foods. *Critical Reviews in Food Science and Nutrition*, **49**, 577–606.

Mohanty, B., Bohidar, H.B. (2003). Systematic of alcohol-induced simple coacervation in aqueous gelatin solutions. *Biomacromolecules*, **4**, 1080–1086.

Mohanty, B., Aswal, V.K., Kohlbrecher, J., Bohidar, H.B. (2004). Small-angle neutron and dynamic light scattering study of gelatin coacervates. *Pramana — Journal of Physics*, **63**, 271–276.

Morris, V.J. (1986). Multicomponent gels. In Phillips, G.O., Wedlock, D.J., Williams, P.A. (Eds). *Gums and Stabilisers for the Food Industry 3*, London: Elsevier Applied Science, pp. 87–99.

Moschakis, T., Murray, B.S., Dickinson, E. (2005). Microstructural evolution of viscoelastic emulsions stabilized by sodium caseinate and xanthan gum. *Journal of Colloid and Interface Science*, **284**, 714–728.

Musampa, R.M., Alves, M.M., Maia, J.M. (2007). Phase separation, rheology and microstructure of pea protein–kappa-carrageenan mixtures. *Food Hydrocolloids*, **21**, 92–99.

Nagasawa, M., Takahashi, A. (1972). Light scattering from polyelectrolyte solutions. In Huglin, M.B. (Ed). *Light Scattering from Polymer Solutions*, London: Academic Press, pp. 671–723.

Nagasawa, K., Ohgata, K., Takahashi, K., Hattori, M. (1996). Role of the polysaccharide content and net charge on the emulsifying properties of β-lactoglobulin–carboxymethyldextran conjugates. *Journal of Agricultural and Food Chemistry*, **44**, 2538–2543.

Neirynck, N., van der Meeren, P., Lukaszewicz-Lausecker, M., Cocquyt, J., Verbeken, D., Dewettinck, K. (2007). Influence of pH and biopolymer ratio on whey protein–pectin interactions in aqueous solutions and in O/W emulsions. *Colloids and Surfaces A: Physicochemical and Engineering Aspects*, **298**, 99–107.

Norton, I.T., Frith, W.J. (2001). Microstructure design in mixed biopolymer composites. *Food Hydrocolloids*, **15**, 543–553.

Ogston, A.G. (1970). On the interaction of solute molecules with porous networks. *Journal of Physical Chemistry*, **74**, 668–669.

Olsson, C., Langton, M., Hermansson, A.-M. (2002). Dynamic measurements of β-lactoglobulin structures during aggregation, gel formation and gel break-up in mixed biopolymer systems. *Food Hydrocolloids*, **16**, 477–488.

Overbeek, J.T.G., Voorn, M.J. (1957). Phase separation in polyelectrolyte solutions: the theory of complex coacervation. *Journal of Cellular and Comparative Physiology*, **49**, 7–26.

Park, J.M., Muhoberac, B.B., Dubin, P.L., Xia, J. (1992). Effects of protein charge heterogeneity in protein–polyelectrolyte complexation. *Macromolecules*, **25**, 290–295.

Pavlovskaya, G., Semenova, M., Tsapkina, E., Tolstoguzov V. (1993). The influence of dextran on the interfacial pressure of adsorbing layers of 11S globulin *Vicia faba* at the planar *n*-decane/aqueous solution interface. *Food Hydrocolloids*, **7**, 1–10.

Peniche, C., Howland, I., Carrillo, O., Zaldívar, C., Argqelles-Monal, W. (2004). Formation and stability of shark liver oil loaded chitosan/calcium alginate capsules. *Food Hydrocolloids*, **18**, 865–871.

Piculell, L., Lindman, B. (1992). Association and segregation in aqueous polymer/polymer, polymer/surfactant, and surfactant/surfactant mixtures: similarities and differences. *Advances in Colloid and Interface Science*, **41**, 149–178.

Piculell, L., Bergfeldt, K., Nilsson, S. (1995). Factors determining phase behaviour of multi-component polymer systems. In Harding, S.E., Hill, S.E., Mitchell, J.R. (Eds). *Biopolymer Mixtures*, Leicestershire, UK: Nottingham University Press, pp. 13–35.

Polyakov, V., Grinberg, V., Tolstoguzov, V. (1997). Thermodynamic incompatibility of proteins. *Food Hydrocolloids*, **11**, 171–180.

Pongsawatmanit, R., Harnsilawat, T., McClements, D.J. (2006). Influence of alginate, pH and ultrasound treatment on palm oil-in-water emulsions stabilized by β-lactoglobulin. *Colloids and Surfaces A: Physicochemical and Engeeniring Aspects*, **287**, 59–67.

Pugnaloni, L.A., Matia-Merino, L., Dickinson, E. (2005). Microstructure of acid-induced caseinate gels containing sucrose: quantification from confocal microscopy and image analysis. *Colloids and Surfaces B: Biointerfaces*, **42**, 211–217.

Quiroga, C.C., Bergenståhl, B. (2008). Phase segregation of amylopectin and β-lactoglobulin in aqueous system. *Carbohydrate Polymers*, **72**, 151–159.

Rottereau, M., Gimel, J.C., Nicolai,T., Durand, D. (2004). Monte Carlo simulation of particle aggregation and gelation. I. Growth, structure and size distribution of the clusters. *European Physical Journal E*, **15**, 133–140.

Sanchez, C., Schmitt, C., Babak, V., Hardy, J. (1997). Rheology of whey protein isolate–xanthan mixed solutions and gels: effect of pH and xanthan concentration. *Nahrung*, **41**, 336–343.

Schmitt, C., Sanchez, C., Desobry-Banon, S., Hardy, J. (1998). Structure and technofunctional properties of protein–polysaccharide complexes. *Critical Reviews in Food Science and Nutrition*, **38**, 689–753.

Schmitt, C., Sanchez, C., Thomas, F., Hardy, J. (1999). Complex coacervation between β-lactoglobulin and acacia gum in aqueous medium. *Food Hydrocolloids*, **13**, 483–496.

Schmitt, C., Sanchez, C., Despond, S., Renard, D., Thomas, F., Hardy, J. (2000). Effect of protein aggregates on the complex coacervation between β-lactoglobulin and acacia gum at pH 4.2. *Food Hydrocolloids*, **14**, 403–413.

Sciortino, F., Bansil, R., Stanley, E.H., Alstrom, P. (1993). Interference of phase separation and gelation: a zeroth-order kinetic model. *Physical Review E*, **47**, 4615–4618.

Scott, R.L. (1949). The thermodynamics of high-polymer solution: phase equilibria in ternary system polymer–polymer–liquid. *Journal of Chemical Physics*, **17**, 279–287.

Sejersen, M.T., Salomonsen, T., Ipsen, R., Clark, R., Rolin, C., Engelsen, S.B. (2007). Zeta potential of pectin-stabilized casein aggregates in acidified milk drinks. *International Dairy Journal*, **17**, 302–307.

Semenova, M.G. (1996). Factors determining the character of biopolymer–biopolymer interactions in multicomponent aqueous solutions modelling food systems. In Parris, N., Kato, A., Creamer, L.K., Pearce, J. (Eds). *Macromolecular Interactions in Food Technology*, ACS Symposium Series No. 650, Washington, DC: American Chemical Society, pp. 37–49.

Semenova, M.G. (2007). Thermodynamic analysis of the impact of molecular interactions on the functionality of food biopolymers in solution and in colloidal systems. *Food Hydrocolloids*, **21**, 23–45.

Semenova, M., Savilova, L. (1998). The role of biopolymer structure in interactions between unlike biopolymers in aqueous medium. *Food Hydrocolloids*, **12**, 65–75.

Semenova, M.G., Bolotina, V.S., Grinberg, V.Ya., Tolstoguzov, V.B. (1990). Thermodynamic incompatibility of the 11S fraction of soybean globulin and pectinate in aqueous medium. *Food Hydrocolloids*, **3**, 447–456.

Semenova, M., Pavlovskaya, G., Tolstoguzov, V. (1991a). Light scattering and thermodynamic phase behaviour of the system 11S globulin–κ-carrageenan–water. *Food Hydrocolloids*, **4**, 469–479.

Semenova, M.G., Bolotina, V.S., Dmitrochenko, A.P., Leontiev, A.L., Polyakov, V.I., Braudo, E.E., Tolstoguzov, V.B. (1991b). The factors affecting compatibility of serum albumin and pectinate in aqueous medium. *Carbohydrate Polymers*, **15**, 367–385.

Semenova, M., Antipova, A., Belyakova, L., Dickinson, E., Brown, R., Pelan, E., Norton, I. (1999a). Effect of pectinate on properties of oil-in-water emulsions stabilized by α_{s1}-casein and β-casein. In Dickinson, E., Rodriguez Patino, J.M. (Eds). *Food Emulsions and Foams: Interfaces, Interactions and Stability*, Cambridge, UK: Royal Society of Chemistry, pp. 163–175.

Semenova, M., Belyakova, L., Antipova, A., Jubanova, M. (1999b). Influence of maltodextrins with different dextrose equivalent on the thermodynamic properties of legumin in a bulk and at the air–water interface. *Colloids and Surfaces B: Biointerfaces*, **12**, 287–297.

Semenova, M.G., Belyakova, L.E., Polikarpov, Yu.N., Antipova, A.S., Dickinson, E. (2009). Light scattering study of sodium caseinate + dextran sulfate in aqueous solution: relationship to emulsion stability. *Food Hydrocolloids*, **23**, 629–639.

Shepherd, R., Robertson, A., Ofman, D. (2000). Dairy glycoconjugate emulsifiers: casein–maltodextrins. *Food Hydrocolloids*, **14**, 281–286.

Singh, S.S., Aswal, V.K., Bohidar, H.B. (2007a). Structural studies of agar–gelatin complex coacervates by small angle neutron scattering, rheology and differential scanning calorimetry. *International Journal of Biological Macromolecules*, **41**, 301–307.

Singh, S.S., Bohidar, H.B., Bandyopadhyay, S. (2007b). Study of gelatin–agar intermolecular aggregates in the supernatant of its coacervate. *Colloids and Surfaces B: Biointerfaces*, **57**, 29–36.

Snoeren, T.H., Payens, T.A.J., Jeunink, J., Both, P. (1975). Electrostatic interaction between κ-carrageenan and κ-casein. *Milchwissenschaft*, **30**, 393–396.

Spagnuolo, P.A., Dalgleish, D.G., Goff, H.D., Morris, E.R. (2005). Kappa-carrageenan interactions in systems containing casein micelles and polysaccharide stabilizers. *Food Hydrocolloids*, **19**, 371–377.

Sperber, B.L.H.M., Schols, H.A., Cohen Stuart, M.A., Norde, W., Voragen, A.G.J. (2009). Influence of the overall charge and local charge density of pectin on the complex formation between pectin and β-lactoglobulin. *Food Hydrocolloids*, **23**, 765–772.

Sriamornsak, P., Wattanakorn, N. (2008). Rheological synergy in aqueous mixtures of pectin and mucin. *Carbohydrate Polymers*, **74**, 458–467.

Sun, C., Gunasekaran, S., Richards, M.P. (2007). Effect of xanthan gum on physico-chemical properties of whey protein isolate stabilized oil-in-water emulsions. *Food Hydrocolloids*, **21**, 555–564.

Surh, J., Decker, E.A., McClements, D.J. (2006). Influence of pH and pectin type on properties and stability of sodium-caseinate stabilized oil-in-water emulsions. *Food Hydrocolloids*, **20**, 607–618.

Tanford, C. (1961). *Physical Chemistry of Macromolecules*, New York: Wiley.

Tolstoguzov, V.B. (1986). Functional properties of protein–polysaccharide mixtures. In Ledward, D.A., Mitchell, J.R. (Eds). *Functional Properties of Food Macromolecules*. London: Elsevier Applied Science, pp. 385–415.

Tolstoguzov, V.B. (1991). Functional properties of food proteins and role of protein–polysaccharide interaction. *Food Hydrocolloids*, **4**, 429–468.

Tolstoguzov, V.B. (1995). Some physicochemical aspects of protein processing in foods. *Food Hydrocolloids*, **9**, 317–332.

Tolstoguzov, V.B. (1996). Structure–property relationships in foods. In Parris, N., Kato, A., Creamer, L.K., Pearce, J. (Eds). *Macromolecular Interactions in Food Technology*, ACS Symposium Series No. 650, Washington, DC: American Chemical Society, pp. 2–14.

Tolstoguzov, V.B. (1997). Protein–polysaccharide interactions. In Damodaran, S., Paraf, A. (Eds). *Food Proteins and their Applications*, New York: Marcel Dekker, pp. 171–198.

Tolstoguzov, V.B. (2000a). Compositions and phase diagrams for aqueous systems based on proteins and polysaccharides. *International Review of Cytology*, **192**, 3–31.

Tolstoguzov, V.B. (2000b). Phase behaviour of macromolecular component in biological and food systems. *Nahrung*, **44**, S299–S308.

Tolstoguzov, V.B. (2003). Some thermodynamic considerations in food formulation. *Food Hydrocolloids*, **17**, 1–23.

Tromp, R.H., de Kruif, C.G., van Eijk, M., Rolin, M.C. (2004). On the mechanism of stabilization of acidified milk drinks by pectin. *Food Hydrocolloids*, **18**, 565–572.

Tsapkina, E., Semenova, M., Pavlovskaya, G., Leontiev, A., Tolstoguzov V. (1992). The influence of incompatibility on the formation of adsorbing layers and dispersion of *n*-decane emulsion droplets in aqueous solution containing a mixture of 11S globulin from *Vicia faba* and dextran. *Food Hydrocolloids*, **6**, 237–251.

Turgeon, S.L., Beaulieu, M., Schmitt, C., Sanchez, C. (2003). Protein–polysaccharide interactions: phase-ordering kinetics, thermodynamics and structural aspects. *Current Opinion in Colloid and Interface Science*, **8**, 401–414.

Turgeon, S.L., Schmitt, C., Sanchez, C. (2007). Protein–polysaccharide complexes and coacervates. *Current Opinion in Colloid and Interface Science*, **12**, 166–178.

van Puyvelde, P., Antonov, Y.A., Moldenaers, P. (2002). Rheo-optical measurement of the interfacial tension of aqueous biopolymer mixtures. *Food Hydrocolloids*, **16**, 395–402.

Vega, C., Goff, H.D. (2005). Phase separation in soft-serve ice cream mixes: rheology and microstructure. *International Dairy Journal*, **15**, 249–254.

Vega, C., Dalgleish, D.G., Goff, H.D. (2005). Effect of κ-carrageenan addition to dairy emulsions containing sodium caseinate and locust bean gum. *Food Hydrocolloids*, **19**, 187–195.

Walstra, P., van Vliet, T., Bremer, L.G.B. (1991). On the fractal nature of particle gels. In Dickinson, E. (Ed.). *Food Polymers, Gels and Colloids*, Cambridge, UK: Royal Society of Chemistry, pp. 369–382.

Wang, Q., Qvist, K.B. (2000). Investigation of the composite system of β-lactoglobulin and pectin in aqueous solutions. *Food Research International*, **33**, 683–690.

Wasserman, L., Semenova, M., Tsapkina, E. (1997). Thermodynamic properties of the 11S globulin of *Vicia faba*–ovalbumin–aqueous solvent system: phase behaviour and light scattering. *Food Hydrocolloids*, **11**, 327–337.

Weinbreck, F., de Kruif, C.G. (2003). Complex coacervation of globular proteins and gum arabic. In Dickinson, E., van Vliet, T. (Eds). *Food Colloids, Biopolymers and Materials*, Cambridge, UK: Royal Society of Chemistry, pp. 337–344.

Weinbreck, F., de Vries, R., Schrooyen, P., de Kruif, C.G. (2003a). Complex coacervation of whey proteins and gum arabic. *Biomacromolecules*, **4**, 293–303.

Weinbreck, F., Nieuwenhuijse, H., Robijn, G.W., de Kruif, C.G. (2003b). Complex formation of whey protein–exocellular polysaccharide EPS B40. *Langmuir*, **19**, 9404–9410.

Weinbreck, F., Nieuwenhuijse, H., Robijn, G.W., de Kruif, C.G. (2004a). Complexation of whey proteins with carrageenan. *Journal of Agricultural and Food Chemistry*, **52**, 3550–3555.

Weinbreck, F., Wientjes, R.H.W., Nieuwenhuijse, H., Robijn, G.W., de Kruif, C.G. (2004b). Rheological properties of whey protein/gum arabic coacervates. *Journal of Rheology*, **48**, 1215–1228

Weinbreck, F., Tromp, R.H., de Kruif, C.G. (2004c). Composition and structure of whey protein/gum arabic coacervates. *Biomacromolecules*, **5**, 1437–1445.

Williams, M.A.K., Fabri, D., Hubbard, C.D., Lundin, L., Foster, T.J., Clark, A.H., Norton, I.T., Loren, N., Hermansson, A.-M. (2001). Kinetics of droplet growth in gelatin/maltodextrin mixtures following thermal quenching. *Langmuir*, **17**, 3412–3418.

Wooster, T.J., Augustin, M.A. (2006). β-Lactoglobulin–dextran Maillard conjugates: their effect on interfacial thickness and emulsion stability. *Journal of Colloid and Interface Science*, **303**, 564–572.

Xing, F., Cheng, G., Yang, B., Ma, L. (2004). Microencapsulation of capsaicin by the complex coacervation of gelatin, acacia and tannins. *Journal of Applied Polymer Science*, **91**, 2669–2675.

Ye, A. (2008). Complexation between milk proteins and polysaccharides via electrostatic interaction: principles and applications – a review. *International Journal of Food Science and Technology*, **43**, 406–415.

Zasypkin, D.V., Braudo, E.E., Tolstoguzov, V.B. (1997). Multicomponent biopolymer gels. *Food Hydrocolloids*, **11**, 159–170.

Zeman, L., Patterson, D. (1972). Effect of the solvent on polymer incompatibility in solution. *Macromolecules*, **5**, 513–516.

Zhang, L., Guo, J., Peng, X., Li, Y. (2004) Preparation and release behaviour of carboxymethylated chitosan/alginate microspheres encapsulating bovine serum albumin. *Journal of Applied Polymer Science*, **92**, 878–882.

PART FOUR

BIOPOLYMER INTERACTIONS AT THE INTERFACES IN FOOD COLLOIDS

INTRODUCTION TO PART FOUR

The interface between two bulk fluid phases is not a mathematical dividing line. Rather it a region of finite thickness containing high-energy solvent molecules. This means that an adsorbed layer of biopolymers or low-molecular-weight surfactant molecules should be viewed as a thin film of highly concentrated solution with properties considerably different from those of the bulk phase (Dickinson, 2004; Damodaran, 2004). The structure and mechanical properties of adsorbed layers are important because they determine the stability behaviour of the oil droplets and the air bubbles in food colloids (Wilde, 2000; Dickinson, 1999, 2001, 2004; Rodriguez Patino *et al.*, 2003, 2007; Murray, 2002; Wilde *et al.*, 2004; Damodaran, 2004). For instance, the thickness and mechanical strength of adsorbed biopolymer layers can be predictive of emulsion and foam stability. Other characteristics of biopolymer adsorbed layers, such as the surface coverage and surface charge density, also determine the nature and strength of the interactions between colloidal particles, which in turn affect the stability behaviour. Owing to the chemical nature of biopolymers and small-molecule surfactants, it is established that interactions within and between adsorbed layers are potentially rather sensitive to variation in the aqueous solution conditions — especially the pH, ionic strength, and calcium ion content (Dalgleish, 1997; Dickinson, 1997, 1998, 2001; Marinova *et al.*, 1997; Makievski *et al.*, 1998).

It is important to understand the characteristic interactions involved at an interface containing each of the main types of surface-active molecules, *i.e.*, biopolymers (proteins, polysaccharides) and low-molecular-weight surfactants (lipids). But that is not the whole story. In real food systems there are almost always mixed ingredients at the interface. So it is necessary to understand what sorts of mixed interfacial structures are formed, and how they are influenced by the intermolecular interactions.

In general, surface activity behaviour in food colloids is dominated by the proteins and the low-molecular-weight surfactants. The competition between proteins and surfactants determines the composition and properties of adsorbed layers at oil–water and air–water interfaces. In the case of mixtures of proteins with non-surface-active polysaccharides, the resulting surface-activity is usually attributed to the adsorption of protein–polysaccharide complexes. By understanding relationships between the protein–protein, protein–surfactant and protein–polysaccharide interactions and the properties of the resulting adsorbed layers, we can aim to establish valid mechanisms to interpret the stability behaviour of food colloidal systems containing mixed ingredients (Dickinson, 2006).

Various experimental techniques have proven useful for studying the interactions and structural organization within interfacial layers (Murray and Dickinson, 1996; Sengupta and Damodaran, 2000; Dickinson, 2001; Bos and van Vliet, 2001; Wen and Frances, 2001; Rodriguez Patino *et al.*, 2003, 2007; McClellan and Frances, 2003; Wilde *et al.*, 2004; Spyropoulos *et al.*, 2008; Jourdain *et al.*, 2009). These methods have their own particular advantages and special features as indicated below.

(i) *Static and dynamic interfacial tension* — Information of this type is particularly sensitive to variations in the adsorption and packing of the surface-active molecules at the interface.
(ii) *Shear and dilatational interfacial rheology* — This is sensitive to surface composition and strength of intermolecular interactions. (Shear measurements are especially sensitive to the type of surface-active species and the adsorbed layer packing density.)
(iii) *Zeta potential of droplets* — This is sensitive to the interactions and distribution of charged molecules at the oil–water interface.
(iv) *Ellipsometry* — This can be useful for determining the adsorbed amount of surface-active molecules and the layer thickness.
(v) *Infrared reflection–adsorption spectroscopy* — This can be used to estimate the composition of protein + surfactant layers, and to detect changes in interfacial protein conformation.
(vi) *Atomic force microscopy (AFM)* — This is a powerful method for imaging surface structural detail on flat substrates at molecular or nanoscale resolution.
(vii) *Brewster angle microscopy (BAM)* — This can provide detailed visualization of surface films containing heterogeneous features (aggregation, phase separation) at micrometre resolution.
(viii) *Fluorescence microscopy* — This is useful for visualization of the phase state of mixed adsorbed layers.
(ix) *Neutron or X-ray reflectivity* — This can be used to give detailed information on the nano/molecular scale about the layer structure and composition perpendicular to the interface.

In addition to these experimental methods, there is also a role for computer simulation and theoretical modelling in providing understanding of structural and mechanical properties of mixed interfacial layers. The techniques of Brownian dynamics simulation and self-consistent-field calculations have, for example, been used to some advantage in this field (Wijmans and Dickinson, 1999; Pugnaloni *et al.*, 2003a,b, 2004, 2005; Parkinson *et al.*, 2005; Ettelaie *et al.*, 2008).

BIBLIOGRAPHY

Bos, M.A., van Vliet, T. (2001). Interfacial rheological properties of adsorbed protein layers and surfactants. *Advances in Colloid and Interface Science*, **91**, 437–471.

Dalgleish, D. (1997). Adsorption of protein and the stability of emulsions. *Trends in Food Science and Technology*, **8**, 1–6.

Damodaran, S. (2004). Adsorbed layers formed from mixtures of proteins. *Current Opinion in Colloid and Interface Science*, **9**, 328–339.

Dickinson, E. (1997). Properties of emulsions stabilized with milk proteins: overview of some recent developments. *Journal of Dairy Science*, **80**, 2607–2619.

Dickinson, E. (1998). Proteins at interfaces and in emulsions: stability, rheology and interactions. *Journal of the Chemical Society, Faraday Transactions*, **94**, 1657–1669.

Dickinson, E. (1999). Adsorbed protein layers at fluid interfaces: interactions, structure and surface rheology. *Colloids and Surfaces B: Biointerfaces*, **15**, 161–176.

Dickinson, E. (2001). Milk protein interfacial layers and the relationship to emulsion stability and rheology. *Colloids and Surfaces B: Biointerfaces*, **20**, 197–210.

Dickinson, E. (2004). Food colloids: the practical application of protein nanoscience in extreme environments. *Current Opinion in Colloid and Interface Science*, **9**, 295–297.

Dickinson, E. (2006). Colloid science of mixed ingredients. *Soft Matter*, **2**, 642–652.

Ettelaie, R., Akinshina, A., Dickinson, E. (2008). Mixed protein–polysaccharide layers: a self consistent field calculation study. *Faraday Discussions*, **139**, 161–178.

Jourdain, L.S., Schmitt, C., Leser, M.E., Murray, B.S., Dickinson, E. (2009). Mixed layers of sodium caseinate + dextran sulfate: influence of order of addition to oil–water interface. *Langmuir*, **25**, 10026–10037.

Makievski, A., Fainerman, V., Bree, M., Wüstneck, R., Krägel, J., Miller, R. (1998). Adsorption of proteins at the liquid air interface. *Journal of Physical Chemistry B*, **102**, 417–425.

Marinova, K., Gurkov, T., Velev, O., Ivanov, I., Campbell, B., Borwankar, R. (1997). The role of additives for the behaviour of thin emulsion films stabilized by proteins. *Colloids and Surfaces A: Physicochemical and Engineering Aspects*, **123–124**, 155–167.

McClellan, S.J., Frances, E.I. (2003). Exclusion of bovine serum albumin from the air–water interface by sodium myristate. *Colloids and Surfaces B: Biointerfaces*, **30**, 1–11.

Murray, B.S. (2002). Interfacial rheology of food emulsifiers and proteins. *Current Opinion in Colloid and Interface Science*, **7**, 426–431.

Murray, B.S., Dickinson, E. (1996). Interfacial rheology and the dynamic properties of adsorbed films of food proteins and surfactants. *Food Science and Technology International* (Japan), **2**, 131–145.

Parkinson, E.L., Ettelaie, R., Dickinson, E. (2005). Using self-consistent-field theory to understand enhanced steric stabilization by casein-like copolymers at low surface coverage in mixed protein layers. *Biomacromolecules*, **6**, 3018–3029.

Pugnaloni, L.A., Ettelaie, R., Dickinson, E. (2003a). Growth and aggregation of surfactant islands during the displacement of an adsorbed protein monolayer: a Brownian dynamics simulation study. *Colloids and Surfaces B: Biointerfaces*, **31**, 149–157.

Pugnaloni, L.A., Ettelaie, R., Dickinson, E. (2003b). Do mixtures of proteins phase separate at interfaces? *Langmuir*, **19**, 1923–1926.

Pugnaloni, L.A., Ettelaie, R., Dickinson, E. (2004). Surface phase separation in complex mixed adsorbing systems: an interface–bulk coupling effect. *Journal of Chemical Physics*, **121**, 3775–3783.

Pugnaloni, L.A., Ettelaie, R., Dickinson, E. (2005). Brownian dynamics simulation of adsorbed layers of interacting particles subjected to large extensional deformation. *Journal of Colloid and Interface Science*, **287**, 401–414.

Rodriguez Patino, J.M., Rodriguez Niño, M.R., Sanchez, C.C. (2003). Protein–emulsifier interactions at the air–water interface. *Current Opinion in Colloid and Interface Science*, **8**, 387–395.

Rodriguez Patino, J.M., Rodriguez Niño, M.R., Sanchez, C.C. (2007). Physico-chemical properties of surfactant and protein films. *Current Opinion in Colloid and Interface Science*, **12**, 187–195.

Sengupta, T., Damodaran, S. (1998). Role of dispersion interactions in the adsorption of proteins at oil–water and air–water interfaces. *Langmuir*, **14**, 6457–6469.

Spyropoulos, F., Ding, P., Frith, W.J., Norton, I.T., Wolf, B., Pacek, A.W. (2008). Interfacial tension in aqueous biopolymer–surfactant mixtures. *Journal of Colloid and Interface Science*, **317**, 604–610.

Wen, X., Frances, E.I. (2001). Adsorption of bovine serum albumin at the air–water interface and its effect on the formation of DPPC surface film. *Colloids and Surfaces A: Physicochemical and Engineering Aspects*, **190**, 319–332.

Wijmans, C.M., Dickinson, E. (1999). Brownian dynamics simulation of the displacement of a protein monolayer by competitive adsorption. *Langmuir*, **15**, 8344–8348.

Wilde, P.J. (2000). Interfaces: their role in foam and emulsion behaviour. *Current Opinion in Colloid and Interface Science*, **5**, 176–181.

Wilde, P., Mackie, A., Husband, F., Gunning, P., Morris, V. (2004). Proteins and emulsifiers at liquid interfaces. *Advances in Colloid and Interface Science*, **108–109**, 63–71.

CHAPTER EIGHT

BIOPOLYMER INTERACTIONS IN ADSORBED LAYERS: RELATIONSHIP TO PROPERTIES OF FOOD COLLOIDS

The amphiphilic character of proteins makes them surface-active at many different types of interfaces. This capacity of proteins to accumulate and form films at phase boundaries (solid–liquid, liquid–liquid, and liquid–vapour) has enormous implications for biological processes such as blood clotting, cell adhesion, lung expansion–contraction, activation of metabolic process, *etc.*. And it also has implications for the technological processes involved in the manufacture of food foams and emulsions (Damodaran, 2004). Hence, for a long time the related subjects of protein adsorption at interfaces and protein interactions with other surface-active molecules (*e.g.*, small-molecule surfactants) have attracted considerable attention. In addition, the phenomenon of electrostatic protein–polysaccharide complexation at the interface has attracted more recent research attention. This is because of its potential use in the nanoscale engineering of delivery vehicles for nutrient encapsulation, and also in the possible protection of adsorbed proteins and emulsified lipids against enzymatic breakdown during digestion (Guzey and McClements, 2006; McClements *et al.*, 2008; Dickinson, 2008; McClements and Decker, 2009).

In order to characterize the properties of adsorbed layers at interfaces, it is imperative to establish the thermodynamic state of the interfacial region. In the first place, since the interfacial region has a finite thickness, it can be regarded as a thin liquid film. Molecular dynamics simulations have shown that the density profile of large water clusters increases from essentially zero at the surface to 1.0 g/cm^3 at a depth of about 1 nm from the surface (Broadskaya *et al.*, 1996; Damodaran, 2004), thereby suggesting that the effective thickness of the interfacial region at the air–water surface is ~ 1 nm. This corresponds to around three layers of water molecules. Neutron reflectivity measurements have revealed a reduced water density up to 4 nm from surfaces of self-assembled monolayers (Schwendel *et al.*, 2003). Therefore, on the basis of various evidence sources, it appears that the air–water interfacial region is at least 1 nm thick. According to Damodaran (2004), this corresponds to an average density of 0.2 g/cm^3 for the water in the interfacial region; the rather high surface tension at the air–water interface arises from the (large) excess standard free energy of this water as compared to that of bulk water.

It is clear that protein adsorption at oil–water or air–water interfaces is basically thermodynamically driven. The thermodynamics of protein–water interactions in the bulk solution, as well as within the interfacial phase, plays a crucial role: that is, during adsorption there is a reduction in the thermodynamic activity (*i.e.*, concentration) of water molecules in the interfacial region (Damodaran, 2004). It is commonly envisaged that the process of forming an adsorbed protein layer involves four main stages, each one contributing significantly to the change in the overall free energy of the system (Graham and Phillips, 1979a,b; Miller *et al.*, 1993; Dickinson and McClements, 1995; Murray, 2002; Damodaran, 2004; Wilde *et al.*, 2004). The four stages are set out below.

(I) Initially there is diffusion from the bulk to the interface. This is not always the predominant or rate-limiting stage. The diffusion rate towards the interface is controlled by the sizes of individual protein molecules (or supramolecular protein particles).

(II) This is the actual adsorption stage. It may also involve an energy barrier, the nature of which has been described in terms of various molecular interactions between the protein and the interface (Sengupta and Damodaran, 1998; Sengupta *et al.*, 1999). For example, in the case of globular proteins, this energy barrier has been estimated as 2–20 kT at the air–water interface (Sengupta *et al.*, 1999). In principle, the energy barrier can be manipulated by altering the physicochemical nature of a protein. For instance, Wierenga *et al.* (2003) have reported that an increase in exposed hydrophobicity of ovalbumin reduces the energy barrier for its adsorption at the air–water interface.

(III) Once the protein is located at the interface, there is reorganization and exposure of the accessible hydrophobic sites on the protein to the hydrophobic phase, followed by changes in the conformation of the protein at the interface. As this takes place, slow macromolecular reorganization via unfolding plays an important role, especially for globular proteins.

(IV) The final stage involves aggregation, cross-linking, and possibly phase separation, caused by the high concentration of protein at the surface. These processes strongly influence the developing mechanical properties of the adsorbed layer.

While these stages I–IV represent generic processes applying to all food proteins, it is significant to note that even small differences in the protein primary structure can sometimes cause major changes in functionality (Wilde, 2000). For example, the genetic variants A and B of β-lactoglobulin differ in terms of just two amino-acid residues. But this difference has been found to cause significant changes in the protein adsorption behaviour (Mackie *et al.*, 1999a) and in the emulsification properties (Euston *et al.*, 1999). The substitution of the aspartic acid residue in variant A for glycine in variant B at residue 64 is apparently responsible for increasing the rate of adsorption, and also for the development of a more elastic interface for variant B (Mackie *et al.*, 1999a). Although the latter has poorer emulsification properties, it is more effective in conferring long-term emulsion stability with respect to coalescence (Euston *et al.*, 1999).

The interfacial functionality of a protein is sensitive to the stability of its native conformation (Dickinson and McClements, 1995). For example, although the rheological properties of equilibrium films formed by ovalbumin and β-lactoglobulin at the air–water interface appear similar, the foaming abilities of the two pure proteins are quite different. The inability of ovalbumin to foam readily has been attributed to its inability to unfold rapidly at the air–water interface (Damodaran, 2004). Another telling example concerns the flexible β-casein, which does not form a gel-like protein network at the air–water interface, although it does readily produce a foam, albeit one of only of moderate stability. Generally speaking, it seems that, in order for a protein to be an effective foaming agent, it should adsorb rapidly and then easily unfold at the air–water interface (Damodaran, 1997, 2004).

During their processing, food protein ingredients are routinely subjected to elevated temperatures, and this has implications for their functional properties. In particular, the mechanical properties of adsorbed protein layers can be influenced by thermal processing. For instance, Roth *et al.* (2000) found that the ageing and heating of films of β-lactoglobulin at the interface can result in considerably higher surface viscosities, depending on the temperature and length of heat treatment.

1. *Protein–Protein Interactions in Adsorbed Layers*

1.1. *Self-Association at Interfaces*

Globular proteins form close-packed monolayers at fluid interfaces. Hence a large contribution to the adsorbed layer viscoelasticity arises from short-range *repulsive* interactions between hard-sphere particles. In addition to, or instead of, this 'glass-like' structure from hard spheres densely packed in two dimensions, many adsorbed proteins can exhibit attractive interactions leading to a more 'gel-like' network structure. Hence the mechanical properties of an adsorbed layer depend on many factors: the surface coverage, the detailed molecular structure of the adsorbed protein, and the balance of the attractive/repulsive interactions between the protein molecules/particles (Wüstneck *et al.*, 1996; Mackie *et al.* 1999a; Bos and van Vliet, 2001; Dickinson, 2006a). These modern molecular interpretations of adsorbed layer properties are consistent with Rehbinder's concept, proposed around 80 years ago, that proteins stabilize foams and emulsions by forming a mechanical barrier. This concept was reviewed by Izmailova *et al.* (1999), including new data showing the importance of the elastic limit of the protein adsorbed layer. Beyond this elastic limit, the surface begins to flow, and the probability of coalescence increases dramatically.

The elasticity of the protein layer structure is supposed to act against the tendency of an emulsion or foam to collapse because it allows the stretching of the interface. This behaviour is most commonly observed for globular proteins, which adsorb, partially unfold, and then develop attractive protein–protein interactions (Dickinson, 1999a; Wilde, 2000; Wilde *et al.*, 2004). The strength of such an adsorbed layer, reflected in the value of the elastic modulus, and the stress at which the structure breaks down, can be successfully correlated with stability of protein-based emulsions and (more especially) protein-based foams (Halling, 1981; Mitchell, 1986; Izmailova *et al.*, 1999; Dickinson, 1999a).

The surface viscoelasticity of the adsorbed protein layer is important in the making and drainage of foams. Prins (1999) used the overflowing cylinder apparatus to demonstrate that the elastic surface formed by an adsorbed protein could slow down the flow of liquid close to the surface. This stagnant layer effect is thought to contribute to the slower drainage rates observed in protein-stabilized foams as compared to those observed for foams stabilized by surfactants. However, what may be applicable to large foam films does not necessarily apply in the same way to small emulsion droplets. The protein β-casein is known to be excellent at stabilizing emulsions by steric repulsion, but it has moderately weak surface mechanical properties (Husband *et al.*, 1997).

Key points relating to the adsorption behaviour of proteins and their influence on colloid stability are summarized below (Dickinson, 1999a).

(i) Proteins adsorb more extensively and less reversibly at hydrophobic surfaces than at hydrophilic surfaces.

(ii) With increasing degree of hydrophobicity of the surface, the ease of exchange of adsorbed protein molecules with the bulk aqueous phase is generally reduced. This difference can be attributed to a greater degree of unfolding at hydrophobic surfaces following protein adsorption, which leads to the development of strong interfacial hydrophobic interactions and the associated displacement of vicinal water molecules from the unfavourable environment of the surface.

(iii) The extent of globular protein unfolding following adsorption is dependent on the protein structure. Little or no structural change following adsorption is expected for 'hard' proteins. Hence, the effective thickness of the globular protein monolayer is often close to the known size of the native protein molecule in solution. The opposite is the case for so-called 'soft' proteins, which show a greater loss of ordered secondary structure (especially α-helix) on adsorption than do the 'hard' proteins.

(iv) The degree of crowding in the adsorbed layer has a considerable influence on the extent of conformational change. The lower is the adsorbed amount, the more space that the protein molecule has to spread out at the surface, and hence the greater is the opportunity for unfolding to minimize the configurational free energy following adsorption.

(v) Differences in the structural properties of adsorbed layers between disordered proteins (*e.g.*, caseins) and globular proteins are most sensitively reflected in the surface shear rheology.

(vi) The mechanical properties of adsorbed protein layers can have a substantial influence on the coalescence stability of oil droplets and (especially) gas bubbles.

There is a long-established body of evidence indicating that the ability to form strong attractive intermolecular interactions and to stabilize air bubbles against coalescence depends on a protein's labile conformation at the interface (Dickinson and McClements, 1995). Thus, it has been well established that, although most globular proteins unfold more slowly at the air–water interface, they tend to form more numerous and stronger intermolecular interactions inside the interfacial layer, in comparison with the flexible caseins (Mitchell, 1986; Dickinson, 1999a).

Immediately following adsorption, the interfacial layer of a globular protein such as β-lactoglobulin can be regarded as being a close-packed monolayer of deformable particles (de Feijter and Benjamins, 1982). Due to the close-range repulsive interactions, this dense layer of mainly unfolded protein molecules has small-deformation viscoelastic properties like those exhibited by a two-dimensional glassy state. As the globular protein molecules unfold, this converts into a two-dimensional gel-like layer (Dickinson, 1999a, 2001; Boerboom *et al.*, 1996; Wijmans and Dickinson, 1998, 1999a) following strengthening of non-bonded physical intermolecular interactions and slow covalent cross-linking (Dickinson and Matsumura, 1991). Whereas an elastic contribution to the rheology is evident with a system of simple repulsive interactions only at small compressive deformations, the presence of a network of attractive interactions provides elasticity (and eventually fracture behaviour) at large deformations. In fact, the rheological and structural properties of such protein adsorbed layers commonly resemble closely those of concentrated globular protein gels (Dickinson, 1999a, 2001). This gel-like character distinguishes the behaviour of adsorbed β-lactoglobulin from that of the adsorbed caseins. The essential structural and rheological features of such a crosslinked gel-like layer have been simulated by the technique of Brownian dynamics (Wijmans and Dickinson, 1998, 1999a). Neutron reflectivity experiments have shown that β-lactoglobulin gives a rather thin adsorbed layer (2–3 nm at neutral pH) (Atkinson *et al.*, 1995). There is also experimental evidence that, at low surface coverage, adsorbed globular proteins exist in more unfolded configurations — possibly even as a phase-separated interfacial state (Erikson *et al.*, 2000).

Let us turn now to a description of adsorbed casein layers. Here a polymer-based model is envisaged, with the flexible chains having some sequences of segments in direct contact with the surface ('trains') and others protruding into the aqueous phase ('loops' and 'tails') (Horne, 1998; Dickinson, 1999b, 2001). In the case of bovine β-casein, its extensive hydrophobic region (160–170 segments) is postulated to be anchored to the interface (trains and small loops), with its hydrophilic tail (40–50 segments) sticking into the aqueous phase, as shown in Figure 8.1b (Dalgleish, 1997; Horne, 1998; Dickinson, 2001). This intuitive representation is supported by modelling based on self-consistent-field (SCF) theory (Leermakers *et al.*, 1996; Dickinson *et al.*, 1997a). That is, the predicted segment density profile at neutral pH and low ionic strength has a dense inner layer (< 2 nm) and an extended outer region, with the polymer segment density reaching just 1 % at a distance 10 nm from the surface. This agrees with inferences from neutron reflectivity data for β-casein adsorbed at both the air–water and oil–water interfaces (Atkinson *et al.*, 1995, 1996; Dickinson *et al.*, 1993a).

In accord with experiments on emulsions (Husband *et al.*, 1997), the molecular configurations deduced from SCF calculations have demonstrated the crucial role of the cluster ('blob') of 5 charged phosphoserine residues in β-casein in maintaining the steric stabilizing layer, whilst also preventing interfacial precipitation (multilayers). The mobility of this 'blob' was demonstrated experimentally by ^{31}P NMR measurements on β-casein-stabilized emulsions (ter Beek *et al.*, 1996). It was inferred that, when the effective charge on the 'blob' is reduced (by dephosphorylation) or screened (by salt addition), the macromolecular 'spring' relaxes back to give a thinner layer (Horne and Leaver, 1995). Specific binding of calcium ions by the phosphoserine residues also causes the 'blob' to move inwards (Dickinson, 2001). The complete neglect of any protein secondary structure in this model is, of course, somewhat unrealistic. While β-casein is undoubtedly a highly disordered protein in bulk solution, it appears that some degree of ordered secondary structure (α-helix) is induced on adsorption (Caessens *et al.*, 1999; Dickinson, 2001).

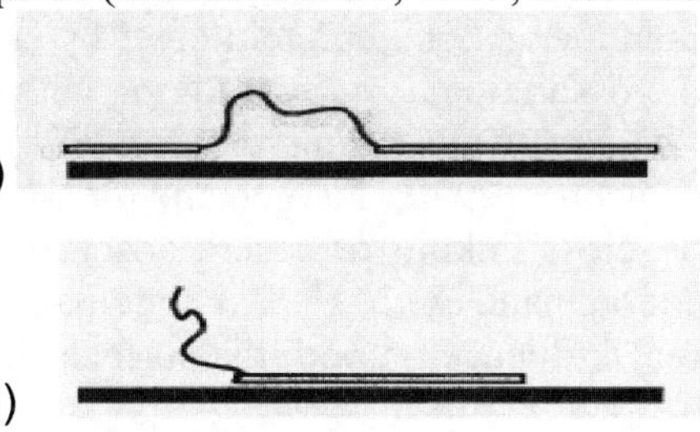

Figure 8.1 Schematic representation of structure of the flexible disordered caseins at a planar hydrophobic interface: (a) α_{s1}-casein; (b) β-casein. The solid bars denote hydrophobic regions of the molecules; they do not imply rigidity. Reproduced from Horne (1998) with permission.

The other major casein monomer in bovine milk is α_{s1}-casein. The SCF theory suggests that a loop-like protein conformation is favoured for adsorbed α_{s1}-casein (see Figure 8.1a) (Dickinson *et al.*, 1997; Horne, 1998). This implies a reduced hydrodynamic thickness of the adsorbed layer for α_{s1}-casein as compared with β-casein.

Differences in intermolecular interactions within adsorbed layers of different kinds of proteins are sensitively reflected in the surface shear rheology (Dickinson *et al.*, 1990; Benjamins and van Voorst Vader, 1992; Murray and Dickinson, 1996). The surface viscosity of α_{s1}-casein (or sodium caseinate) is rather higher than that of β-casein, but still much lower than that of α-lactalbumin or β-lactoglobulin (Murray and Dickinson, 1996). At the hydrocarbon–water interface, the surface shear viscosity is some 10^3–10^4 times larger for β-lactoglobulin than for β-casein (Murray and Dickinson, 1996; Dickinson, 1999a). The strongly viscoelastic character of adsorbed β-lactoglobulin can be attributed to a high packing density and strong protein–protein interactions, compared with the rather loose packing and weak protein–protein interactions of casein monolayers.

Dilatational surface rheology is a less discriminating experimental technique. At air–water and sunflower oil–water interfaces, it is found (Lucassen-Reynders and Benjamins, 1999) that both disordered β-casein and globular BSA show purely elastic behaviour in terms of the surface dilatational modulus over a considerable range of surface pressures. As the surface packing density is higher near the isoelectric point (and the attractive protein–protein interactions are correspondingly stronger), it is not surprising that the monolayer elasticity of β-casein is higher at pH = 5 than at pH = 7 (Rodriguez Nino *et al.*, 1999).

One potential strategy for influencing foam or emulsion stability is to attempt to strengthen the adsorbed protein layer by increasing the degree of protein–protein crosslinking, *e.g.*, by bridging with calcium ions (Hunt *et al.*, 1993), or enzymatically using the cross-linking enzyme transglutaminase (Dickinson, 1997b; Færgemand *et al.*, 1997; Dickinson, 1999a). Enzymatic cross-linking of casein adsorbed at fluid interfaces has been shown (Færgemand *et al.*, 1997; Færgemand and Murray, 1998) to have a substantial influence on both the surface shear and the surface dilatational properties.

1.2. *Segregation in Mixed Protein Layers*

On the basis that an adsorbed protein film can be viewed as a thin layer of highly concentrated solution of low local water activity, the excluded volume interactions should be especially significant (Dickinson, 2004). Therefore it is not unreasonable to expect phase separation in binary protein films. On the other hand, the process of phase separation within an adsorbed layer is determined not just by equilibrium thermodynamics, but also by the kinetics. That is, it is possible that the highly viscoelastic character of an adsorbed globular protein layer could 'trap' thermodynamically incompatible polymers into a notionally homogeneous (metastable) adsorbed state — one that is structurally either glass-like or gel-like (Dickinson, 2004; Damodaran, 2004).

Modelling or simulation studies can be useful tools for predicting the conditions under which segregative phase separation might occur over normal experimental timescales. Adopting such an approach, Brownian dynamics simulations were carried out by Pugnaloni *et al.* (2003b) on mixtures of two types of model adsorbing particles with properties mimicking those of globular proteins. The aim was to explore the conditions under which phase separation might take place in binary protein films. The simulations indicated that phase separation could occur in a binary particle mixture under two kinds of conditions: (i) if the particles of one of the components can exhibit reversible self-association, or (ii) if there is an effective repulsive interaction between the unlike particles. Figure 8.2 shows a set of pictures of the simulated interfacial structure. When the particles of either component can form irreversible bonds, the system becomes kinetically trapped. It is therefore suggested that phase separation in a protein mixture at a fluid interface might occur not because of direct thermodynamic incompatibility between the two different species, but rather as a consequence of the self-association of one of the species (Pugnaloni *et al.*, 2003b). In an alternative interpretation of the problem, Damodaran (2004) has suggested that the preferential interaction of one of the protein components with water could be the mechanistic origin of the competitive adsorption and phase separation behaviour in mixed protein films.

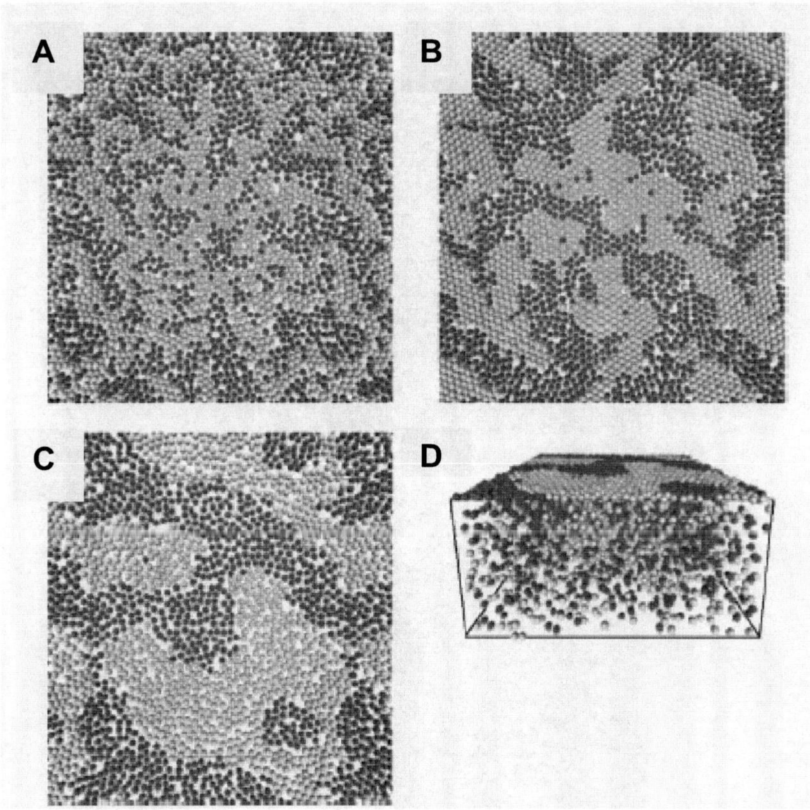

Figure 8.2 Phase separation in binary mixtures of model spherical particles at a planar interface generated by Brownian dynamics simulation. The three 2-D images refer to systems in which (A) light particles form irreversible bonds, (B) light particles form reversible bonds, and (C) neither dark nor light particles form bonds, but they repel each other. Picture D shows a 3-D representation. Reproduced from Pugnaloni *et al.* (2003b) with permission.

Interfacial phase separation has been postulated for several combinations of adsorbed food proteins: β-casein + soy 11S globulin, β-casein + bovine serum albumin (BSA), β-casein + lysozyme, β-casein + α_{s1}-casein, β-casein + α-lactalbumin, *etc.* (Xu and Damodaran, 1994; Anand and Damodaran, 1995, 1996; Cao and Damodaran, 1995; Razumovsky and Damodaran, 1999; Sengupta and Damodaran, 2000; Damodaran, 2004). Figure 8.3 shows visual evidence from a fluorescence microscopy study of a mixed monolayer film revealing phase-separated BSA-rich and β-casein-rich regions coexisting with inhomogeneous mixed regions (Sengupta and Damodaran, 2000).

(a)

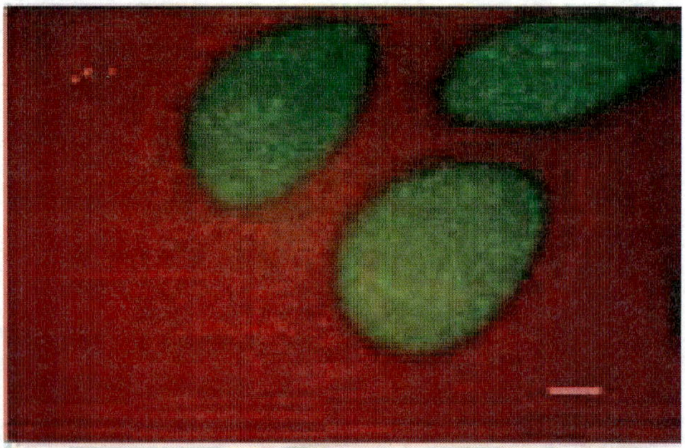

(b)

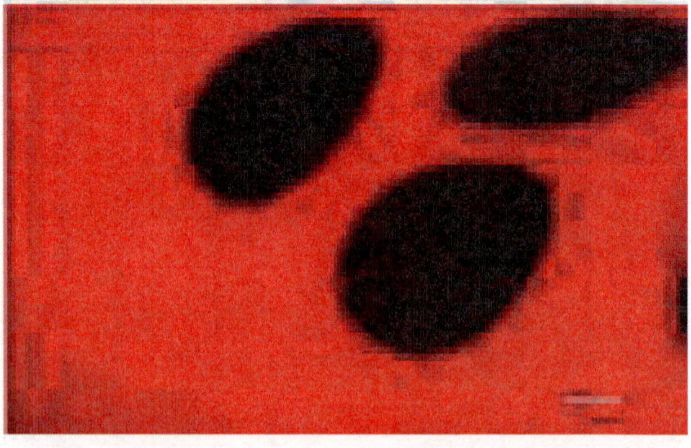

Figure 8.3 Fluorescence microscope images of a BSA + β-casein film at the air–water interface having a surface composition of 0.18 mg m^{-2} of BSA and 1.05 mg m^{-2} of β-casein. (a) Image taken with mutual filter. (b) Image of same system taken with red filter. Scale bar = 100 μm. Reproduced from Sengupta and Damodaran (2000) with permission.

The complementary images in Figure 8.3 refer to a BSA + β-casein mixed film transferred from the air–water interface after 4 days of adsorption and ageing at a well-defined ratio of the bulk protein concentrations. The β-casein-rich phase (red background) forms the continuous phase of the mixed film while the BSA-rich phase (green domains in Figure 8.3a) exists as segregated patches dispersed within the continuous phase. This preferred morphology was attributed to the weaker attractive protein–protein interactions for β-casein within the layer (Sengupta and Damodaran, 2000). The authors also suggested that, because the free energy at the boundaries between the phase-separated regions is expected to be locally greater than in other parts of the film, these boundaries may act as zones of instability in protein-stabilized foams and emulsions.

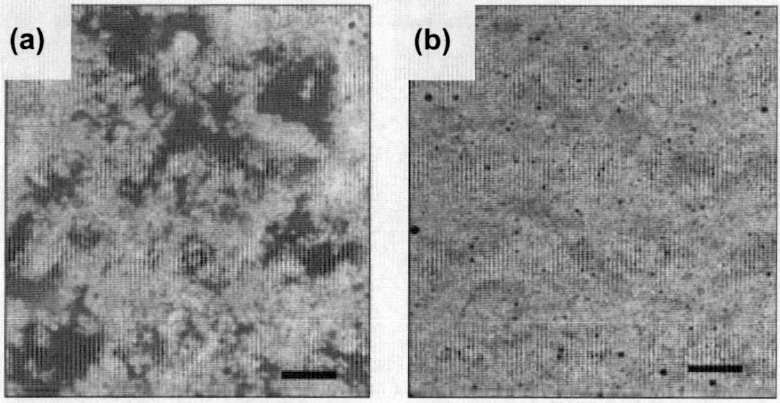

Figure 8.4 Effect of replacing small fraction of β-lactoglobulin by sodium caseinate on microstructure of concentrated oil-in-water emulsion (45 vol% oil, 3 wt% total protein, pH = 6.8, ionic strength = 0.03 M) heated for 6 min at 90 °C. Confocal micrographs were obtained with Rhodamine B as fluorescent protein stain: (a) emulsion contains 3 wt% β-lactoglobulin; (b) emulsion contains 2.85 wt% β-lactoglobulin + 0.15 wt% caseinate. Scale bar = 20 μm. Reproduced from Parkinson and Dickinson (2004) with permission.

As well as *lateral* heterogeneity in mixed protein layers, there is also the possibility of segregation of biopolymer components *perpendicular* to the interface, *i.e.*, bilayer formation (Dickinson, 1995, 2009). Let us consider the case of an interface containing casein and whey protein in an emulsion system. The images in Figure 8.4 are confocal micrographs of an emulsion for which the incorporation of a very low surface coverage of sodium caseinate was found to have the capability to convert the heat-sensitive globular protein-stabilized interface into one that remained stable to heat treatment (Parkinson and Dickinson, 2004). In this context 'stability' means the absence of droplet flocculation as detected by light scattering, viscometry and confocal microscopy (see Figure 8.4).

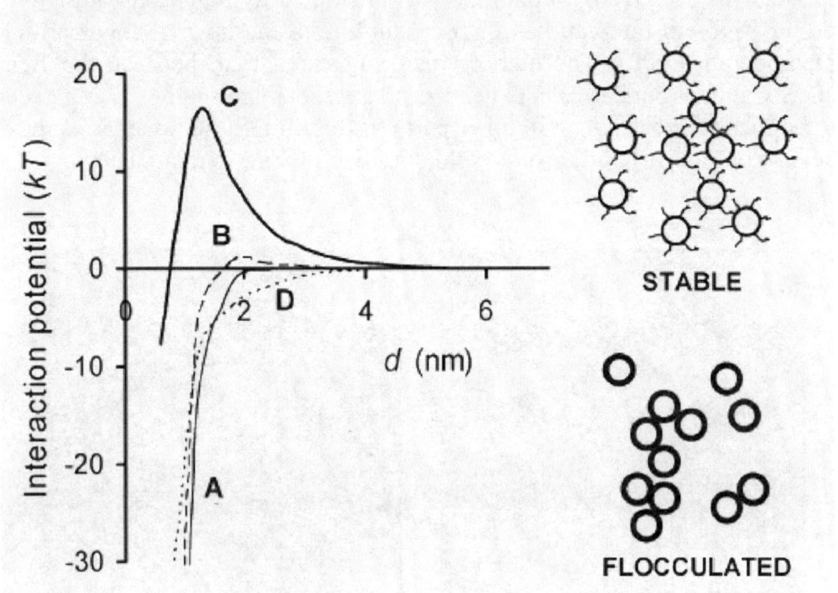

Figure 8.5 Interaction potential for model whey protein layer consisting of densely packed brush-like tethered chains with small a fraction of the whey protein replaced by β-casein chains as represented by a copolymer model. The energy $A(d)$ calculated from SCF theory is plotted as a function of surface–surface separation d: A, no β-casein; B, 2.5% β-casein; C, 5% β-casein; D, 5% β-casein alone (without whey protein layer). Potentials A, B and D imply that the emulsion system is flocculated; potential C implies a stable emulsion state. Reproduced from Dickinson (2006b) with permission.

Calculations from SCF theory of the mixed layer structure, and of the interaction potential for a pair of mixed layers as a function of interlayer separation, suggest that the mixed layer has a heterogeneous morphology perpendicular to the interface (Parkinson *et al.*, 2005). This localized segregation arises from the excluded volume interaction between spaced-out casein chains and the dense brush-like layer that was invoked in the simple SCF model to represent the β-lactoglobulin adsorbed monolayer.

Using a horticultural analogy, the structure of the mixed layer has been compared with an 'overgrown garden' in which there competition for space and light between the 'grass-like' whey protein layer and the partly segregated 'weed-like' casein molecules. A crowding effect increases the extension of the casein molecules from the surface, thereby greatly increasing the steric stabilizing capacity of the mixed ingredient interface.

Figure 8.5 shows some results of SCF calculations of the interaction potential for this mixed system of milk proteins (Parkinson *et al.*, 2005; Dickinson, 2006b). For interfaces coated with β-lactoglobulin (no casein present), the theory generates an attraction of range $d \approx 2$ nm (curve A). Similarly, for only a low density of grafted β-casein chains (2.5 % replacement), in the absence of any whey protein, the theory also gives an attraction, but now of a longer range (3–4 nm), due to casein polymer bridging (curve D). But when the whey protein and β-casein (2.5 % replacement) are present together, the theory predicts a stabilizing inter-layer repulsion and a maximum in the interaction potential at $d \approx 1$–2 nm (curve B). The height of the predicted 'primary' maximum increases from $A_{max} \approx 1\ kT$ for 2.5 % β-lactoglobulin substitution (curve B) to $A_{max} > 15\ kT$ for 5 % substitution (curve C). Therefore we see that this simple model predicts that the replacement of just a few per cent of the whey protein by β-casein tails can produce enough steric repulsion to transform the system from 'flocculated' to 'stable'. Enhanced steric stabilization is also predicted for limited replacement of whey protein by α_{s1}-casein, but to a lesser extent (Parkinson *et al.*, 2005; Dickinson, 2006b).

The full implications of segregation of components in mixed protein layers are as yet not fully understood. To address the problem properly, it would seem that some further fundamental research on the nanoscience of mixed protein layers is required (Dickinson, 2004; Damodaran, 2004).

2. *Proteins + Surfactants at Interfaces*

Low-molecular-weight surfactants ('emulsifiers') are important ingredients in food products. The types of surfactants most commonly studied in food colloids research are: phospholipids (lecithin), mono/diglycerides (particularly glycerol monostearate), polysorbates (Tweens), sorbitan monostearate or monooleate (Spans), and sucrose esters. These small lipid-based amphiphiles can typically lower the interfacial tension to a greater extent than the macromolecular amphiphiles such as proteins and certain gums (Bos and van Vliet, 2001).

Surfactants adsorb reversibly during the making of foams and emulsions, and the resulting surfactant-coated interface remains fluid-like (in the absence of adsorbed protein). The dynamic stretching of an interface containing mobile surfactants (or polar lipids) around newly formed emulsion droplets (or air bubbles) leads to concentration gradients, and the resulting rapid diffusion of the molecules to regions of reduced surfactant concentration leads to a stabilizing effect known as the Gibbs–Marangoni mechanism (Dickinson, 1992, 1994).

A situation that commonly occurs with food foams and emulsions is that there is a mixture of protein and low-molecular-weight surfactant available for adsorption at the interface. The composition and structure of the developing adsorbed layer are therefore strongly influenced by dynamic aspects of the competitive adsorption between protein and surfactant. This competitive adsorption in turn is influenced by the nature of the interfacial protein–protein and protein–surfactant interactions. At the most basic level, what drives this competition is that the surfactant–surface interaction is stronger than the interaction of the surface with the protein (or protein–surfactant complex) (Dickinson, 1998; Goff, 1997; Rodriguez Patino *et al.*, 2007; Miller *et al.*, 2008; Kotsmar *et al.*, 2009).

The dynamic adsorption behaviour is therefore complex: it involves adsorption, interaction and rearrangement of a mixed protein + surfactant system. Amongst the many contributory factors, the behaviour depends most of all on the composition of the interfacial film (Miller *et al.*, 2000a, 2008; Rodriguez Patino *et al.*, 2003, 2007). Experimental curves of dynamic surface tension for the adsorption of protein + emulsifier mixtures are commonly considered to consist of two sections. The first (at short adsorption times) corresponds to rapid adsorption of emulsifier, and the second (at long times) corresponds to relatively slow adsorption of protein (Rodriguez Nino and Rodriguez Patino, 1998; Miller *et al.*, 2000b; Rodriguez Patino *et al.*, 2003, 2007; Kotsmar *et al.*, 2009).

At sufficiently high surfactant concentrations, it is generally observed that the protein plays little part in the surface properties. This is due to the total displacement of the protein from the interface by the excess emulsifier (de Feijter *et al*, 1987; Courthaudon *et al.*, 1991a,b; Dickinson and Tanai, 1992; Chen and Dickinson, 1995; Dickinson, 2001). Thermodynamically speaking, we say that the surfactant lowers the surface free energy more than the protein at these high surfactant concentrations. The process of protein displacement from the air–water interface has been demonstrated directly through neutron reflectivity measurements (Horne *et al.*, 1998). Figure 8.6 shows such data for the gradual and systematic competitive displacement of β-lactoglobulin from the interface by a non-ionic or anionic surfactant (Dickinson, 2001).

In a mixed protein + surfactant layer, the presence of surfactant tends to weaken the mechanical properties (MacRitchie, 1978; Courthaudon *et al.*, 1991a; Wüstneck *et al.*, 1996; Bos and van Vliet, 2001). There is a dramatic drop from high values of interfacial shear viscosity and interfacial shear modulus observed for most proteins to the low values which are characteristic of adsorbed layers of surfactants (Krägel *et al.*, 2008). For example, on displacing β-lactoglobulin from the oil–water interface by Tween 20, a reduction in surface shear viscosity from 500 mN s m^{-1} to 0.1 mN s m^{-1} was recorded by Clark *et al.* (1993). For the same mixed system, Coke *et al.* (1990) reported a decrease in film thickness from 30 to 15 nm as the protein was displaced. Under conditions where both protein and surfactant coexist at the interface, it has been generally observed that the foam or emulsion is less stable than the equivalent one containing the pure surfactant or pure protein (Coke *et al.*, 1990; Dickinson *et al.*, 1993b; Cornec *et al.*, 1998; Wilde, 2000). It has also been observed that a water-soluble surfactant is generally more effective than an oil-soluble surfactant at destabilizing the adsorbed layer in a protein-based emulsion (Cornec *et al.*, 1996; Dickinson, 2001).

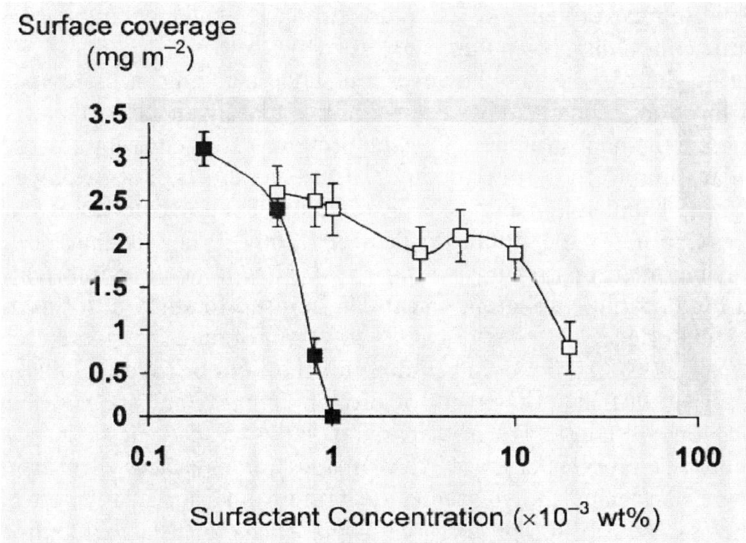

Figure 8.6 Comparison of the influence of non-ionic $C_{12}E_6$ (hexaoxyethylene *n*-dodecyl ether) or anionic SDS (sodium dodecyl sulfate) on adsorbed amount of β-lactoglobulin at the air–water interface (0.1 wt% protein, pH = 6, ionic strength = 0.02 M, 25 °C) as determined by neutron reflectivity measurements. Protein surface concentration is plotted against the aqueous phase surfactant concentration: (■) $C_{12}E_6$; (□) SDS. Reproduced from Dickinson (2001) with permission.

2.1. *The Competitive Displacement Mechanism*

Experimental information on the competitive adsorption of proteins and surfactants has been gradually accumulating in the literature over many years (MacRitchie, 1978, Dickinson and Woskett, 1989; Coke *et al.*, 1990; Courthaudon *et al.*, 1991a,b; Dickinson *et al.*, 1993b; Cornec *et al.*, 1998; Wilde, 2000; Wilde *et al.*, 2004; Rodriguez Patino *et al.*, 2007). Thus, it is established that hydrophobic proteins are generally more resistant to the destabilizing effects of surfactants (Brierley *et al.*, 1996), and notably that β-lactoglobulin is more resistant than β-casein (Cornec *et al.*, 1996, 1998; Clark *et al.*, 1994). An essential difference between the two latter proteins is that β-lactoglobulin forms a much more elastic adsorbed layer than does β-casein. The observed effect of protein hydrophobicity can be readily interpreted along the lines of relating the ease of displacement to the surface activity. However, until recently, it was not entirely clear why a globular protein, which forms a more elastic interface, should be more resistant to competitive displacement by surfactant. The supposition was that the interfacial interactions between globular protein molecules are stronger, hence making the layer more difficult to disrupt (Wilde *et al.*, 2004).

During the past decade, however, our understanding of the competitive adsorption mechanism has been significantly advanced by the ability to visualize the nanostructure of mixed protein + surfactant interfaces. In particular, atomic force microscopy (AFM) has emerged as a powerful experimental tool for imaging surface detail on flat substrates at molecular resolution (Wilde, 2000; Wilde *et al.*, 2004). The advantage of the AFM technique is that it can also impart interfacial thickness information with a high spatial resolution, and it can be used to study adsorption at both solid surfaces (McMaster *et al.*, 2000) and liquid interfaces (Johnson *et al.*, 2000). An important finding for the field of food colloids was the observation that most layers of protein + surfactant are not homogeneous, but are laterally phase-separated on the mesoscopic scale.

Figure 8.7 presents a set of AFM images for interfaces of protein + non-ionic surfactant showing separate protein and surfactant domains. The images indicate the progressive displacement of β-lactoglobulin by Tween 20 as the surface pressure is increased by sequential additions of surfactant (Wilde *et al.*, 2004). It is assumed that, in the early stages of the displacement process, the surfactant adsorbs into small defects in the protein film (Mackie *et al.*, 2000). This leads to the formation of discrete surfactant domains which apply an increasing surface pressure against the surrounding protein regions, hence effectively 'squeezing' the protein from the interface.

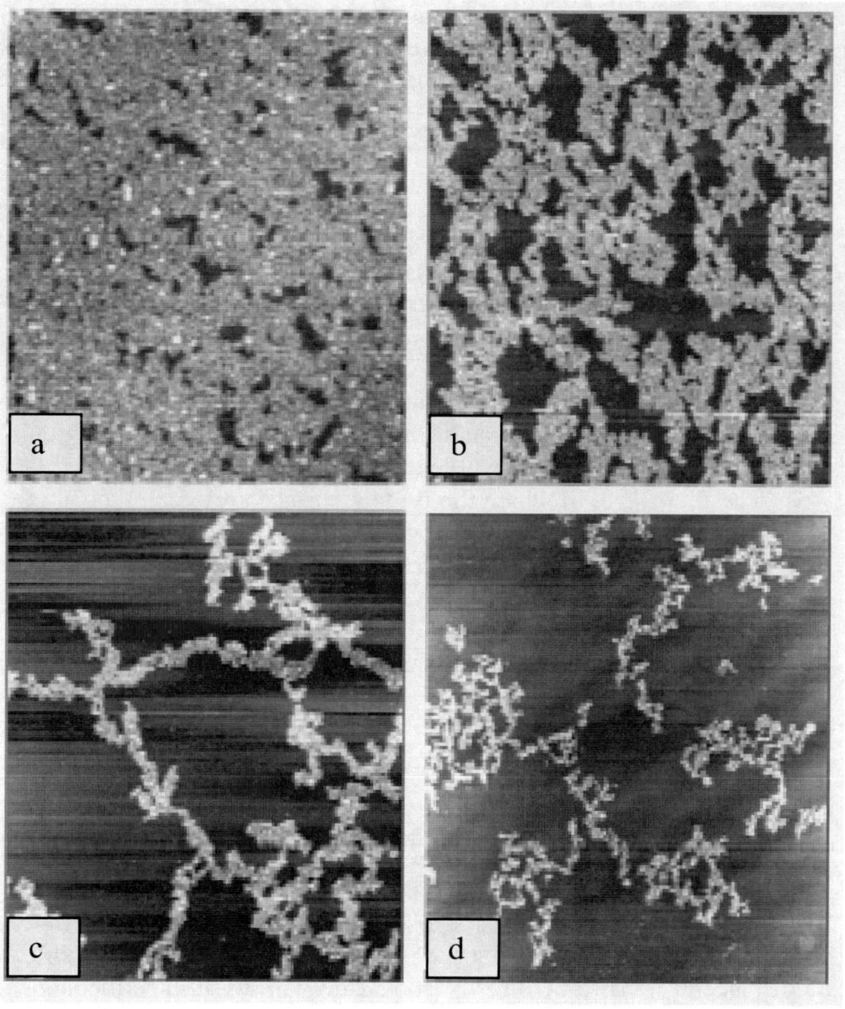

Figure 8.7 AFM images of a mixed β-lactoglobulin + Tween 20 interface. Light regions represent the protein network, and dark regions represent surfactant domains. At low surface pressures, the surfactant adsorbs into the protein network, forming small domains (a). As more surfactant is added, the surface pressure in the domains increases and the domains expand (b). The protein is progressively compressed into narrow regions (c). Finally the protein network fails, and the protein is displaced as aggregates (d). Image sizes: (a) 1 μm; (b) 3.2 μm; (c) 6 μm ; (d) 8 μm. Reproduced from Wilde *et al*. (2004) with permission.

On theoretical grounds, the phenomenon of two-phase coexistence at a liquid interface in a complex mixed adsorbing system is certainly not unexpected. So long as the various molecular interactions are of the normally expected form, statistical thermodynamics indicates that interfacial demixing can readily lead to a more favoured stable equilibrium state (Pugnaloni *et al.*, 2004).

Based on experimental evidence from AFM, the physical mechanism of 'orogenic' displacement was proposed (Mackie *et al.*, 1999b, 2000, 2003). A crucial aspect of the mechanism is that the surfactant domains exert a lateral surface pressure which compresses the protein layer. The AFM data obtained by Mackie and co-workers (1999b) provided direct visual evidence, for the first time, of a *gel-like* protein network at the air–water interface, as well as a structural explanation of protein displacement by small-molecule surfactant. In essence, the process of 'orogenic' displacement is assumed to involve three stages:

- The first stage, in which the protein surface area decreases but the film thickness does not change, involves simple compression of the film. This compression presumably reflects an improved packing of the protein within the monolayer network.
- The second phase is a regime in which the protein film is no longer compressible, but it compensates for the decrease in surface area through a corresponding increase in film thickness. This causes a buckling of the protein layer, including dissociation of at least some of the protein from the air–water interface, but not from the protein film itself.
- Finally, at a sufficiently high surface pressure, the protein network breaks down completely, thereby allowing outright desorption of the protein in the form of aggregates.

The general mechanism appears to be similar for proteins of different structure and conformation, but the ease by which a specific protein can be displaced seems to be directly correlated with its surface rheological properties (Wilde, 2000). Thus, for example, some major differences were found between AFM images of the mixed surfaces of β-casein and those of the globular whey proteins, α-lactalbumin and β-lactoglobulin, each in the presence of Tween 20 at the air–water interface (Mackie *et al.*, 1999b). Figure 8.8 shows the more rounded shape of the surfactant domains in the β-casein film (image A) as compared to the highly irregular nature of the surfactant domains in the globular protein film (image B). The differences in domain shape are characteristic of a more uniform propagation of stress in the low-viscosity β-casein films, in contrast to large local stress gradients in the viscoelastic films of β-lactoglobulin and α-lactalbumin (Mackie *et al.*, 1999b). With some small variations in the detailed behaviour, similar results were obtained at the oil–water interface. And so the generality of the orogenic mechanism would appear to be convincingly validated (Mackie *et al.*, 2000).

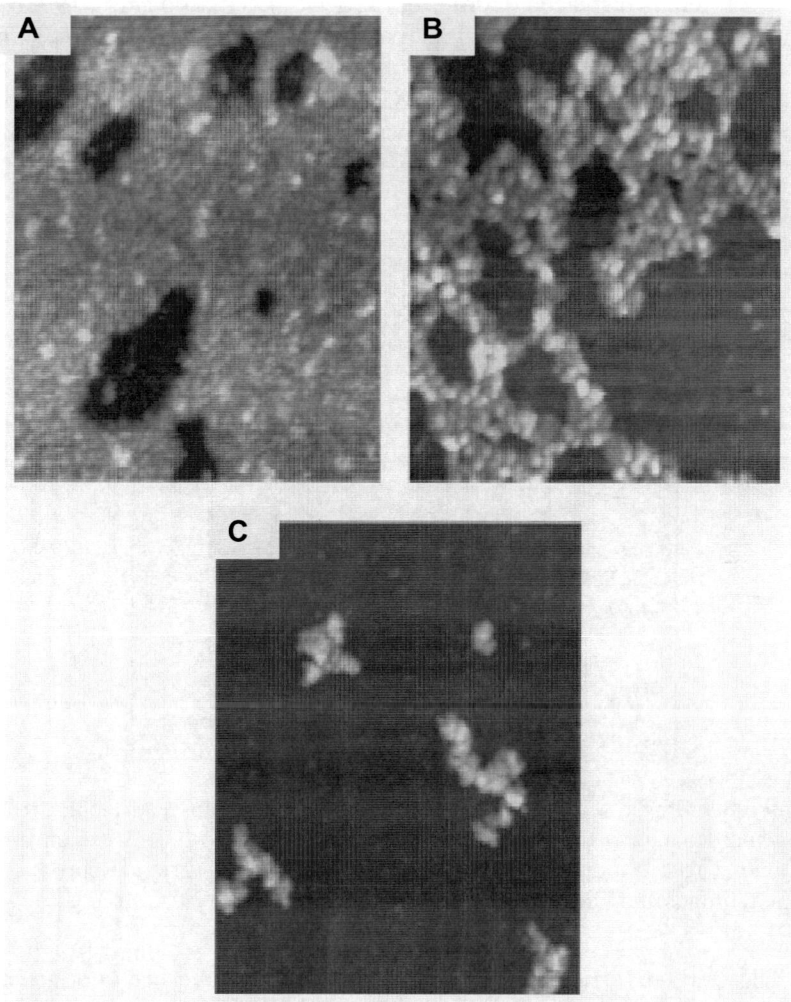

Figure. 8.8 Comparison of high resolution AFM images of Langmuir–Blodgett films obtained from the air–water interface for three protein + Tween 20 systems: (A) β-casein, 0.6 × 0.6 μm, surface pressure π = 20.6 mN/m; (B) β-lactoglobulin, 0.8 × 0.8 μm, π = 26.1 mN/m; (C) α-lactalbumin, 0.7 × 0.7 μm, π = 28.4 mN/m. Reproduced from Mackie *et al.* (1999b) with permission.

These detailed AFM studies of the structure of surfactant + protein layers have helped to explain the followings observations:

(i) why the stability of protein-based foams and emulsions can often be affected at very low concentrations of surfactant, before any protein is displaced (Cornec *et al.*, 1998; Dickinson *et al.*, 1993b);
(ii) why there is an observed increase in the thickness or surface concentration of an adsorbed protein layer following incorporation of surfactant (Dickinson, 1999a; Cornec and Narsimhan, 2000); and
(iii) why proteins that exhibit higher viscoelastic moduli are generally more difficult to displace (Cornec *et al.*, 1998).

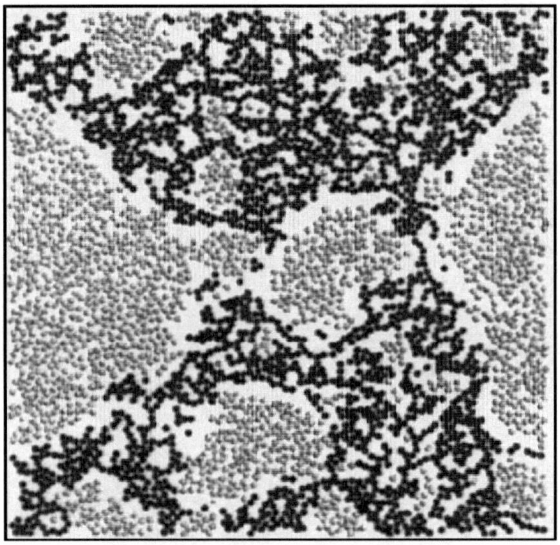

Figure 8.9 Phase separation in a mixed layer of protein + surfactant from Brownian dynamic simulation. In the picture are cross-linked protein-like particles (black) and surfactant-like displacer particles (grey). Reproduced from Wijmans and Dickinson (1999b) with permission.

The qualitative features of the orogenic model have been confirmed by computer simulation. Using the Brownian dynamics approach, we have simulated the competitive displacement process between protein-like and surfactant-like particles (Wijmans and Dickinson, 1999b; Pugnaloni *et al.*, 2003a). First and foremost, the computed images show a qualitative structural similarity to the experimental observations made using AFM (compare Figure 8.9 with Figure 8.7). In addition, the simulations have demonstrated the importance of the strength of the protein network in the displacement process. It was also demonstrated that the presence of repulsive forces between the adsorbed protein and surfactant enhances the phase separation. More generally, it seems that computer modelling is a powerful tool for investigating and understanding the role of specific intermolecular interactions in the development of interfacial structure, including the behaviour under large deformation conditions (Pugnaloni *et al.*, 2005).

When it comes to understanding the detailed behaviour of mixtures of proteins with specific emulsifiers, a range of experimental interfacial techniques is required. Monoglycerides are an especially important class of food emulsifiers. In a systematic study of surface pressure *versus* area isotherms of protein + monoglyceride monolayers, Rodriguez Patino *et al.* (2001a,b, 2003, 2007) have found that, at a macroscopic level, these mixtures form a practically immiscible monolayer at the air–water interface at surface pressures lower than that required for protein collapse. At higher surface pressures the collapsed protein layer is displaced from the interface by the emulsifier (monoglyceride) through the orogenic mechanism. The existence of rather weak protein–protein interactions in the mixtures involving disordered proteins (β-casein or sodium caseinate) is assumed to facilitate protein displacement from the air–water interface by the monoglycerides (Rodriguez Patino *et al.*, 2003).

In situ measurements of large-scale lateral structural features at fluid interfaces have been obtained using Brewster angle microscopy (BAM). Studies by Rodriguez Patino *et al.* (1999a,b, 2003, 2007) have shown that it is possible to image interfaces containing both proteins and surfactants. Figure 8.10 shows a set of BAM images for films of whey protein mixed with monoolein or monopalmitin (Rodriguez Patino *et al.*, 2003). The surfactants have a high refractive index compared to the proteins, allowing easy discrimination between the two species at the interface. These BAM experiments have demonstrated that the AFM-based orogenic mechanism can be considered valid and reproducible: the BAM measurements were performed *in situ* on a fluid interface, and essentially the same process of phase separation and thickening of the interfacial layers was observed. So, although Brewster angle microscopy has much lower resolution than AFM, it appears that the two approaches are quite complementary, with BAM having the advantage of being entirely a non-invasive technique (Wilde, 2000; Rodriguez Patino *et al.*, 2003, 2007). A recent BAM study by Murray *et al.* (2009) has investigated the effects of heat processing and large-scale deformation on milk protein films. The experimental observations of wrinkling and fracture of β-lactoglobulin films under large deformation were found to be qualitatively similar to those generated by computer simulation (Pugnaloni *et al.*, 2005).

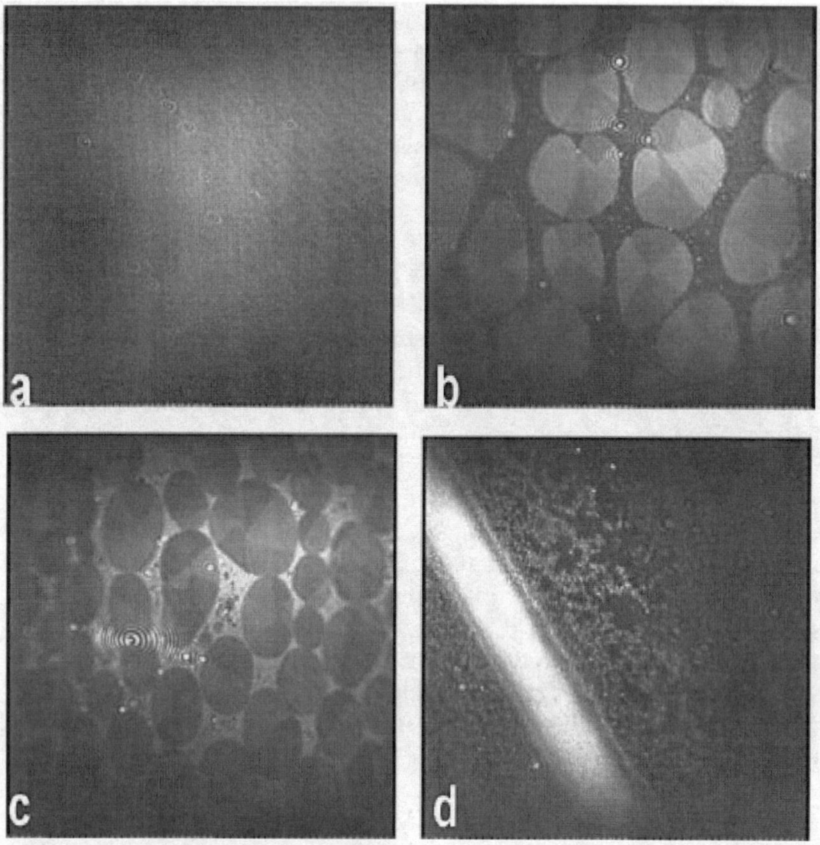

Figure 8.10 Visualization by Brewster angle microscopy (BAM) of pure and mixed protein + emulsifier films. (a) Image of homogeneous pure or mixed film with gaseous or liquid-expanded structure (*i.e.*, monoolein at every surface pressure π or monopalmitin at $\pi < 5$ mN/m). (b) Coexistence of heterogeneous liquid-condensed (circular) domains and homogeneous liquid-expanded domains (dark background) of emulsifier (*i.e.*, monopalmitin at $5 < \pi < 30$ mN/m). This image resembles the mixed protein + emulsifier film (protein + monopalmitin) before protein collapse. (c) Typical squeezing out phenomenon of protein by emulsifier after protein collapse (*i.e.*, at $\pi >$ 20 mN/m). The emulsifier domains (dark circles) float over a sub-layer of collapsed protein (white background). (d) Small islands of collapsed protein in protein + emulsifier film at higher surface pressures indicate that the protein is not completely displaced. Reproduced from Rodriguez Patino *et al.* (2003) with permission.

An alternative interpretation of desorption of proteins in the presence of surfactants was suggested by Damodaran (2004). It was proposed that the displacement of protein by the surfactant may not be because of the lateral pressure exerted by the surfactant film, but rather as a result of a reduction in interfacial water activity and consequent changes in protein–water interactions in the interfacial region. This suggestion was largely based on a correlation between the elastic modulus in the adsorbed layer and the calculated interfacial water activity (a_w^s) (Damodaran, 2004). Thus the data in Figure 8.11 show that, for various food protein films, the elastic modulus reaches a minimum value at $a_w^s \sim 0.15$–0.25. This range of water activity has been interpreted to have a special significance in the water sorption behaviour of proteins. It has been stated that, when proteins are exposed to an equilibrium relative humidity at $a_w \approx 0.2$, the ionic groups on the protein's surface become hydrated first; and then, as a_w is increased further, the other hydrophilic groups and the hydrophobic surfaces, in that order, become sequentially hydrated.

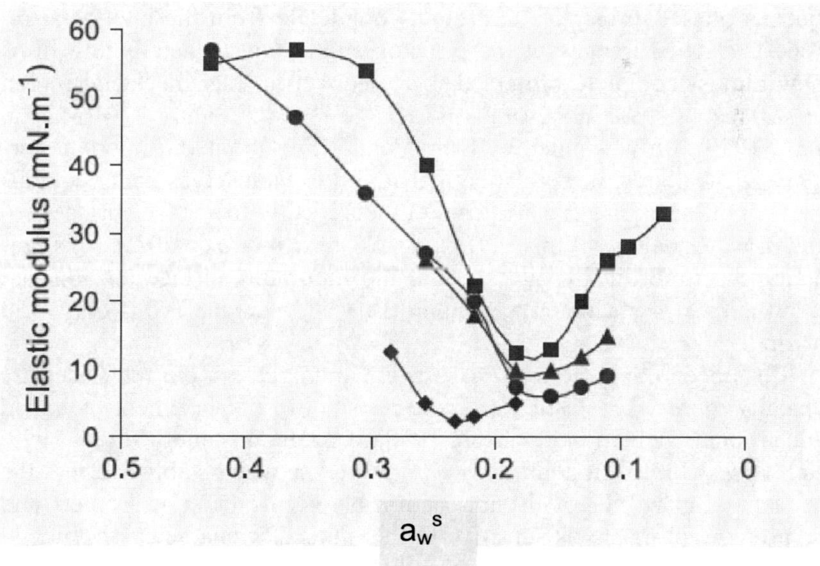

Figure 8.11 Elastic modulus *versus* interfacial water activity a_w^s of protein layers at the air–water interface: (◆) adsorbed β-casein; (▲) adsorbed α-lactalbumin; (●) adsorbed β-lactoglobulin; (■) spread β-lactoglobulin. Reproduced from Damodaran (2004) with permission.

When the water activity is reduced from $a_w = 0.8$ to $a_w = 0.3$ in a bulk protein solution, the 'monolayer' water associated with the hydrophobic patches around the protein's surface dissociates first (Rupley and Careri, 1991). As a_w is further reduced to below 0.3, water molecules associated with polar and ionic groups then begin to dissociate. It is therefore not unreasonable to suppose that a similar dehydration process also might occur in the air–water interfacial region (Damodaran, 2004). The lowering of surface tension as a result of surfactant adsorption is fundamentally due to the displacement of water molecules from the interfacial region, *i.e.*, the reduction in the interfacial water activity. Therefore, as surfactant molecules adsorb to fill up small spaces in the protein film, they progressively lower the interfacial water activity, first locally and then globally. This results in a progressive dehydration of the adsorbed protein molecules.

According to Damodaran (2004), the gradual decrease in the elastic modulus of the film (Figure 8.11) could be due to a progressive decrease in the plasticizing effect of water on the protein in the interfacial region. Such a dehydration would tend to relocate the protein into the bulk aqueous phase, so displacing protein completely from the interfacial region. The AFM images of the β-lactoglobulin and α-lactalbumin films show almost complete protein displacement from the interfacial region for surface pressure $\pi > 24$ mN/m ($a_w^s < 0.19$) (see Figure 8.8) (Mackie *et al.*, 1999b). It was noted by Damodaran (2004) that an important corollary of the above interpretation is that, left to themselves under equilibrium conditions, pure proteins cannot theoretically lower the surface tension of water below 49 mN/m (*i.e.*, $\pi = 23$ mN/m or $a_w^s = 0.2$). This prediction is borne out by the fact that the minimum surface tension that even the most surface-active protein (β-casein) can achieve is only ~ 50 mN/m.

It has been observed that heat-treated and aged protein interfaces are typically more viscoelastic and more resistant to displacement by added emulsifiers (Chen and Dickinson, 1999a; Garofalakis and Murray, 1999; Roth *et al.*, 2000). In addition, it is now rather well established that the mechanical character of the interaction between emulsion droplets and the protein gel matrix is sensitive to the molecular character of protein–surfactant interactions (Dickinson and Hong, 1997; Chen and Dickinson, 1998; 1999b). In particular, it has been demonstrated that, by displacing protein from the surface of the emulsion droplets, the viscoelasticity of an emulsion gel can be significantly reduced. What is happening here is that, as the surfactant content increases, the emulsion droplets are being gradually transformed from 'active' filler particles (protein-coated) to 'inactive' filler particles (mainly surfactant-coated).

2.2. *Protein–Surfactant Complexation*

It is well established that surfactants can associate with proteins to form molecular complexes involving various kinds of attractive interactions (Coke *et al.*, 1990; Bos *et al.*, 1997; Bos and van Vliet, 2001; Kelley and McClements, 2003; Semenova *et al.*, 2003, 2006; Il'in *et al.*, 2004, 2005; Malhotra and Coupland, 2004; Istarova *et al.*, 2005; Semenova, 2007). The role of protein–surfactant complexes is a more important issue with charged surfactants. As a consequence of complexation, a mixed system of protein + surfactant may have completely different surface activity and rheological properties from those of the pure protein or the pure surfactant (see section 2 in chapter six for more details).

The existence of protein–surfactant complexation can induce destabilization of a colloidal system by removing the adsorbed protein from the solid or liquid surface by means of a solubilization mechanism (Dickinson, 1998). That is, the desorption of protein from the interface is considered to arise as a result of protein solubilization into the bulk aqueous phase through the formation of a more hydrophilic protein–surfactant complex (Dickinson, 1997a, 1998; Fang and Dalgleish, 1997; Wasserman and Semenova, 1997). This complexation is largely driven by the attractive interactions between the hydrophobic domains on the two kinds of molecules. It is likely that the presence of interfacial complexes accounts for the frequent occurrence of states of incomplete displacement of protein from the interface in the high surfactant concentration regime (Cornec *et al.*, 1998; Chen and Dickinson, 1998).

In many systems we may suppose that the actual mechanism of protein displacement by surfactant involves a combination of solubilization and direct replacement mechanisms. Furthermore, following the formation of protein–surfactant attractive interactions, a change in protein conformation will probably take place, which may contribute further to protein removal from the interface (Dickinson, 1997a, 1998; Goff, 1997; Dalgleish, 1997; Fang and Dalgleish, 1997; Cornec *et al.*, 1998; Rodriguez Patino *et al.*, 1997; Le Meste *et al.*, 1997).

3. *Proteins + Polysaccharides at Interfaces*

In the case of a protein + polysaccharide mixture, whether the adsorption process is competitive or cooperative in character will depend on the concentration and surface activity of each adsorbed biopolymer species, and on the nature and strength of the protein–polysaccharide interactions (Murray, 2002; Baeza *et al.*, 2005; Dickinson, 2008). In addition, the structure and the colloidal stabilizing properties of any mixed protein + polysaccharide layer will depend on the order in which the two biopolymers become introduced to the interface (Guzey and McClements, 2006; McClements *et al.*, 2008; Dickinson, 2008; Jourdain *et al.*, 2008, 2009) (as described in section 2.4.5 of chapter seven).

3.1. *Complexation and Associative Interactions*

In a mixed system in which protein and ionic polysaccharide are introduced simultaneously, the formation of an electrostatic complex in the bulk aqueous medium can be expected to affect all the stages of protein adsorption — diffusion, adsorption, reorganization, and interactions within the adsorbed layer. As might be expected intuitively, the diffusive transport stage occurs more slowly, as observed for complexes of BSA with dextran sulfate adsorbing at the oil–water interface (Gurov and Nuss, 1986) and β-lactoglobulin + low-methoxy pectin at the air–water interface (Ganzevles, 2006). This slowing down is attributable to the greater hydrodynamic radius of complexes as compared to pure protein. Also there may be a kinetic barrier associated with the protein–polysaccharide complex adsorption as a consequence of the lower diffusion rate of protein through the complex layer.

When complexes form spontaneously in the bulk aqueous phase, there is a lowering of the chemical potential of protein in bulk solution, and consequently also at the interface, which implies a reduction in the surface activity (see chapters three and six for more details) (Galazka *et al.*, 2000; Semenova, 2007; Jourdain *et al.*, 2009). The molecular origin of this effect is an increase in average hydrophilicity of the protein surface, and the associated blocking of those hydrophobic patches on the protein surface which are important for adsorption at the liquid interface. The adsorption of complexes of protein + highly charged polysaccharide generally leads to formation of a more highly charged and thicker interfacial layer than with the protein alone. Hence there is enhancement in the strength and range of the steric and electrostatic repulsive interactions between the biopolymer-coated colloidal particles (Khalloufi *et al.*, 2009; Turgeon *et al.*, 2007; Jourdain *et al.*, 2008, 2009).

Protein–polysaccharide complexation affects the surface viscoelastic properties of the protein interfacial layer. Surface shear rheology is especially sensitive to the strength of the interfacial protein–polysaccharide interactions. Experimental data on BSA+ dextran sulfate (Dickinson and Galazka, 1992), α_{s1}-casein + high-methoxy pectin (Dickinson *et al.*, 1998), β-lactoglobulin + low-methoxy pectin (Ganzevles *et al.*, 2006), and β-lactoglobulin + acacia gum (Schmitt *et al.*, 2005) have all demonstrated an enhanced interfacial shear viscosity for the mixed biopolymer layer as compared to that for the corresponding pure protein layer.

Figure 8.12 illustrates the effect of complex formation between protein and polysaccharide on the time-dependent surface shear viscosity at the oil–water interface for the system BSA + dextran sulfate (DS) at pH = 7 and ionic strength = 50 mM. The film adsorbed from the 10^{-3} wt % solution of pure protein has a surface viscosity of $\eta_S > 200$ mPa s after 24 h. As the polysaccharide is not itself surface-active, it exhibited no measurable surface viscosity ($\eta_S < 1$ mPa s). But, when 10^{-3} wt% DS was introduced into the aqueous phase below the 24-hour-old BSA film, the surface viscosity showed an increase (after a further 24 h) to a value around twice that for the original protein film. Hence, in this case, the new protein–polysaccharide interactions induced at the oil–water interface were sufficiently strong to influence considerably the viscoelastic properties of the adsorbed biopolymer layer.

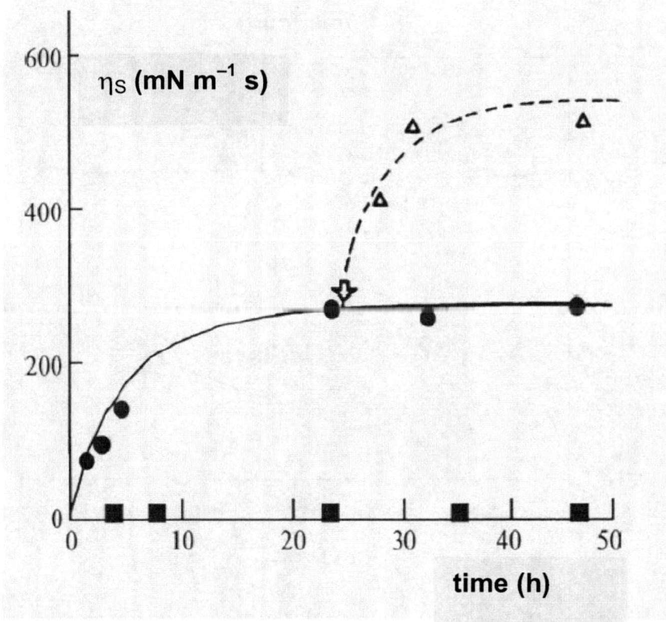

Figure 8.12 Time-dependent surface shear viscosity η_S of bovine serum albumin (BSA) + dextran sulfate (DS) at the *n*-tetradecane–water interface (pH = 7, ionic strength = 50 mM, 25 °C): (■) 10^{-3} wt% DS; (●) 10^{-3} wt% BSA; (△) 10^{-3} wt% DS added (⇓) to aqueous sub-phase below the BSA film after 24 h. Reproduced from Dickinson (1995) with permission.

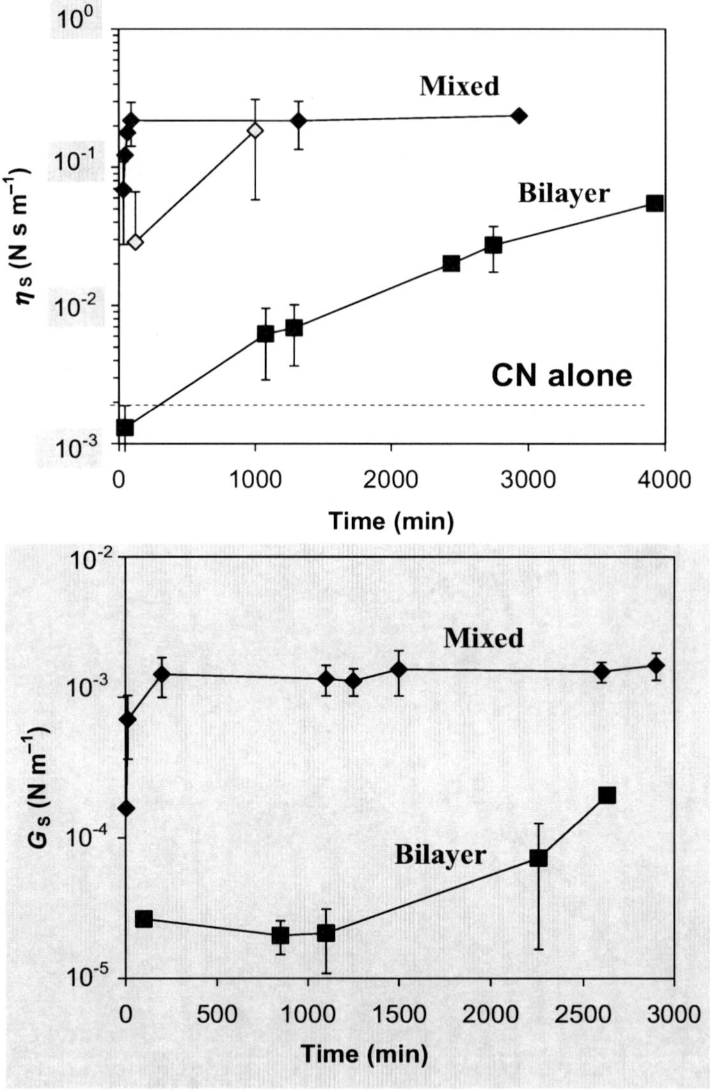

Figure 8.13 Effect of the method of preparation of the complexes of sodium caseinate (CN) + dextran sulfate (DS) on the time dependence of interfacial shear viscosity, η_S, and interfacial shear elasticity, G_S (frequency 0.1 s^{-1}): (◆) CN + DS, mixed layer, freshly prepared complexes; (◇) CN + DS, mixed layer, 24-h-old complexes; (■) CN + DS, bilayer. Aqueous solutions contained 0.5 wt% CN and 1 wt% DS in 20 mM imidazole buffer at pH = 6. Reproduced from Jourdain *et al.* (2009) with permission.

Figure 8.13 demonstrates the effect of the method of preparation of the oil–water interface ('bilayer' or 'mixed') on the time dependence of the interfacial shear viscosity (η_S) and the interfacial shear elasticity (G_S) (Jourdain *et al.*, 2009). For the film formed from adsorption of the mixed biopolymers (sodium caseinate + dextran sulfate), the shear viscosity was found to increase rapidly, reaching a limiting maximum value within 1 h. The interfacial shear elasticity was also very high initially. These data can be explained in terms of the presence of strongly bound electrostatic complexes, which inhibit the mobility of the casein molecules at the oil–water interface, leading to the formation of a thick rigid layer, and generating a mixed biopolymer network within the adsorbed layer. In contrast, for the 'bilayer' system, in which the caseinate was adsorbed first to the oil–water interface, the interfacial shear viscosity and elastic modulus increased rather slowly with time when the dextran sulfate was subsequently introduced into the aqueous sub-phase. The results suggest that, although the polysaccharide initially comes into close proximity with the interface, there is a requirement for extensive conformational rearrangement prior to strong complex formation and the creation of a final rigid interfacial structure.

3.2. Phase Separation and Segregative Interactions

We have previously referred to the notion that an adsorbed biopolymer film could be viewed like a thin layer of highly concentrated solution of low local water activity. On this basis, the presence of excluded volume interactions (thermodynamic incompatibility) between protein and surface-active polysaccharide would be expected to cause local segregation of the biopolymer species, and even possibly interfacial phase separation. Some polysaccharides like sugar beet pectin and gum arabic are themselves used as effective emulsifying agents (Dickinson, 2003). They can adsorb to oil–water (or air–water) interfaces and exhibit significant surface activity as evidenced from the reduction in the interfacial tension (Bergenståhl *et al.*, 1986; Randall *et al.*, 1988; Gaonkar, 1991; Garti, 1999; Huang *et al.*, 2001; Dickinson, 2009). However, it is important to note that the surface activity of most of these hydrocolloids is actually due to the presence of a small fraction of bound protein. For instance, gum arabic has been shown to contain about 10 % of its total mass as a hydroxyproline-rich glycoprotein (gum arabic glycoprotein, GAGP) with 90 % carbohydrate content and 10 % protein content (Qi *et al.*, 1991).

It is to be anticipated that thermodynamic incompatibility between a protein and a polysaccharide in an adsorbed film around the oil droplets or air bubbles in an emulsion or foam would have an influence on the stability properties. In competitive adsorption experiments involving the simultaneous adsorption of protein and gum arabic from a bulk mixture, it was reported that β-casein dynamically displaces the hydrocolloid (GAGP) from the air–water interface (Damodaran and Razumovsky, 2003). At a β-casein to GAGP bulk concentration ratio of 8%, the presence of the β-casein was found to effectively inhibit adsorption of GAGP to the air–water interface. However, the ability of β-casein to displace the gum arabic was found to be greatly reduced when the latter was already present as a pre-formed film. This behaviour could be attributed to the thermodynamic incompatibility of the mixing of β-casein into the polysaccharide film. The unfavorable free energy of mixing between GAGP and β-casein at the air–water interface was estimated to be around 5 kJ mol^{-1} at 25 °C (Damodaran and Razumovsky, 2003).

4. *Particles at the Interface of a Phase-Separated Mixed Biopolymer System*

The possibility of the accumulation and structuring of colloidal particles at the liquid–liquid interface of a phase-separating mixed biopolymer solution has recently been investigated (Firoozmand *et al.*, 2009). The system studied was an aqueous sugar solution of gelatin + oxidized (non-gelling) starch. Figure 8.14 shows a time sequence of images from confocal microscopy demonstrating the predominant accumulation of the polystyrene latex particles at the liquid–liquid interface and in the bulk of the gelatin-rich phase. The bright features indicate places where the particle density is high, and dark domains indicate regions depleted of the colloidal particles. From these and other images, some general features of particle behaviour could be inferred (Firoozmand *et al.*, 2009):

(i) the particles are preferentially associated with gelatin-rich regions;
(ii) the particles tend to be accumulating at the developing interface between the gelatin-rich and the starch-rich phases; and
(iii) the particles do not prevent growth of the bicontinuous phase-separated structure, but their presence at the interface does appear to influence (reduce) its rate of development.

It is postulated that the main thermodynamic driving force for particle adsorption at the liquid–liquid interface is the osmotic repulsion between the colloidal particles and hydrophilic starch polymer molecules. This leads to an effective depletion flocculation of particles at the boundaries of the starch-rich regions. At the same time, the gelatin has a strong tendency to adsorb at the hydrophobic surface of the polystyrene particles, thereby conferring upon them some degree of thermodynamic compatibility with the biopolymer environment of the gelatin-rich phase regions. The microstructure of the aged two-phase samples shows large variability in the local curvature of the particle-rich interfaces. This is a clear signature of surface viscoelasticity (Clegg, 2008).

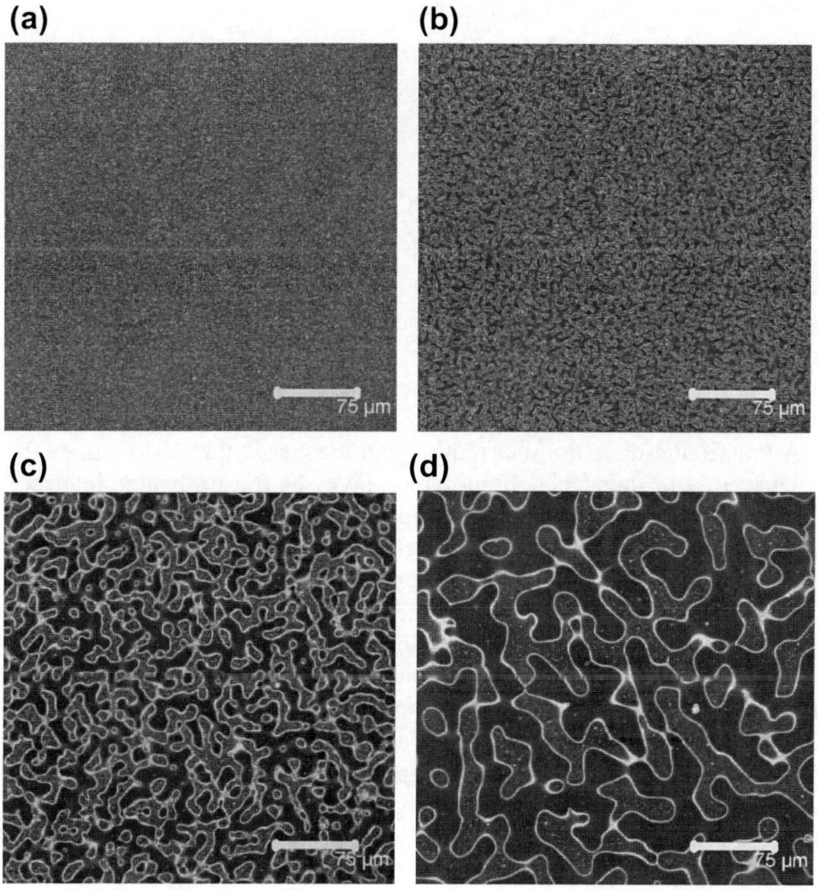

Figure 8.14 CLSM images showing the initial development of the microstructure of a phase-separated mixed biopolymer system (25.5 wt% sugar, 31.4 wt% glucose syrup, 7 wt% gelatin, and 4 wt% oxidized starch; pH = 5.2, low ionic strength) containing 0.7 wt% polystyrene latex particles (d_{32} = 0.3 μm). The sample was quenched from 90 to 1 °C, held at 1 °C for 10 min, heated to 40 °C at 6 °C min^{-1}, and observed at 40 °C for various times: (a) 2 min, (b) 4 min, (c) 8 min, and (d) 16 min. White regions are rich in colloidal particles. Reproduced from Firoozmand *et al.* (2009) with permission.

Incorporation of the appropriate kinds of (nano)particles into heterogeneous mixed biopolymer systems may offer new opportunities for the generation of novel structures having interesting physicochemical properties. In particular, the concept of particle structuring at liquid–liquid interfaces may have future possibilities in food-based biopolymer-based delivery and encapsulation technology.

5. *Enzymatic Hydrolysis in the Presence of Biopolymer Adsorbed Layers*

The biochemical stability of food colloids is now attracting considerable research interest because of its obvious relevance to the delivery and bioavailability of nutrients and nutraceuticals *in vivo*. In particular, the processes of enzymatic hydrolysis occurring at the triglyceride–water interface appear important because most dietary lipids are present in the human stomach at some stage in the form of emulsified droplets (size 20–40 μm) (Armand *et al.*, 1994; McClements *et al.*, 2008; Dickinson, 2008; Singh *et al.*, 2009; McClements and Decker, 2009).

A tantalizing question that requires an answer is this: "How does the structure of the interfacial biopolymer layer at the oil–water interface affect accessibility of enzymes to emulsified lipids and adsorbed proteins during digestion processes?" It is well recognized that the biochemical breakdown of proteins and lipids depends on the availability of the digestive enzymes at the appropriate molecular sites. Therefore the activities of proteases and lipases in the gastrointestinal tract depends *inter alia* on the extent to which gel particles and interfacial layers around lipid droplets inhibit the accessibility of enzymes to their substrates (Dickinson, 2008). The system contains a complicated mixture of fat, protein, polysaccharide and surfactant (including phospholipids, bile salts, *etc.*). The possible structural arrangement is (greatly) oversimplified by the cartoon in Figure 8.15.

There is currently little understanding of the influence of interfacial composition and (nano)structure on the kinetics of enzymatic hydrolysis of biopolymers and lipids. However, a few preliminary studies are beginning to emerge (McClements *et al.*, 2008; Dickinson, 2008). Thus, for example, Jourdain *et al.* (2009) have shown recently that, in a 'mixed' sodium caseinate + dextran sulfate system, the measured interfacial viscosity increased from η_S = 220 mN s m^{-1} without enzyme to η_S = 950 mN s m^{-1} with trypsin present. At the same time, the interfacial elasticity was initially slightly reduced from G_S = 1.6 mN m^{-1} to G_S = 0.7 mN m^{-1}, although it later returned to close to its original value. Conversely, in the equivalent 'bilayer' system, with other conditions remaining the same, large increases in the interfacial shear viscosity (~ one order of magnitude) and shear elasticity (~ two orders of magnitude) were observed upon addition of trypsin. The full implications of these data for protease activity in emulsions containing the same biopolymers are complicated by the interaction of the enzyme itself with the protein–polysaccharide adsorbed layer (Jourdain *et al.*, 2009).

In another set of studies, it has been reported that the *in vitro* digestibility of lipid droplets by pancreatic lipase is significantly affected by emulsifier type (Mun *et al.*, 2006, 2007; Park *et al.*, 2007). Intuitively, one might expect that a thick dense layer of strongly bound protein–polysaccharide complex at the oil–water interface would reduce considerably the *in vivo* accessibility of lipases, and hence would reduce the rate of human metabolism of fats. Establishment of the validity of this hypothesis must still await consolidation of a substantial body of detailed results from independent systematic studies on a broad range of mixed biopolymer systems.

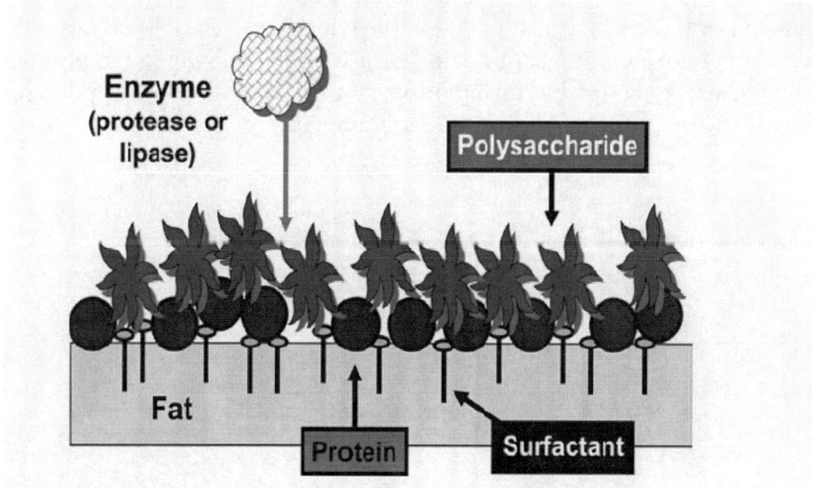

Figure 8.15 Cartoon showing how proteins, polysaccharides and surfactants (emulsifiers) might be distributed at the triglyceride–water interface. Interfacial complexation *in vivo* between adsorbed protein and charged polysaccharide in the gastrointestinal tract could affect digestion of protein and fat by forming structures that inhibit the accessibility and activity of enzymes (proteases and lipases). Reproduced from Dickinson (2008) with permission.

Concluding Remarks

A major objective for the future is to understand better the nanoscale organization of mixed systems of low-molecular-weight surfactants and biopolymers at macroscopic surfaces and in dispersed systems. Such information is a requirement for influencing and controlling the various interfacial parameters which may be regarded as important contributory factors to foam and emulsion behaviour. One question that still remains open is the extent to which it really is possible to make a direct quantitative connection between the formation, stability, texture, and taste of a food system, on the one hand, and the complete set of interactions on the molecular and colloidal scales, on the other. Any solution to this complex problem will undoubtedly require data from experimental and simulation studies using the wide range of emerging nanoscience techniques.

Finally, it is clear that much more research needs to be performed to connect the disciplines of colloid science and nutrition, especially in the application areas of functional foods and delivery systems. The combination of information from both *in vitro* and *in vivo* studies is evidently required in order to be able to understand properly the complex physicochemical processes that can occur during the digestion of food colloids. Our ultimate objective must be to make use of this knowledge to produce effective strategies for the formulation of a healthier diet with optimized bioavailability of the most desirable ingredients.

BIBLIOGRAPHY

Anand, K., Damodaran, S. (1995). Kinetics of adsorption of lysozyme and bovine serum albumin at the air/water interface from a binary mixture. *Journal of Colloid and Interface Science*, **176**, 63–73.

Anand, K., Damodaran, S. (1996). Dynamics of exchange between α_{s1}-casein and β-casein during adsorption at air–water interface. *Journal of Agricultural and Food Chemistry*, **44**, 1022–1028.

Armand, M., Borel, P., Dubois, C., Senft, M., Peyrot, J., Salducci, J., Lafont, H., Lairon, D. (1994). Characterization of emulsions and lipolysis of dietary lipids in the human stomach. *American Journal of Physiology — Gastrointestinal and Liver Physiology*, **266**, G372–G381.

Atkinson, P.J., Dickinson, E., Horne, D.S., Richardson, R.M. (1995). Neutron reflectivity of adsorbed β-casein and β-lactoglobulin at the air–water interface. *Journal of the Chemical Society, Faraday Transactions*, **91**, 2847–2854.

Atkinson, P.J., Dickinson, E., Horne, D.S., Leermakers, F.A.M., Richardson, R.M. (1996). Theoretical and experimental investigations of adsorbed protein structure at a fluid interface. *Berichte der Bunsen-Gesellschaft für Physikalische Chemie*, **100**, 994–998.

Baeza, R., Sanchez, C.C., Pilosof, A.M.R., Rodriguez Patino, J.M. (2005). Interactions of polysaccharides with β-lactoglobulin adsorbed films at the air–water interface. *Food Hydrocolloids*, **19**, 239–248.

Benjamins, J., van Voorst Vader, F. (1992). The determination of the surface shear properties of adsorbed protein layers. *Colloids and Surfaces*, **65**, 161–174.

Bergenståhl, B., Fogler, S., Stenius, P. (1986). The influence of gums on the stability of dispersions. In Phillips, G.O., Wedlock, D.J., Williams, P.A. (Eds), *Gums and Stabilisers for the Food Industry 3*, London: Elsevier Applied Science, pp. 285–293.

Boerboom, F.J.G., de Groot-Mostert, A.E.A., Prins, A., van Vliet, T. (1996). Bulk and surface rheological behaviour of aqueous milk protein solutions: a comparison. *Netherlands Milk and Dairy Journal*, **50**, 183–198.

Bos, M.A., van Vliet, T. (2001). Interfacial rheological properties of adsorbed protein layers and surfactants. *Advances in Colloid and Interface Science*, **91**, 437–471.

Bos, M., Nylander, T., Arnebrant, T., Clark, D.C. (1997). Protein/emulsifier interactions. In Hasenhuettl, G.L., Hartel, R.W. (Eds). *Food Emulsions and their Applications*, New York: Chapman & Hall, pp. 95–146.

Brierley, E.R., Wilde, P.J., Onishi, A., Hughes, P.J., Simpson, W.J., Clark, D.C. (1996). The influence of ethanol on the foaming properties of beer protein fractions: a comparison of Rudin and microconductivity methods of foam assessment. *Journal of the Science of Food and Agriculture*, **70**, 531–537.

Broadskaya, E.N., Eriksson, J.C., Laaksonen, A., Rusanov, A.I. (1996). Local structure and work of formation of water clusters studied by molecular dynamics simulations. *Journal of Colloid and Interface Science*, **180**, 86–97.

Caessens, P.W.J.R., de Jongh, H.H.J., Norde, W., Gruppen, H. (1999). The adsorption-induced secondary structure of β-casein and of distinct parts of its sequence in relation to foam and emulsion properties. *Biochimica et Biophysica Acta*, **1430**, 73–83.

Cao, Y., Damodaran, S. (1995). Coadsorption of β-casein and bovine serum albumin at the air–water interface from a binary mixture. *Journal of Agricultural and Food Chemistry*, **43**, 2567–2573.

Chen, J., Dickinson, E. (1995). Protein–surfactant interfacial interactions. Part 3. Competitive adsorption of protein + surfactant in emulsions. *Colloids and Surfaces A: Physicochemical and Engineering Aspects*, **101**, 77–85.

Chen, J., Dickinson, E. (1998). Viscoelastic properties of protein-stabilized emulsions: effect of protein–surfactant interactions. *Journal of Agricultural and Food Chemistry*, **46**, 91–97.

Chen, J., Dickinson, E. (1999a). Interfacial ageing effect on the rheology of heat-set protein emulsion gel. *Food Hydrocolloids*, **13**, 363–369.

Chen, J., Dickinson, E. (1999b). Effect of monoglycerides and diglycerol esters on viscoelasticity of heat-set whey protein emulsion gels. *International Journal of Food Science and Technology*, **34**, 493–501.

Clark, D.C., Mackie, A.R., Wilde, P.J., Wilson, D.R. (1993). In Schwenke, K., Mothers, R. (Eds). *Food Proteins: Structure and Functionality*, Weinheim: VCH, pp. 263–269.

Clark, D.C., Mackie, A.R., Wilde, P.J., Wilson, D.R. (1994). Differences in the structure and dynamics of the adsorbed layers in protein-stabilized model foams and emulsions. *Faraday Discussions*, **98**, 253–262.

Clegg, P.S. (2008). Fluid-bicontinuous gels stabilized by interfacial colloids: low and high molecular weight fluids. *Journal of Physics:Condensed Matter*, **20**, 113101.

Coke, M., Wilde, P.J., Russell, E.J., Clark, D.C. (1990). The influence of surface composition and molecular diffusion on the stability of foams formed from protein detergent mixtures. *Journal of Colloid and Interface Science*, **138**, 489–504.

Cornec, M., Narsimhan, G. (2000). Adsorption and exchange of β-lactoglobulin onto spread monoglyceride monolayers at the air–water interface. *Langmuir*, **16**, 1216–1225.

Cornec, M., Mackie, A.R., Wilde, P.J., Clark, D.C. (1996). Competitive adsorption of β-lactoglobulin and β-casein with Span 80 at the oil–water interface and the effect on emulsion behaviour. *Colloids and Surfaces A: Physicochemical and Engineering Aspects*, **114**, 237–244.

Cornec, M., Wilde, P.J., Gunning, P.A., Mackie, A.R., Husband, F.A., Parker, M.L., Clark, D.C. (1998). Emulsion stability as affected by competitive adsorption between an oil-soluble emulsifier and milk proteins at the interface. *Journal of Food Science*, **63**, 39–43.

Courthaudon, J.-L., Dickinson, E., Matsumura, Y., Clark, D.C. (1991a). Competitive adsorption of β-lactoglobulin + Tween 20 at the oil–water interface. *Colloids and Surfaces*, **56**, 293–300.

Courthaudon, J.-L., Dickinson, E., Matsumura, Y., Williams, A. (1991b). Influence of emulsifier on the competitive adsorption of whey proteins in emulsions. *Food Structure*, **10**, 109–115.

Dalgleish, D. (1997). Adsorption of protein and the stability of emulsions. *Trends in Food Science and Technology*, **8**, 1–6.

Damodaran, S. (1997). *Food Proteins and their Applications*, New York: Marcel Dekker.

Damodaran, S. (2004). Adsorbed layers formed from mixtures of proteins. *Current Opinion in Colloid and Interface Science*, **9**, 328–339.

Damodaran, S., Razumovsky, L. (2003). Competitive adsorption and thermodynamic incompatibility of mixing of β-casein and gum arabic at the air–water interface. *Food Hydrocolloids*, **17**, 355–363.

de Feijter, J.A., Benjamins, J. (1982). Soft-particle model of compact macromolecules at interfaces. *Journal of Colloid and Interface Science*, **90**, 289–292.

de Feijter, J.A., Benjamins, J., Tamboer, M. (1987). Adsorption displacement of proteins by surfactants in oil-in-water emulsions. *Colloids and Surfaces*, **27**, 243–266.
Dickinson, E. (1992). *An Introduction to Food Colloids*, Oxford: Oxford University Press.
Dickinson, E. (1994). Emulsions and droplet size control. In Wedlock, D.J. (Ed.). *Controlled Particle, Droplet and Bubble Formation*, Oxford: Butterworth, pp. 191–216.
Dickinson, E. (1995). Mixed biopolymers at interfaces. In Harding, S.E., Hill, S.E., Mitchell, J.R. (Eds). *Biopolymer Mixtures*, Leicestershire, UK: Nottingham University Press, pp. 349–372.
Dickinson, E. (1997a). Properties of emulsions stabilized with milk proteins: overview of some recent developments. *Journal of Dairy Science*, **80**, 2607–2619.
Dickinson, E. (1997b). Enzymic crosslinking as a tool for food colloid rheology control and interfacial stabilization. *Trends in Food Science and Technology*, **8**, 334–339.
Dickinson, E. (1998). Proteins at interfaces and in emulsions: stability, rheology and interactions. *Journal of the Chemical Society, Faraday Transactions*, **94**, 1657–1669.
Dickinson, E. (1999a). Adsorbed protein layers at fluid interfaces: interactions, structure and surface rheology. *Colloids and Surfaces B: Biointerfaces*, **15**, 161–176.
Dickinson, E. (1999b). Caseins in emulsions: interfacial properties and interactions. *International Dairy Journal*, **9**, 305–312.
Dickinson, E. (2001). Milk protein interfacial layers and the relationship to emulsion stability and rheology. *Colloids and Surfaces B: Biointerfaces*, **20**, 197–210.
Dickinson, E. (2003). Hydrocolloids at interfaces and the influence on the properties of dispersed systems. *Food Hydrocolloids*, **17**, 25–39.
Dickinson, E. (2004). Food colloids: the practical application of protein nanoscience in extreme environments. *Current Opinion in Colloid and Interface Science*, **9**, 295–297.
Dickinson, E. (2006a). Structure formation in casein–based gels, foams, and emulsions. *Colloids and Surfaces A: Physicochemical and Engineering Aspects*, **288**, 3–11.
Dickinson, E. (2006b). Colloid science of mixed ingredients. *Soft Matter*, **2**, 642–652.
Dickinson, E. (2008). Interfacial structure and stability of food emulsions as affected by protein–polysaccharide interactions. *Soft Matter*, **4**, 932–942.
Dickinson, E. (2009). Hydrocolloids as emulsifiers and emulsion stabilizers. *Food Hydrocolloids*, **23**, 1473–1482.
Dickinson, E., Galazka, V.B. (1992). Emulsion stabilization by protein–polysaccharide complexes. In Phillips, G.O., Wedlock, D.J., Williams, P.A. (Eds). *Gums and Stabilisers for the Food Industry 6*, Oxford: IRL Press, pp. 351–362.
Dickinson, E., Hong, S.-T. (1997). Influence of an anionic surfactant on the rheology of heat-set β-lactoglobulin-stabilized emulsion gels. *Colloids and Surfaces A: Physicochemical and Engineering Aspects*, **127**, 1–10.
Dickinson, E., Matsumura, Y. (1991). Time-dependent polymerization of β-lactoglobulin through disulphide bonds at the oil–water interface in emulsions. *International Journal of Biological Macromolecules*, **13**, 26–30.
Dickinson, E., McClements, D.J. (1995). *Advances in Food Colloids*, Glasgow: Blackie, pp. 27–80.
Dickinson, E., Tanai, S. (1992). Protein displacement from the emulsion droplet surface by oil-soluble and water-soluble surfactants. *Journal of Agricultural and Food Chemistry*, **40**, 179–183.
Dickinson, E., Woskett, C.M. (1989). Competitive adsorption between proteins and small-molecule surfactants in food emulsions. In Bee, R.D., Richmond, P., Mingins, J. (Eds). *Food Colloids*, Cambridge, UK: Royal Society of Chemistry, pp. 74–96.
Dickinson, E., Horne, D.S., Phipps, J.S., Richardson, R.M. (1993a). A neutron reflectivity study of the adsorption of β-casein at fluid interfaces. *Langmuir*, **9**, 242–248.

Dickinson, E., Owusu, R.K., Williams, A. (1993b). Orthokinetic destabilization of a protein-stabilized emulsion by a water soluble surfactant. *Journal of the Chemical Society, Faraday Transactions*, **89**, 865–866.

Dickinson, E., Horne, D.S., Pinfield, V.J., Leermakers, F.A.M. (1997). Self-consistent-field modelling of casein adsorption. Comparison of results for α_{s1}-casein and β-casein. *Journal of the Chemical Society, Faraday Transactions*, **93**, 425–432.

Dickinson, E., Rolfe, S.E., Dalgleish, D.G. (1990). Surface shear viscometry as a probe of protein–protein interactions in mixed milk protein films adsorbed at the oil–water interface. *International Journal of Biological Macromolecules*, **12**, 189–194.

Dickinson, E., Semenova, M.G., Antipova, A.S., Pelan, E. (1998). Effect of high-methoxy pectin on properties of casein-stabilized emulsions. *Food Hydrocolloids*, **12**, 425–432.

Erikson, J.S., Sundaram, S., Stebe, K.J. (2000). Evidence that the induction time in the surface pressure evolution of lysozyme solutions is caused by a surface phase transition. *Langmuir*, **16**, 5072–5078.

Ettelaie, R., Akinshina, A., Dickinson, E. (2008). Mixed protein–polysaccharide layers: a self consistent field calculation study. *Faraday Discussions*, **139**, 161–178.

Euston, S.R., Hirst, R.L., Hill, J.P. (1999). The emulsifying properties of β-lactoglobulin genetic variants A, B and C. *Colloids and Surfaces B: Biointerfaces*, **12**, 193–202.

Færgemand, M., Murray, B.S. (1998). Interfacial dilatational properties of milk proteins cross-linked by transglutaminase. *Journal of Agricultural and Food Chemistry*, **46**, 884–890.

Færgemand, M., Murray, B.S., Dickinson, E. (1997). Cross-linking of milk proteins with transglutaminase at the oil–water interface. *Journal of Agricultural and Food Chemistry*, **45**, 2514–2519.

Fang, Y., Dalgleish, D. (1997). Interactions between sodium caseinate and dioleoylphosphatidylcholine on oil–water interfaces and in solution. In Dickinson, E., Bergenståhl, B. (Eds). *Food Colloids: Proteins, Lipids and Polysaccharides*, Cambridge, UK: Royal Society of Chemistry, pp. 67–76.

Firoozmand, H., Murray, B.S., Dickinson, E. (2009). Interfacial structuring in a phase-separating mixed biopolymer solution containing colloidal particles. *Langmuir*, **25**, 1300–1305.

Galazka, V.B., Dickinson, E., Ledward, D.A. (2000). Emulsifying properties of ovalbumin in mixtures with sulfated polysaccharides: effects of pH, ionic strength, heat and high-pressure treatment. *Journal of the Science of Food and Agriculture*, **80**, 1219–1229.

Ganzevles, R.A. (2006). PhD Thesis, Wageningen University, the Netherlands.

Ganzevles, R.A., Zinoviadou, K., van Vliet, T., Cohen Stuart, M.A., de Jongh, H.H.J. (2006). Modulating surface rheology by electrostatic protein–polysaccharide interactions. *Langmuir*, **22**, 10089–10096.

Gaonkar, A.G. (1991). Surface and interfacial activities and emulsion characteristics of some food hydrocolloids. *Food Hydrocolloids*, **5**, 329–337.

Garofalakis, G., Murray, B.S. (1999). Effect of film ageing on the surface properties of lactoglobulin and lactoglobulin + sucrose stearate monolayers. *Colloids and Surfaces B: Biointerfaces*, **12**, 231–237.

Garti, N. (1999). Hydrocolloids as emulsifying agents for oil-in-water emulsion. *Journal of Dispersion Science and Technology*, **20**, 327–355.

Goff, H. (1997). Colloidal aspects of ice cream — a review. *International Dairy Journal*, **7**, 363–373.

Graham, D.E., Phillips, M.C. (1979a). Proteins at liquid interfaces. I. Kinetics of adsorption and surface denaturation. *Journal of Colloid and Interface Science*, **70**, 403–414.

Graham, D.E., Phillips, M.C. (1979b). Proteins at liquid interfaces. II. Adsorption isotherms. *Journal of Colloid and Interface Science*, **70**, 415–426.

Gurov, A.N., Nuss, P.V. (1986). Protein–polysaccharide complexes as surfactants. *Nahrung*, **30**, 349–353.

Guzey, D., McClements, D.J. (2006). Formation, stability and properties of multilayer emulsions for application in the food industry. *Advances in Colloid and Interface Science*, **128–130**, 227–248.

Halling, P.J. (1981). Protein-stabilized foams and emulsions. *Critical Reviews in Food Science and Nutrition*, **15**, 155–203.

Horne, D. S. (1998). Casein interactions: casting light on the black boxes, the structure in dairy products. *International Dairy Journal*, **8**, 171–177.

Horne, D.S., Leaver, J. (1995). Milk proteins on surfaces. *Food Hydrocolloids*, **9**, 91–95.

Horne, D.S., Atkinson, P.J., Dickinson, E., Pinfield, V.J., Richardson, R.M. (1998). Neutron reflectivity study of competitive adsorption of β-lactoglobulin and non-ionic surfactant at the air–water interface. *International Dairy Journal*, **8**, 73–77.

Huang, X., Kakuda, Y., Cui, W. (2001). Hydrocolloids in emulsions: particle size distribution and interfacial activity. *Food Hydrocolloids*, **15**, 533–542.

Hunt, J.A., Dickinson, E., Horne, D.S. (1993). Competitive displacement of proteins in oil-in-water emulsions containing calcium ions. *Colloids and Surfaces A: Physicochemical and Engineering Aspects*, **71**, 197–203.

Husband, F.A., Wilde, P.J., Mackie, A.R., Garrood, M.J. (1997). A comparison of the functional and interfacial properties of β-casein and dephosphorylated β-casein. *Journal of Colloid and Interface Science*, **195**, 77–85.

Il'in, M.M., Semenova, M.G., Belyakova, L.E., Antipova, A.S., Polikarpov, Yu.N. (2004). Thermodynamic and functional properties of legumin in the presence of small-molecule surfactants: effect of temperature and pH. *Journal of Colloid and Interface Science*, **278**, 71–80.

Il'in, M.M., Anokhina, M.S., Semenova, M.G., Belyakova, L.E., Polikarpov, Yu.N. (2005). Calorimetric study of the interactions between small-molecule surfactants and sodium caseinate with reference to the surface activity of their binary mixtures. *Food Hydrocolloids*, **19**, 441–453.

Istarova, T.A., Semenova, M.G., Sorokoumova, G.M., Selishcheva, A.A., Belyakova, L.E., Polikarpov, Yu.N. (2005). Effect of pH on caseinate interactions with soy phospholipids in relation to surface activity of their mixtures. *Food Hydrocolloids*, **19**, 429–440.

Izmailova, V.N., Yampolskaya, G.P., Tulovskaya, Z.D. (1999). Development of Rehbinder's concept on structure-mechanical barrier in stability of dispersions stabilized with proteins. *Colloids and Surfaces A: Physicochemical and Engineering Aspects*, **160**, 89–106.

Johnson, C.A., Yuan, Y., Lenhoff, A.M. (2000). Adsorbed layers of ferritin at solid and fluid interfaces studied by atomic force microscopy. *Journal of Colloid and Interface Science*, **223**, 261–272.

Jourdain, L., Leser, M.E., Schmitt, C., Michel, M., Dickinson, E. (2008). Stability of emulsions containing sodium caseinate and dextran sulfate: relationship to complexation in solution. *Food Hydrocolloids*, **22**, 647–659.

Jourdain, L.S., Schmitt, C., Leser, M.E., Murray, B.S., Dickinson, E. (2009). Mixed layers of sodium caseinate + dextran sulfate: influence of order of addition to oil–water interface. *Langmuir*, **25**, 10026–10037.

Kelley, D., McClements, D.J. (2003). Interactions of bovine serum albumin with ionic surfactants in aqueous solutions. *Food Hydrocolloids*, **17**, 73–85.

Khalloufi, S., Corredig, M., Goff, H.D., Alexander, M. (2009). Flaxseed gums and their adsorption on whey protein-stabilized oil-in-water emulsions. *Food Hydrocolloids*, **23**, 616–618.

Kotsmar, Cs., Pradines, V., Alahverdjieva, V.S., Aksenenko, E.V., Fainerman, V.B., Kovalchuk, V.I., Krägel, J., Leser, M.E., Noskov, B.A., Miller, R. (2009). Thermodynamics, adsorption kinetics and rheology of mixed protein–surfactant interfacial layers. *Advances in Colloid and Interface Science*, **150**, 41–54.

Krägel, J., Derkatch, S.R., Miller, R. (2008). Interfacial shear rheology of protein–surfactant layers. *Advances in Colloid and Interface Science*, **144**, 38–53.

Leermakers, F.A.M., Atkinson, P.J., Dickinson, E., Horne, D.S. (1996). Self-consistent-field modeling of adsorbed β-casein: effects of pH and ionic strength on surface coverage and density profile. *Journal of Colloid and Interface Science*, **178**, 681–693.

Le Meste, M., Tainturier, P., Gelin, J.-L. (1997). Lipid–protein interactions — consequences for surface activity in food emulsions. In Dickinson, E., Bergenståhl, B. (Eds). *Food Colloids: Proteins, Lipids and Polysaccharides*, Cambridge, UK: Royal Society of Chemistry, pp. 185–200.

Lucassen-Reynders, E.H., Benjamins, J. (1999). Dilational rheology of proteins adsorbed at fluid interfaces. In Dickinson, E., Rodriguez Patino, J.M. (Eds). *Food Emulsions and Foams: Interfaces, Interactions and Stability*, Cambridge, UK: Royal Society of Chemistry, pp. 195–206.

Mackie, A.R., Husband, F.A., Holt, C., Wilde, P.J. (1999a). Adsorption of β-lactoglobulin variants A and B to the air–water interface. *International Journal of Food Science and Technology*, **34**, 509–516.

Mackie, A.R., Gunning A.P., Wilde, P.J., Morris, V.J. (1999b). The orogenic displacement of protein from the air–water interface by surfactant. *Journal of Colloid and Interface Science*, **210**, 157–166.

Mackie, A.R., Gunning A.P., Wilde, P.J., Morris, V.J. (2000). Orogenic displacement of protein from the oil–water interface. *Langmuir*, **16**, 2243–2247.

Mackie, A.R., Gunning, A.P., Pugnaloni, L.A., Dickinson, E., Wilde, P.J., Morris, V.J. (2003). Growth of surfactant domains in protein films. *Langmuir*, **19**, 6032–6038.

MacRitchie, F. (1978). Proteins at interfaces. In Anfinsen, C.B., Edsall, J.T., Richards, F.M. (Eds). *Advances in Protein Chemistry*, New York: Academic Press, vol. 32, pp. 283–326.

Malhotra, A., Coupland, J.N. (2004). The effect of surfactants on the solubility, zeta potential, and viscosity of soy protein isolates. *Food Hydrocolloids*, **18**, 101–108.

McClements, D.J., Decker, E.A. (2009). Controlling lipid bioavailability using emulsion-based delivery systems. In McClements, D.J., Decker, E.A. (Eds). *Designing Functional Foods*, Boca Raton, FL: CRC Press, pp. 502–546.

McClements, D.J., Decker, E.A., Park, Y., Weiss, J. (2008). Designing food structure to control stability, digestion, release and adsorption of lipophilic food components. *Food Biophysics*, **3**, 219–228.

McMaster, T.J., Miles, M.J., Shewry, P.R., Tatham, A.S. (2000). *In situ* surface adsorption of the protein C hordein using atomic force microscopy. *Langmuir*, **16**, 1463–1468.

Miller, R., Policova, Z., Sedev, R., Neumann, A.W. (1993). Relaxation behaviour of human albumin adsorbed at the solution–air interface. *Colloids and Surfaces A: Physicochemical and Engineering Aspects*, **76**, 179–185.

Miller, R., Alahverdjieva, V.S., Fainerman, V.B. (2008). Thermodynamics and rheology of mixed protein–surfactant adsorption layers. *Soft Matter*, **4**, 1141–1146.

Miller, R., Fainerman, V.B., Makievski, A.V., Krägel, J., Grigoriev, D.O., Kazakov, V.N., Sinyachenko, O.V. (2000a). Dynamics of protein and mixed protein + surfactant adsorption layers at the water–fluid interface. *Advances in Colloid and Interface Science*, **86**, 39–82.

Miller, R., Fainerman, V.B., Makievski, A.V., Krägel, J., Wüstneck, R. (2000b). Adsorption characteristics of mixed monolayers of a globular protein and a non-ionic surfactant. *Colloids and Surfaces A: Physicochemical and Engineering Aspects*, **161**, 151–157.

Mitchell, J.R. (1986). Foaming and emulsifying properties of proteins. In Hudson, B.J.F. (Ed.). *Developments in Food Proteins 4*, London: Elsevier Applied Science, pp. 291–338.

Mun, S., Decker, E.A., Park, Y., Weiss, J., McClements, D.J. (2006). Influence of interfacial composition on *in vitro* digestibility of emulsified lipids: potential mechanism for chitosan's ability to inhibit fat digestion. *Food Biophysics*, **1**, 21–29.

Mun, S., Decker, E.A., McClements, D.J. (2007). Influence of emulsifier type on *in vitro* digestibility of lipid droplets by pancreatic lipase. *Food Research International*, **40**, 770–781.

Murray, B.S. (2002). Interfacial rheology of food emulsifiers and proteins. *Current Opinion in Colloid and Interface Science*, **7**, 426–431.

Murray, B.S., Dickinson, E. (1996). Interfacial rheology and the dynamic properties of adsorbed films of food proteins and surfactants. *Food Science and Technology International* (Japan), **2**, 131–145.

Murray, B.S., Xu, R., Dickinson, E. (2009). Brewster angle microscopy of adsorbed protein films at air–water interfaces after compression, expansion and heat processing. *Langmuir*, **23**, 1190–1197.

Park, G.Y., Mun, S., Park, Y., Rhee, S., Decker, E.A., Weiss, J., McClements, D.J., Park, Y. (2007). Influence of encapsulation of emulsified lipids with chitosan on their *in vivo* digestibility. *Food Chemistry*, **104**, 761–767.

Parkinson, E.L., Dickinson, E. (2004). Inhibition of heat-induced aggregation of a β-lactoglobulin-stabilized emulsion by very small additions of casein. *Colloids and Surfaces B: Biointerfaces*, **39**, 23–30.

Parkinson, E.L., Ettelaie, R., Dickinson, E. (2005). Using self-consistent-field theory to understand enhanced steric stabilization by casein-like copolymers at low surface coverage in mixed protein layers. *Biomacromolecules*, **6**, 3018–3029.

Prins, A. (1999). Stagnant surface behaviour and its effect on foam and film stability. *Colloids and Surfaces A: Physicochemical and Engineering Aspects*, **149**, 467–473.

Pugnaloni, L.A., Ettelaie, R., Dickinson, E. (2003a). Growth and aggregation of surfactant islands during the displacement of an adsorbed protein monolayer: a Brownian dynamics simulation study. *Colloids and Surfaces B: Biointerfaces*, **31**, 149–157.

Pugnaloni, L.A., Ettelaie, R., Dickinson, E. (2003b). Do mixtures of proteins phase separate at interfaces? *Langmuir*, **19**, 1923–1926.

Pugnaloni, L.A., Ettelaie, R., Dickinson, E. (2004). Surface phase separation in complex mixed adsorbing systems: an interface–bulk coupling effect. *Journal of Chemical Physics*, **121**, 3775–3783.

Pugnaloni, L.A., Ettelaie, R., Dickinson, E. (2005). Brownian dynamics simulation of adsorbed layers of interacting particles subjected to large extensional deformation. *Journal of Colloid and Interface Science*, **287**, 401–414.

Qi, W., Fong, C., Lamport, D.T.A. (1991). Gum arabic glycoprotein is a twisted hairy rope: a new model based on O-galactosylhydeoxyproline as the polysaccharide attachment site. *Plant Physiology*, **96**, 848–855.

Randall, C., Phillips, G.O., Williams, P.A. (1988). The role of the proteinaceous component on the emulsifying properties of gum arabic. *Food Hydrocolloids*, **2**, 131–140.

Razumovsky, L., Damodaran, S. (1999). Thermodynamic incompatibility of proteins at the air–water interface? *Colloids and Surfaces B: Biointerfaces*, **13**, 251–261.

Rodriguez Niño, M.R., Rodriguez Patino, J.M. (1998). Surface tension of bovine serum albumin and Tween 20 at the air–aqueous phase interface. *Journal of the American Oil Chemists Society*, **75**, 1241–1248.

Rodriguez Niño, M.R., Sanchez, C.C., Rodriguez Patino, J.M. (1999). Interfacial characteristics of β-casein spread films at the air–water interface. *Colloids and Surfaces B: Biointerfaces*, **12**, 161–173.

Rodriguez Patino, J.M., Rodriguez Niño, M.R., Gomez, J. (1997). Interfacial and foaming characteristics of protein–lipid systems. *Food Hydrocolloids*, **11**, 49–58.

Rodriguez Patino, J.M., Sanchez, C.C., Rodriguez Niño, M.R. (1999a). Is Brewster angle microscopy a useful technique to distinguish between isotropic domains in β-casein-monoolein mixed monolayers at the air–water interface? *Langmuir*, **15**, 4777–4788.

Rodriguez Patino, J.M., Sanchez, C.C., Rodriguez Niño, M.R. (1999b). Analysis of β-casein-monopalmitin mixed films at the air–water interface. *Journal of Agricultural and Food Chemistry*, **47**, 4998–5008.

Rodriguez Patino, J.M., Rodriguez Niño, M.R., Carrera, C., Cejudo, M. (2001a). The effect of pH on monoglyceride–caseinate mixed monolayers at the air–water interface. *Journal of Colloid and Interface Science*, 240, 113–126.

Rodriguez Patino, J.M., Rodriguez Niño, M.R., Carrera, C., Cejudo, M. (2001b). Whey protein isolate–monoglyceride mixed monolayers at the air–water interface: structure, morphology, and interactions. *Langmuir*, **17**, 7545–7553.

Rodriguez Patino, J.M., Rodriguez Niño, M.R., Sanchez, C.C. (2003). Protein–emulsifier interactions at the air–water interface. *Current Opinion in Colloid and Interface Science*, **8**, 387–395.

Rodriguez Patino, J.M., Rodriguez Niño, M.R., Sanchez, C.C. (2007). Physico-chemical properties of surfactant and protein films. *Current Opinion in Colloid and Interface Science*, **12**, 187–195.

Roth, S., Murray, B.S., Dickinson, E. (2000). Interfacial shear rheology of aged and heat-treated β-lactoglobulin films: displacement by nonionic surfactant. *Journal of Agricultural and Food Chemistry*, **48**, 1491–1497.

Rupley, J.A., Careri, G. (1991). Protein hydration and function. *Advances in Protein Chemistry*, **41**, 37–172.

Schmitt, C., Kolodziejczyk, E., Leser, M.E. (2005). Interfacial and foam stabilization properties of β-lactoglobulin + acacia gum electrostatic complexes. In Dickinson, E. (Ed.). *Food Colloids: Interactions, Microstructure and Processing*, Cambridge, UK: Royal Society of Chemistry, pp. 284–300.

Schwendel, D., Hayashi, T., Dahint, R., Pertsin, A., Grunze, M., Steitz, R., Schreiber, F. (2003). Interaction of water with self-assembled monolayers: neutron reflectivity measurements of the water density in the interface region. *Langmuir*, **19**, 2284–2293.

Semenova, M.G. (2007). Thermodynamic analysis of the impact of molecular interactions on the functionality of food biopolymers in solution and in colloidal systems. *Food Hydrocolloids*, **21**, 23–45.

Semenova, M.G., Il'in, M.M., Belyakova, L.E., Antipova, A.S. (2003). Protein + small-molecule surfactant mixtures: thermodynamics of interactions and functionality. In

Dickinson, E., van Vliet, T. (Eds). *Food Colloids, Biopolymers and Materials*, Cambridge, UK: Royal Society of Chemistry, pp. 377–387.

Semenova, M.G., Belyakova, L.E., Polikarpov, Yu.N., Il'in, M.M., Istarova, T.A., Anokhina, M.S., Tsapkina, E.N. (2006). Thermodynamic analysis of the impact of the surfactant–protein interactions on the molecular parameters and surface behaviour of food proteins. *Biomacromolecules*, **7**, 101–113.

Sengupta, T., Damodaran, S. (1998). Role of dispersion interactions in the adsorption of proteins at oil–water and air–water interfaces. *Langmuir*, **14**, 6457–6469.

Sengupta, T., Damodaran, S. (2000). Incompatibility and phase separation in a bovine serum albumin/β-casein/water ternary film at the air–water interface. *Journal of Colloid and Interface Science*, **229**, 21–28.

Sengupta, T., Razumovsky, L., Damodaran, S. (1999). Energetics of protein–interface interactions and its effect on protein adsorption. *Langmuir*, **15**, 6991–7001.

Singh, H., Ye, A., Horne, D. (2009). Structuring food emulsions in the gastrointestinal tract to modify lipid digestion. *Progress in Lipid Research*, **48**, 92–100.

ter Beek, L.C., Ketelaars, M., McCain, D.C., Smulders, P.E.A., Walstra, P., Hemminga, M.A. (1996). Nuclear magnetic resonance study of the conformation and dynamics of β-casein at the oil–water interface in emulsions. *Biophysical Journal*, **70**, 2396–2402.

Turgeon, S.L., Schmitt, C., Sanchez, C. (2007). Protein–polysaccharide complexes and coacervates. *Current Opinion in Colloid and Interface Science*, **12**, 166–178.

Wasserman, L., Semenova, M. (1997). Effect of lipophilic molecules on food protein surface activity at the air–water interface. In Dickinson, E., Bergenståhl, B. (Eds). *Food Colloids: Proteins, Lipids and Polysaccharides*, Cambridge, UK: Royal Society of Chemistry, pp. 77–91.

Wierenga, P.A., Meinders, M.B.J., Egmond, M.R., de Voragen, F.A.G.J., de Jongh, H.H.J. (2003). Protein exposed hydrophobicity reduces the kinetic barrier for adsorption of ovalbumin to the air–water interface. *Langmuir*, **19**, 8964–8970.

Wijmans, C.M., Dickinson, E. (1998). Simulation of interfacial shear and dilatational rheology of an adsorbed protein monolayer modeled as a network of spherical particles. *Langmuir*, **14**, 7278–7286.

Wijmans, C.M., Dickinson, E. (1999a). Brownian dynamics simulation of a bonded network of reversibly adsorbed particles: towards a model of protein adsorbed layers. *Physical Chemistry Chemical Physics*, **1**, 2141–2147.

Wijmans, C.M., Dickinson, E. (1999b). Brownian dynamics simulation of the displacement of a protein monolayer by competitive adsorption. *Langmuir*, **15**, 8344–8348.

Wilde, P.J. (2000). Interfaces: their role in foam and emulsion behaviour. *Current Opinion in Colloid and Interface Science*, **5**, 176–181.

Wilde, P., Mackie, A., Husband, F., Gunning, P., Morris, V. (2004). Proteins and emulsifiers at liquid interfaces. *Advances in Colloid and Interface Science*, **108–109**, 63–71.

Wüstneck, R., Krägel, J., Miller, R., Wilde, P.J., Clark, D.C. (1996). Adsorption of surface-active complexes between β-casein, β-lactoglobulin and ionic surfactants and their shear rheological behaviour. *Colloids and Surfaces A: Physicochemical and Engineering Aspects*, **114**, 255–265.

Xu, S., Damodaran, S. (1994). Kinetics of adsorption of proteins at the air–water interface from a binary mixture. *Langmuir*, **10**, 472–480.

Index